Workbook to Accompany Saunders Interactive General Chemistry CD-ROM

Version 2.5 with ActivChemistry™

John C. Kotz

State University of New York
College at Oneonta

William J. Vining

The University of Massachusetts, Amherst

with ActivChemistry Lessons by
Justin Fermann
The University of Massachusetts, Amherst

SAUNDERS COLLEGE PUBLISHING

Harcourt Brace College Publishers

Ft. Worth Philadelphia San Diego New York
Orlando Austin San Antonio Montreal
London Sydney Tokyo

Portions of this work were published in previous editions.

Text typeface: Sabon

Saunders College Publishing

Publisher: John Vondeling
Acquisitions Editor: Kent Porter Hamann
Developmental Editor: Christine Benedetto / Alexandra Buczek
Text Designer: Ana Aguilar
Cover Designer: Cara Castiglio
Production Manager: Doris Bruey
Product Manager: Pauline Mula

CD-ROM Production

Archipelago:
Ana Aguilar, Darin Eriksen, Brian Griffith, Bill Gudmundson, Charles Hamper, Patrick Harman, Bruce Hoffman, Mark Keller, Gary Lopez, Birgit Maddox, Beth Pickett, Ruth Rominger, Brian Rowlett, Nicole Taylor

Oneonta:
Stephanie C. Codden, Kathrine Kotz, Andrew L. Maillet, John Pape, Charles D. Winters, Susan M. Young

Narration:
Sylvia J. Aimerito, Lloyd Sherr

Computer Programming:
Ganesa Media Laboratories, Inc.

Printed in the United States of America

ISBN 0-03-021483-1

90 021 9876543

Introduction to the Saunders General Chemistry CD-ROM

The *Saunders General Chemistry CD-ROM* for Macintosh and Windows is an interactive presentation of general chemistry for college and university students embarking on careers in the sciences and health professions. Our goals in producing the CD-ROM are

- to emphasize the concepts of chemistry.
- to illustrate the images and sounds of chemistry.
- to illustrate the connection between the macroscopic world and the submicroscopic world of atoms and molecules.
- to provide some important tools for learning and doing chemistry.

We want students to view the world as chemists do, that is, to think about molecules. We also want students to gain a feel for the way data and observations are used to generate concepts, in particular the concepts usually presented in general chemistry.

The focus of the discs is an *Interactive Presentation* of general chemistry. The presentation is based on the introductory chapter and the first 21 numbered chapters (of 24 chapters) in *Chemistry & Chemical Reactivity*, 3rd edition, by John C. Kotz and Paul M. Treichel. The discs and this *Workbook* are intended to be used with this book, with the 4th edition of *Chemistry & Chemical Reactivity*, or with any other general chemistry text. The discs and *Workbook* could also be used alone and, as such, constitute nearly an entire chemistry course.

The core of Version 2.5 of the CD-ROM is identical to Versions 2.0 and 2.1. No significant alterations have been made in the material. However, *ActivChemistry* has been added. This software package, which allows the student to design and perform simulated laboratory experiments in chemistry, has been integrated into the screens of the previous version.

The CD-ROM can also be used with the 4th edition of *Chemistry & Chemical Reactivity*. The new edition largely follows the organization of the 3rd edition, with the exception of a rearrangement of topics in Chapters 2-5. Icons in the margin of the 4th edition correlate the book with the CD-ROM.

FEATURES OF THE CD-ROM

While the *Interactive Presentation* makes up the majority of the material on these discs, other items are

- ActivChemistry™ Software
- CAChe Visualizer for Education, a tool for visualizing molecules and their properties
- Interactive periodic table database
- Database of Common Chemical Compounds
- Plotting tool
- Molecular weight and molarity calculators

Each is described briefly below with reference to the location of more detailed instructions for its use.

Interactive Presentation

Information is presented as "screens" within chapters, the organization of which follow *Chemistry & Chemical Reactivity*. Each screen in each chapter is meant to introduce a key concept in that chapter. The ideas that stem from that concept are then presented in sidebars such as "A Closer Look" screen, a problem screen, a biography of a scientist, or an environmental note. Complete instructions for using this presentation are in the section of this *Workbook* titled *Using the CD-ROM* and in the instructions that accompany the discs.

CAChe Visualizer for Education

CAChe Scientific/Oxford Molecular has licensed the *CAChe Visualizer for Education*, a tool for viewing and manipulating molecules. All of the molecules featured in *Chemistry & Chemical Reactivity*, 4th edition, are included on the CD-ROM plus many others. The user can view these molecules and crystal structures and measure bond distances and bond angles. In addition, energy maps, molecular orbitals, and electron density surfaces can be visualized. In the Macintosh version it is possible to view calculated infrared spectra. Brief instructions for using the software are in Appendix A of the *Workbook*.

Tools

Interactive Periodic Table

The *Interactive Periodic Table* includes a database of 18 element properties, a tool to view three-dimensional representations of periodic trends, definitions of terms, and an interactive tool that gives the electron configuration of any element. The database is found by clicking on the "Tools" button in the *Interactive Presentation*.

Database of Chemical Compounds

Over 80 compounds are contained in a database that can be searched by name. Information available includes a model of the molecule or crystal lattice, its common and systematic name, density, melting point and boiling point, thermodynamic data, and, for weak acids and bases, ionization constants. Models of virtually all these molecules are included in the files that accompany the *CAChe Visualizer for Education*.

Plotting Tool

A general purpose plotting program is included that allows plotting and manipulation of x-y type data. It allows students to do a least-squares analysis of data and will return a slope, intercept, and correlation coefficient. Directions for using the tool are in Appendix A of this *Workbook*.

Tools

Molecular Weight and Molarity Calculators

These programs allow the user to calculate quickly the molecular weight of any compound and to calculate the quantity of a compound required to prepare a certain volume of a solution with a given concentration. They are located in the "Tools" button in the *Interactive Presentation*.

Important Tables

For your reference, some important tables of information in the *Interactive Presentation* are

Screen	Table Title
6.17	Enthalpies of Formation
8.10	Atomic Size
8.11	Ion Sizes
8.12	Ionization Energy
8.13	Electron Affinity
9.9	Bond Length
9.9	Bond Energies
9.11	Electronegativity
17.6	Weak Acids and Bases
19.4	Solubility Products
21.6	Standard Potentials

WORKBOOK

The *Workbook* is organized around the main "concept" screens, each of which is numbered by the chapter and screen. The questions in the *Workbook* can be answered after reading the text on the screen (and related screens), observing the videos, photos, and animations, and working through associated problems.

The *Workbook* questions are often not the usual questions found in a textbook. Instead, they ask you to make observations from videos or animations of reactions and draw conclusions from those observations. Or, you may be asked to derive data from your observations, treat those data using a mathematical tool, and then draw conclusions from the outcome. Other questions ask you to observe molecules using the *CAChe Visualizer for Education* and to use those models to derive your own understanding of relevant chemical phenomena. Still another possibility is to use one of the "interactive tools." An example is the one on Coulomb's law (Chapter 3), where you can vary the charge on two ions and the distance between them to better understand the consequences of chemical bonding involving ions. Finally, you may be asked to search the chemical databases to find data on elements and then to find the relationships among those data.

Because the CD-ROM presentation cannot include some of the details of problem solving that are given in the usual textbook, the questions for each screen in the CD-ROM also include appropriate Exercises; Study Questions are found at the end of each chapter. (All are taken from *Chemistry & Chemical Reactivity*; answers to these questions are found in Appendices B and C, respectively, of this *Workbook*.)

ACKNOWLEDGEMENTS

Fashioning this interactive presentation has been extremely challenging and time-consuming. In addition to the two of us, and many people at Archipelago Productions, there are others who worked very hard to bring this project to a successful conclusion, and they deserve special mention.

Charlie Winters did much of the photography and videography for this project as well as for *Chemistry & Chemical Reactivity*. Susan Young built many of the models that accompany the CAChe software, wrote and edited text, drafted the documentation for the CAChe Visualizer, assembled and checked data, helped to set up photos and videos, served as a model when needed, and performed innumerable other tasks. Katie Kotz managed the flow of media between New York, Philadelphia, and San Diego. Stephanie Codden at Hartwick College helped in the initial design of chapters, in writing directions, and in checking screens. Andy Maillet and Christine Nelson wrote prototypes of the interactive"tools." Flick Coleman at Wellesley College checked all of the screens for scientific accuracy.

Many at Saunders College Publishing were important in bringing this project to fruition. Carol Field acted as the Development Editor for the original project and provided liaison between Saunders and Archipelago. Angus McDonald, Nathalie Cunningham, and Christine Benedetto were extremely helpful in testing and marketing. Finally, we especially thank John Vondeling, our publisher at Saunders, for his support and encouragement.

At Archipelago we worked with many talented people. In particular we wish to thank Gary Lopez for his leadership, Nicole Taylor for her organizational skills, Pat Harman and Bill Gudmundson for their creativity, Brian Rowlett for his technical expertise, and Bruce Hoffman for his writing skills.

We also wish to acknowledge John and Betty Moore, Kathy Christoph, Jon Holmes, Paul Schatz, and Paul Treichel for their contributions at a design meeting at the University of Wisconsin in 1992.

This *Workbook* was prepared by us using QuarkXpress™, Adobe Photoshop™, and ChemDraw™. Titration curves were done using Acid-Base, a program from *JCE: Software*. The *Workbook* was built around a design by Ana Aguilar of Archipelago Productions.

Finally, we wish to thank the Chemistry Departments at SUNY-Oneonta and Hartwick College and our colleagues and students in those departments for their support.

FINALLY ...

And finally, as a personal note we wish to make it very clear that these discs are not a product solely of our labors and ideas. The development of the CD-ROM has truly been a joint project with Archipelago Productions. The design and programming of the discs and many of the ideas for animations and presentations were theirs. This project simply could not have been done without their collaboration. We wish to express our deepest gratitude for their creativity and friendship. We also want to thank our wives—Katie and Kathy—for their constant help, support, and encouragement over the several years that it took to develop these materials.

We are very excited about this CD-ROM and the opportunity it offers students as a new way to learn chemistry and its importance in our world. We and the people at Saunders College Publishing and at Archipelago Productions thoroughly enjoyed producing the discs. We hope you enjoy them—and learn from them.

John C. Kotz (KotzJC@Oneonta.edu)

William Vining (Vining@chem.umass.edu)

August, 1998

Using the CD-ROM

The *Saunders Interactive General Chemistry CD-ROM* is a complete multimedia presentation of college-level introductory chemistry. The presentation follows the structure of the Saunders College Publishing textbook *Chemistry & Chemical Reactivity*, 3rd and 4th editions, by Kotz and Treichel.

The two-disc set includes

- The Interactive Presentation, with the Introduction and Chapters 1-9 on Disc 1, and Chapters 10-21 on Disc 2.
- ActivChemistry™ Software and 52 lessons coordinated with each chapter.
- Database of Chemical Compounds
- CAChe™ Visualizer for education (molecular modeling software) and library of molecular models.
- Tool for plotting graphs
- Interactive Periodic Table

MINIMUM SYSTEM REQUIREMENTS

Windows™ installation requires:

Computer/OS:
MPC Level II compliant computer
486DX-66 computer running Windows 3.1 or greater
(note: ActivChemistry will not run on Windows NT)

RAM: 8 megabytes for Windows 3.1, 16 megabytes for Windows 95 or NT

Hard Disk Space: 15 MB

Other:

256 color capable VGA video board that is MPC Level II compliant, and a color VGA display (minimum)
SoundBlaster(TM) compatible audio card, and speakers or a headset
Double speed CD-ROM drive (minimum)
Mouse

Macintosh installation requires:

Computer/OS:
Macintosh, 68020 or greater, Power Mac (Acrobat Reader) running Mac OS 7.1 or greater

RAM: 16 MB

Hard Disk Space: 20 MB

Other:
256-color display monitor at 640 x 480 pixels (minimum)
Double speed CD-ROM drive (minimum)
QuickTime™ version 2.1 software
Mouse

Running the CAChe Visualizer for Education or ActivChemistry software requires a Floating-Point-Unit (FPU)

INSTALLATION INSTRUCTIONS

Installation can be performed from either disc in the two-disc set. Only one installation is needed. Each disc contains both Macintosh and Windows versions of the presentation.

Installation Procedure for the Macintosh

Put either disc in the CD-ROM drive and double-click the Chemistry Disc icon on the desktop to open the CD-ROM window.

In the CD-ROM window, open the Folder named *Install* by double-clicking on it.

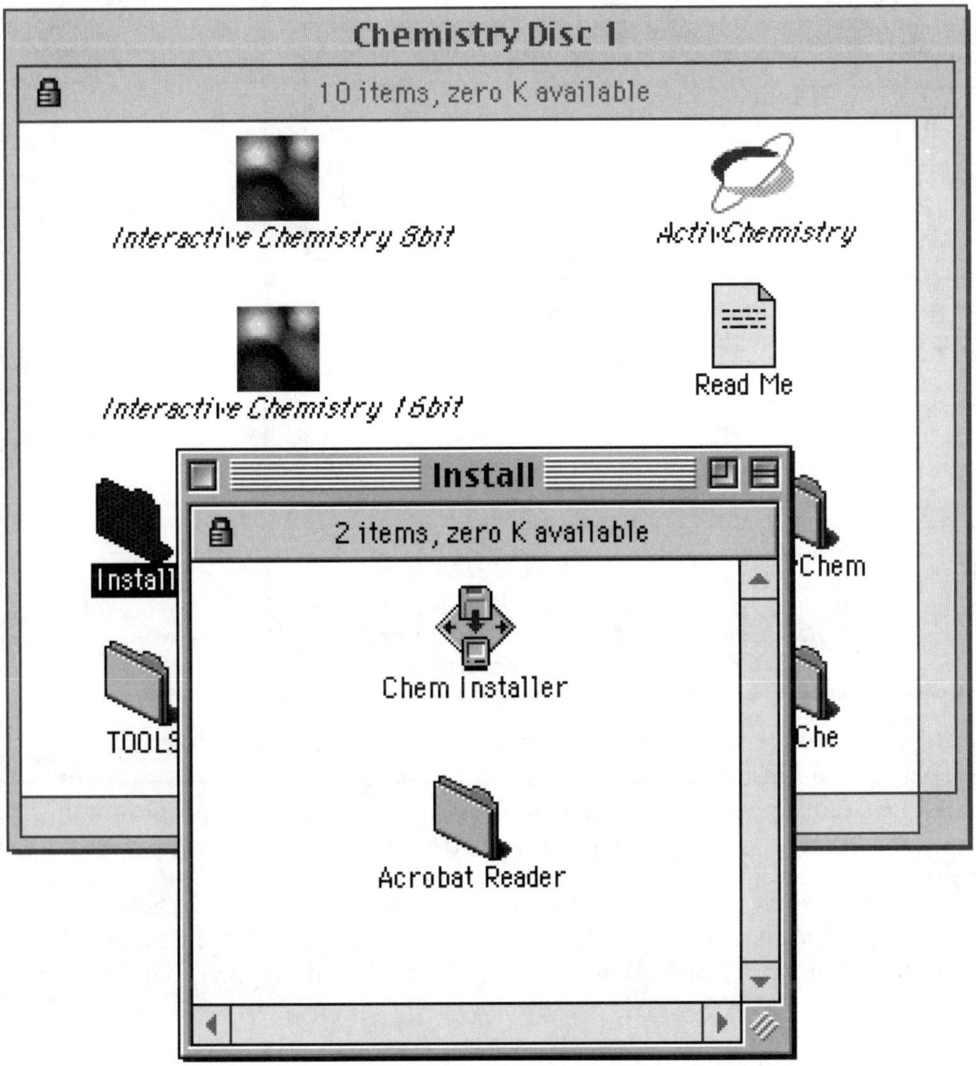

Chemistry Disc 1
10 items, zero K available
Interactive Chemistry 8bit
ActivChemistry
Interactive Chemistry 16bit
Read Me
Install
TOOLS
Chem
Che
Install
2 items, zero K available
Chem Installer
Acrobat Reader

Double-click on the *Chem Installer* icon.

The installer start-up screen opens. Click *"Continue."*

The End-User License Agreement screen appears. Please read the agreement carefully. If you do not agree with the terms of the license agreement, click *"Disagree"* and return the CD-ROM to Saunders College Publishing for a full refund. If you agree, click *"Agree."*

The installation window appears with the "Standard Installation" option pre-selected.

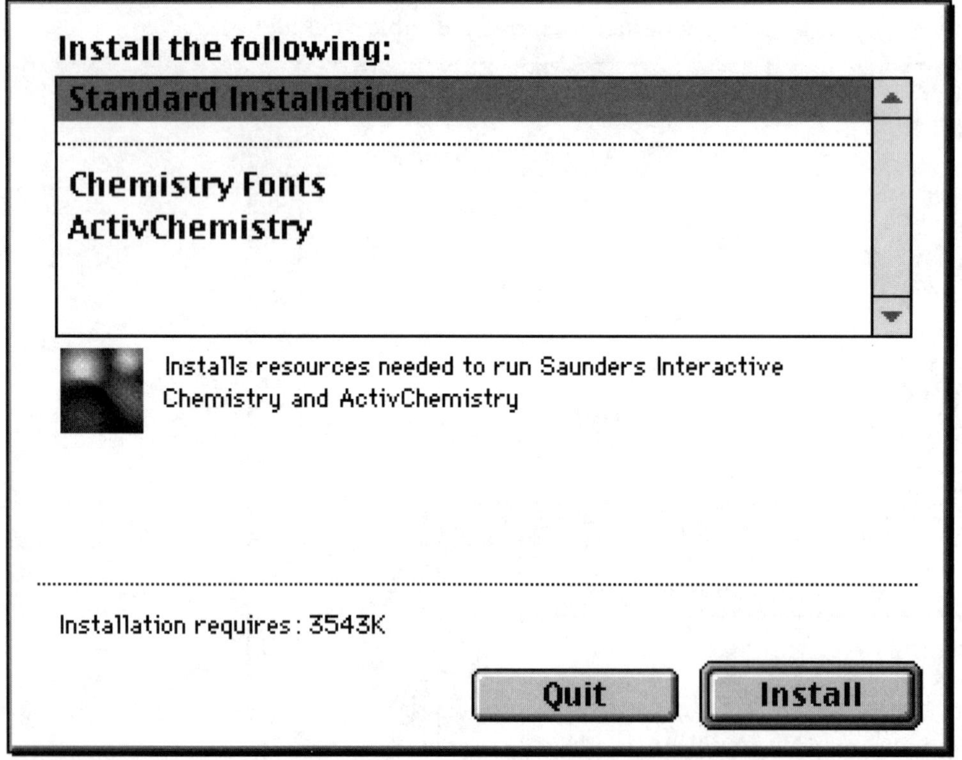

Note: Interactive Chemistry uses specially designed fonts that must be installed for the application to run properly. In addition, it requires that you have QuickTime version 2.1 or greater already installed on your computer. (QuickTime™ is Apple Computer's programming "architecture" for motion pictures, and it comes pre-installed with most Apple computers.)

Click the *"Install"* button. The fonts will be placed in the Fonts folder, and ActivChemistry resources will be placed in the Extensions folder, both of which are in your System folder. If you wish to install only the Chemistry Fonts or ActivChemistry, select the appropriate option and click *"Install."*

A message appears telling you that installation was successful. Click *"Quit."*

Several files on the CD-ROM require the installation of a separate application, the Adobe™ Acrobat™ Reader, to view the files properly. These files include:

- the Interactive General Chemistry CD-ROM workbook file, which is found in the workbook folder, on the desktop (top level of the CD-ROM).
- the ActivChemistry User's Guide, the list of image sources for the CD-ROM, and a compilation of the tables found in the CD-ROM. All these files are found in the DOCS folder on the desktop.
- correlation guides that correlate the screens of the CD-ROM to five of Saunders's most popular chemistry textbooks. These files are found in the COR_GUID folder that is found in the DOCS folder on the desktop.
- instructions for use and indexes for the CAChe molecular models found on the CD. These files are found in the DOCS folder that is located in the CAChe folder on the desktop.
- a database of chemical compounds, found in the Tools folder on the desktop.

To install this program, open the Acrobat Reader folder, double-click on Reader 3.01 Installer, and follow the directions.

Reader 3.01 Installer

Following successful installation of the Acrobat software, you will be prompted to restart your computer so it can recognize the new additions to your System folder. If you wish to run the presentation immediately, choose "Restart."

Installation is now complete.

To run the Interactive Chemistry presentation, one of the discs must be in the CD-ROM drive.

Choose *Interactive Chemistry 8bit* if you are running the program on a computer whose monitor is set to 8-bit (256-color) resolution.

Interactive Chemistry 8bit

Choose *Interactive Chemistry 16bit* if you are running the program on a computer whose monitor is set to 16-bit (thousands of colors) or higher resolution.

Interactive Chemistry 16bit

Note: Your Macintosh will display the following message if you attempt to run the 16-bit program on a monitor with 8-bit resolution.

INSTALLATION PRODEDURE FOR WINDOWS

From within Windows™ 3.1

Put either disc in the CD-ROM drive.

In the Program Manager window, select Run from the file menu.

Type "D:\setup.exe" and choose OK.

Note: If your CD-ROM drive is not accessed through the drive letter "D," then substitute the appropriate letter in the setup command.

Follow the steps listed under General Windows Installation Procedures below.

From within Windows™ 95

Put either disc in the CD-ROM drive.

Double-click on the "Shortcut to CD-ROM Disc" icon, or open "My Computer" and double-click on the CD-ROM icon.

Double-click on the "Setup" icon.

In the dialog box, click "Full - Install all files."

The next dialog box asks if you want setup to create Program Manager groups. Click *"Yes."*

The third dialog box asks if you would like the installer to add lines to AUTOEXEC.BAT. Click *"Yes."*

The last dialog box indicates that to complete installation, you should run the QTW and Acrobat installers. Click *"OK."*

To finish installation, follow the General Windows Installation Procedures below.

General Windows Installation Procedures

In the new Program Manager group, double-click on Install QTW and follow the installation instructions that appear in the dialog boxes.

When QuickTime has finished installing, close the QuickTime for Windows Program Manager group to return to the Saunders Chemistry 2.5 Program Manager group.

Several files on the CD-ROM require the installation of a separate application, the Adobe™ Acrobat™ Reader, to view the files properly. These files include:

- the Interactive General Chemistry CD-ROM workbook file, which is found on the top level of the CD-ROM.
- the ActivChemistry User's Guide, the list of image sources for the CD-ROM, a database of chemical compounds, and a compilation of the tables found in the CD-ROM. All these files are found in the DOCS folder on the desktop.

- correlation guides that correlate the screens of the CD-ROM to five of Saunders's most popular chemistry textbooks. These files are found in the COR_GUID folder that is found in the DOCS folder on the desktop.
- instructions for use and indexes for the CAChe molecular models found on the CD. These files are found in the DOCS folder that is located in the CAChe folder on the desktop.

There are two options available for installing Acrobat Reader: 16 bit (for those running Windows 3.x) and 32 bit (for those running Windows 95). Double-click on the version that matches your computer's operating system and follow the instructions in the dialog boxes that appear.

STARTING THE PRESENTATION

Once the program is properly installed and the desired disc is in the CD-ROM drive, open the CD-ROM by clicking on its icon.

Interactive Chemistry To begin the presentation, double-click on the Interactive Chemistry icon.

A Table of Contents appears. The presentation is divided into chapters, with the Introduction and Chapters 1–9 on Disc 1 and Chapters 10–21 on Disc 2.

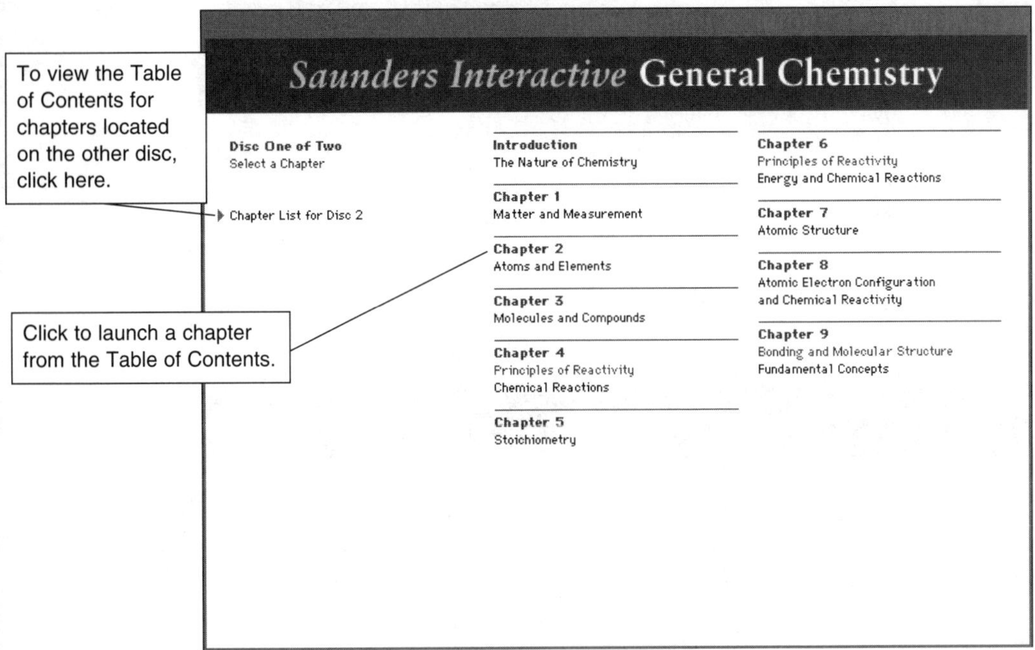

Notes

The mouse is used for all navigation. A pointing finger cursor indicates an active, clickable area. Inactive screen areas are indicated by the arrow cursor . Navigation within the presentation is accomplished by a single click of the mouse.

Notes

Each chapter exists as a separate application. When you change chapters, the chapter in use closes and the selected chapter opens.

USING THE PRESENTATION

Chapter Opener

The first screen of every chapter (except in the Introductory chapter) is a Chapter Opener.

Each chapter is organized into a series of Main Screens, which address a single topic or a group of closely related topics. The Chapter Opener screen provides a list of that chapter's Main Screens.

In addition to a Table of Contents, each Chapter Opener screen includes an Introduction, Chapter Map, and Chemical Puzzler, all indicated by red arrows.

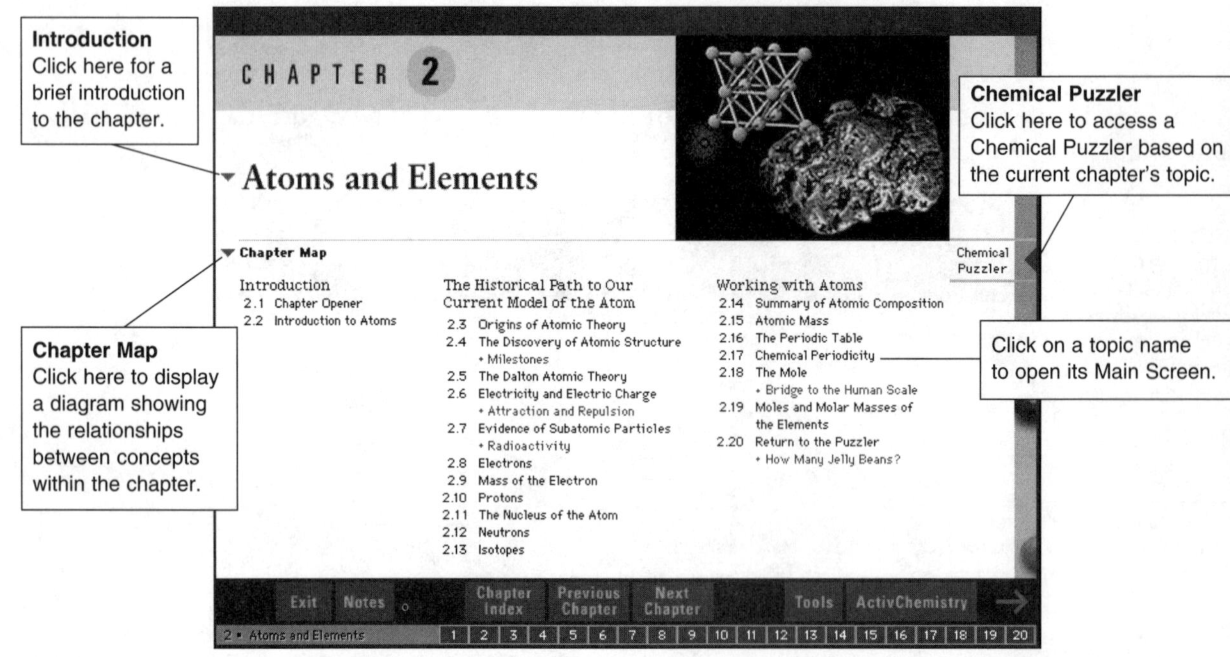

Navigation Bar Operation

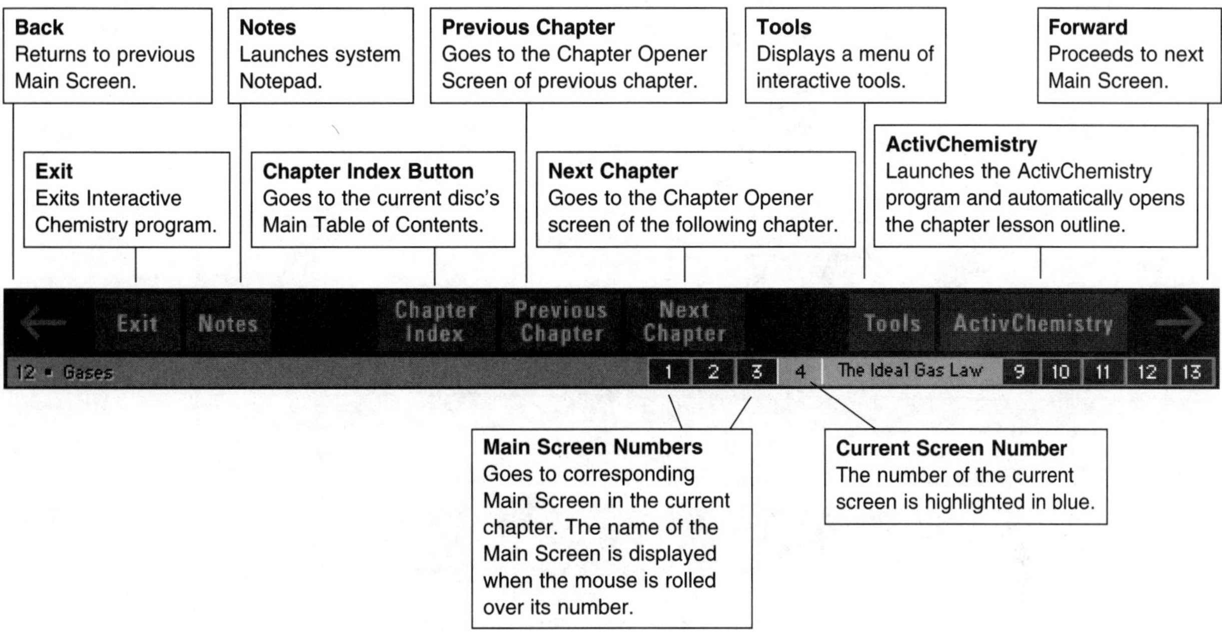

At the bottom of the screen is the navigation bar. It is used to move between screens and chapters, to access additional features, and to exit the program. The navigation bar also provides access to interactive tools, ActivChemistry, and the notebook.

Main Screens

Main Screens are accessed either from the Table of Contents on the Chapter Opener Screen or from the Navigation Bar at the bottom of each screen in a chapter. Each Main Screen includes several features such as video or audio clips, problems, sidebars, tables, or animated simulations that provide more information about the current topic. Most features can be accessed and navigated by clicking on their corresponding red arrows.

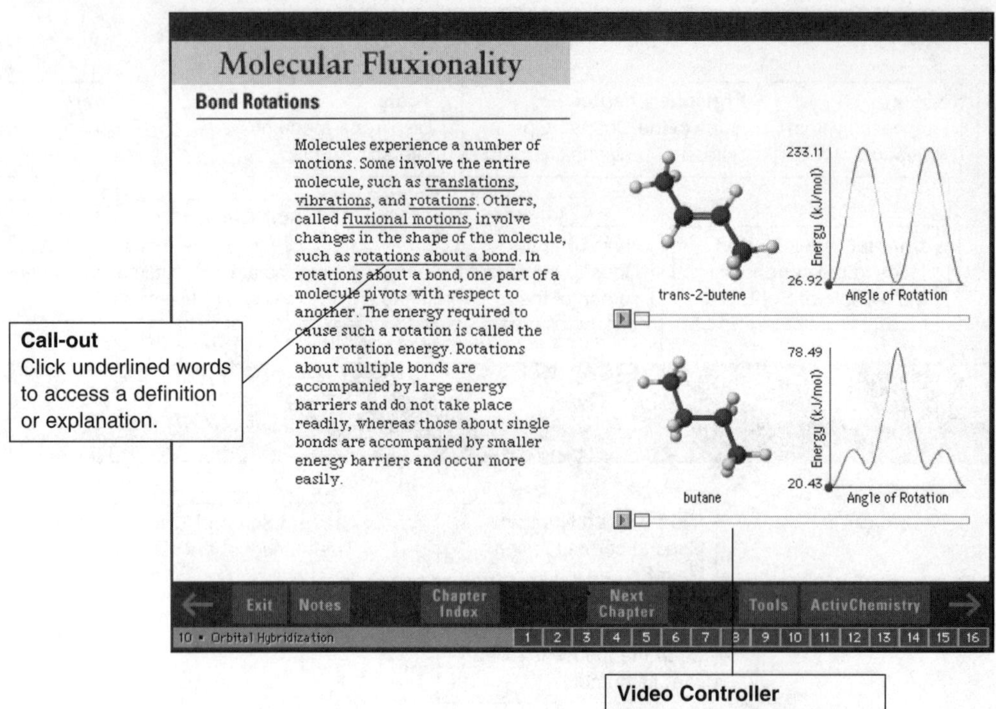
Molecular Fluxionality
Bond Rotations
Molecules experience a number of motions. Some involve the entire molecule, such as translations, vibrations, and rotations. Others, called fluxional motions, involve changes in the shape of the molecule, such as rotations about a bond. In rotations about a bond, one part of a molecule pivots with respect to another. The energy required to cause such a rotation is called the bond rotation energy. Rotations about multiple bonds are accompanied by large energy barriers and do not take place readily, whereas those about single bonds are accompanied by smaller energy barriers and occur more easily.
233.11
Energy (kJ/mol)
26.92
Angle of Rotation
trans-2-butene
78.49
Energy (kJ/mol)
20.43
Angle of Rotation
butane
Exit
Notes
Chapter Index
Next Chapter
Tools
ActivChemistry
10 • Orbital Hybridization
1 2 3 4 5 6 7 8 9 10 11 12 13 14 15 16
Call-out
Click underlined words to access a definition or explanation.
Video Controller
Push the Play/Pause button to see the video with narration. Slide the controller to move forward or backward through the clip.

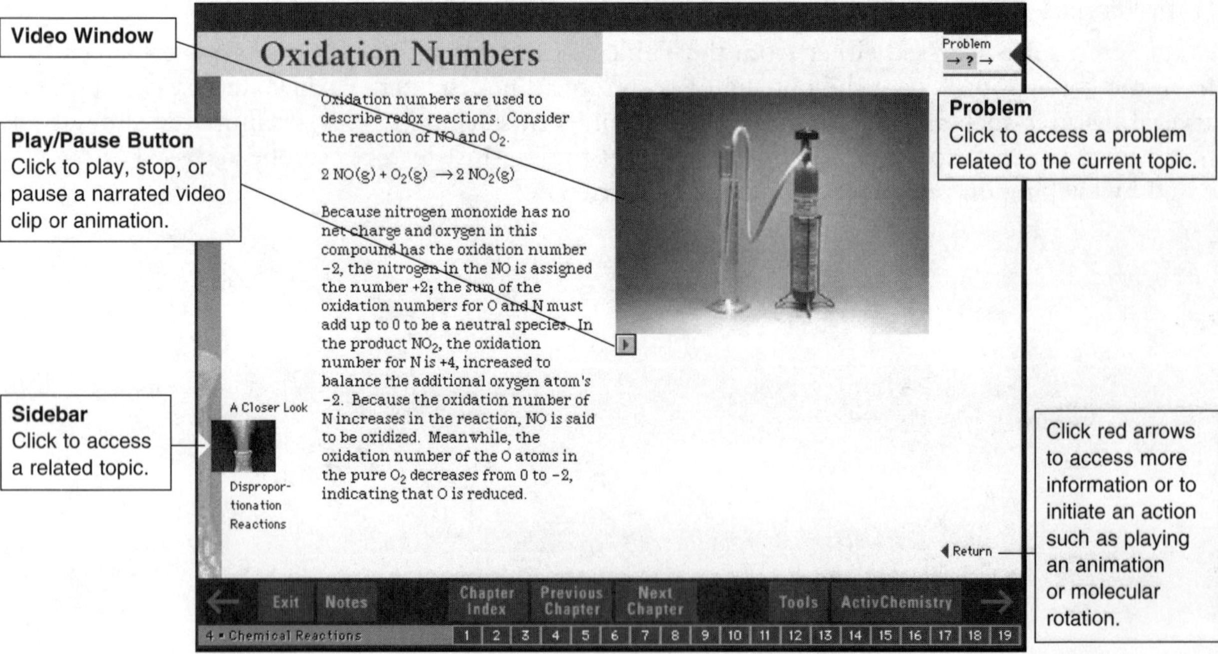
Oxidation Numbers
Problem
→ ? →
Oxidation numbers are used to describe redox reactions. Consider the reaction of NO and O2.
2 NO(g) + O2(g) → 2 NO2(g)
Because nitrogen monoxide has no net charge and oxygen in this compound has the oxidation number −2, the nitrogen in the NO is assigned the number +2; the sum of the oxidation numbers for O and N must add up to 0 to be a neutral species. In the product NO2, the oxidation number for N is +4, increased to balance the additional oxygen atom's −2. Because the oxidation number of N increases in the reaction, NO is said to be oxidized. Meanwhile, the oxidation number of the O atoms in the pure O2 decreases from 0 to −2, indicating that O is reduced.
A Closer Look
Dispropor-tionation Reactions
Return
Exit
Notes
Chapter Index
Previous Chapter
Next Chapter
Tools
ActivChemistry
4 • Chemical Reactions
1 2 3 4 5 6 7 8 9 10 11 12 13 14 15 16 17 18 19
Video Window
Play/Pause Button
Click to play, stop, or pause a narrated video clip or animation.
Sidebar
Click to access a related topic.
Problem
Click to access a problem related to the current topic.
Click red arrows to access more information or to initiate an action such as playing an animation or molecular rotation.

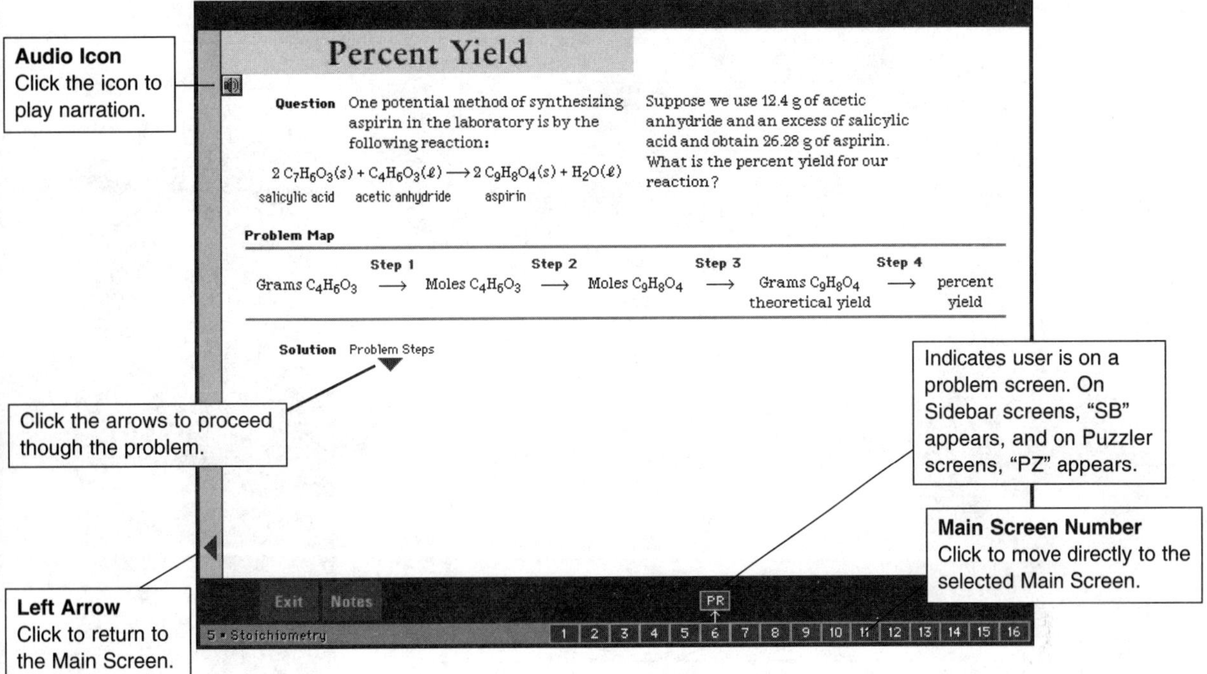

Accessing the Tools

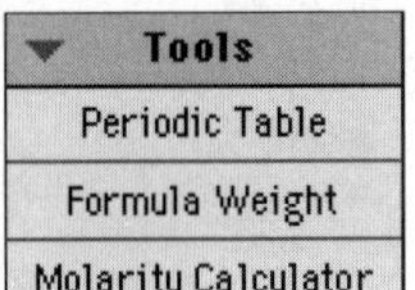

Three interactive tools are available from every Main Screen, accessed from the Tools button in the Navigation Bar. Clicking on the icon displays a pop-up menu of the tools: the Periodic Table of the Elements, the Formula Weight Calculator, and the Molarity Calculator. To access a tool, click on its name. Click on the red arrow at the top of the menu to close it.

To return to the Main Screen from the tool, click on the red arrow at the left side of the screen.

Accessing Adobe Acrobat File

Several files on the CD-ROM are saved in Adobe Acrobat format. These files include:

- the Interactive General Chemistry CD-ROM workbook file, which is found in the workbook folder, on the desktop (top level of the CD-ROM).
- The ActivChemistry User's Guide, the list of image sources for the CD-ROM, a database of chemical compounds, and a compilation of the tables found in the CD-ROM. All these files are found in the DOCS folder on the desktop.
- correlation guides that correlate the screens of the CD-ROM to five of Saunders's most popular chemistry textbooks. These files are found in the COR_GUID folder that is found in the DOCS folder on the desktop.
- instructions for use and indexes for the CAChe molecular models found on the CD. These files are found in the DOCS folder that is located in the CAChe folder on the desktop.

Accessing these files launches Adobe Acrobat Reader, a separate application with its own navigational features.

Sample page in Adobe Acrobat Reader:

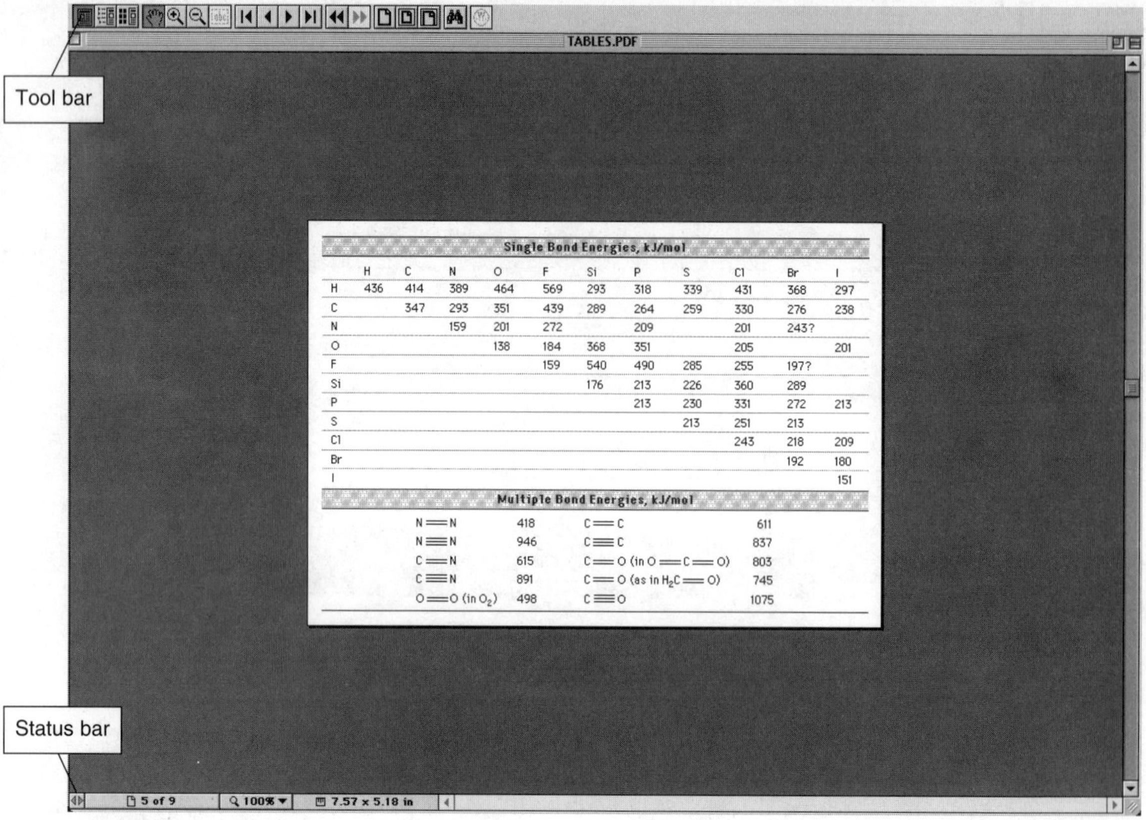

TABLES.PDF

Tool bar

Single Bond Energies, kJ/mol

	H	C	N	O	F	Si	P	S	Cl	Br	I
H	436	414	389	464	569	293	318	339	431	368	297
C		347	293	351	439	289	264	259	330	276	238
N			159	201	272		209		201	243?	
O				138	184	368	351		205		201
F					159	540	490	285	255	197?	
Si						176	213	226	360	289	
P							213	230	331	272	213
S								213	251	213	
Cl									243	218	209
Br										192	180
I											151

Multiple Bond Energies, kJ/mol

N=N	418	C=C	611
N≡N	946	C≡C	837
C=N	615	C=O (in O=C=O)	803
C≡N	891	C=O (as in H_2C=O)	745
O=O (in O_2)	498	C≡O	1075

Status bar

5 of 9 | 100% | 7.57 x 5.18 in

Tool bar:

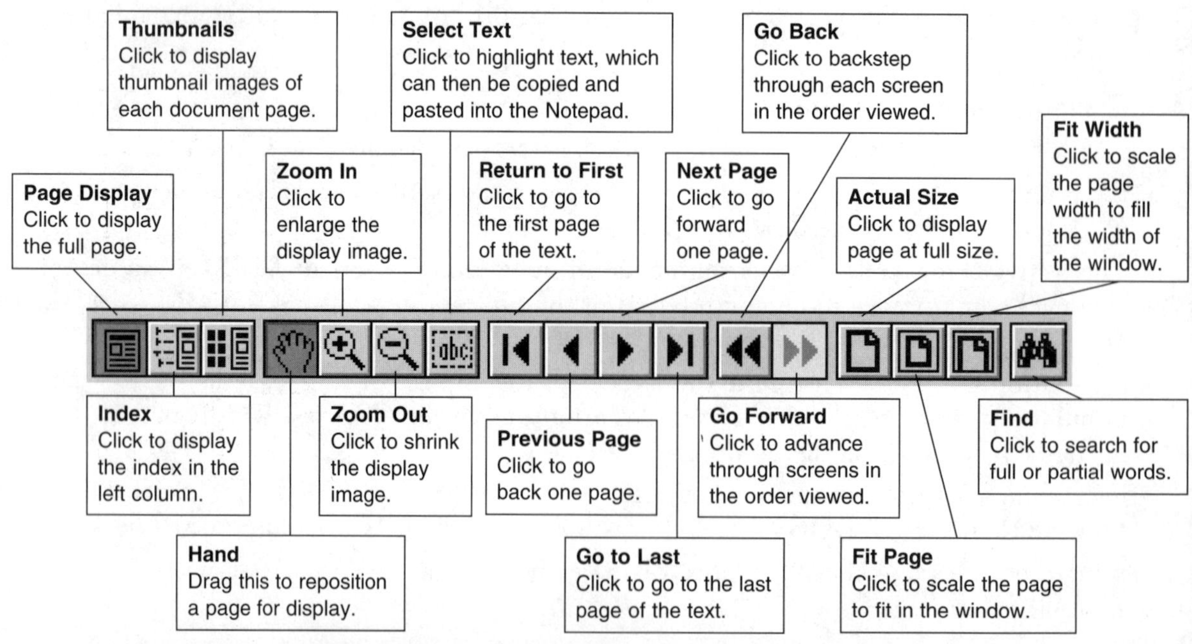

At the bottom of each screen is a Status Bar.

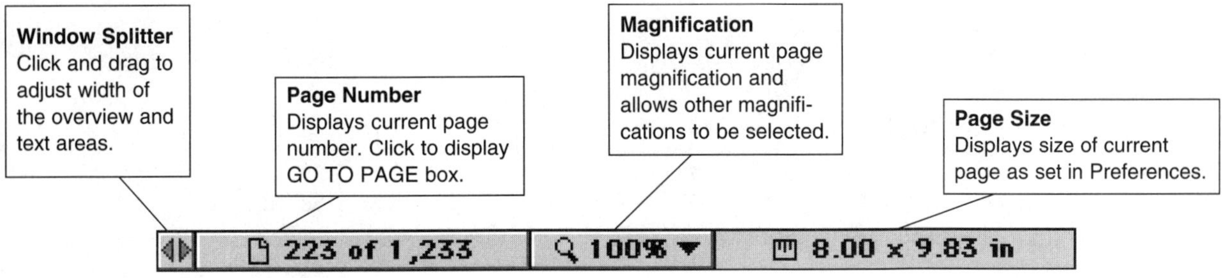

REFERENCE SECTION AND TROUBLESHOOTING GUIDE

This section is organized by

- — Navigation Functions
- — Media Access Functions
- — Utilities

NAVIGATION FUNCTIONS

Forward/Backward Arrows

Forward and reverse arrows allow the user to move to the next or previous Main Screen.

Notes

The Chapter Opener contains no reverse arrow, and no forward arrow appears on the last screen of a chapter.

Next Chapter/Previous Chapter Icons

The Next Chapter button moves the user to the Chapter Opener screen of the next chapter on the current disc.

The Previous Chapter button allows the user to move to the Chapter Opener screen of the previous chapter.

Notes

Chapters exist as separate applications. When moving between chapters, the program exits from the current chapter and launches the selected chapter. Chapter Navigation icons apply only to the current disc; for example, there is no icon to move from Chapter 9, the last chapter on Disc 1, to Chapter 10, the first chapter on Disc 2.

Chapter Index Button

The Chapter Index button allows the user to return to the Main Table of Contents for the current disc.

Notes

Disc 1 and Disc 2 have separate Tables of Contents screens listing the chapters available on the discs. The Table of Contents for chapters of the disc not in use can be viewed by clicking on the arrow at the left side of the screen.

Main Screen Numbers

8 9 10 11 Ion Sizes At the bottom of every Chapter screen is a sequence of numbers corresponding to the Main Screens in the Chapter. The current screen is indicated in blue. Moving the mouse over these numbers displays the name of each Main Screen. Clicking on a number takes the user to that Main Screen.

Notes
When the user is on a Chapter Puzzler, Problem, or Sidebar, the Main Screen Number also displays one of the following, as appropriate:

PZ Puzzler

PR Problem

SB Sidebar

Exit Icon

Exit The Exit Icon quits the presentation.

Notes
The Interactive Chemistry program can also be terminated by choosing Quit from the file menu.

MEDIA ACCESS FUNCTIONS

Red Arrows

Small red arrows, found throughout the Interactive Chemistry presentation, initiate some action such as displaying more text, rotating a molecular model, or starting an animation or video.

Video Buttons

Play

Stop

Pause

Video Controller

Video buttons allow the user to play, stop, pause, and replay video or animation clips. Some clips have a sliding controller that allows users to advance or reverse the clip.

Notes
If the video plays poorly, drops frames, or the audio tends to "cut-out," make sure you have quit any open applications to free up additional RAM. If the problem persists, try running the application on a computer with more memory or a faster CD-ROM drive.

Audio Button

Play

Audio buttons allow the user to play, stop, pause, and replay audio clips.

Notes
If your system does not play audio or the audio levels are too low or high, adjust the volume levels in your sound control panel (Macintosh) or Sound Card Software (Windows). Also check the volume level of your external speakers or headphones. If the audio tends to "skip" or "cut-out," make sure you have quit any other open applications to free up additional RAM. If the problem persists, try running the application on a computer with more memory or a faster CD-ROM drive.

UTILITIES

The Navigation Bar includes the following utilities.

Tools Button

The Tools button allows access to the Periodic Table of Elements, Formula Weight Calculator, and Molarity Calculator tools from any Main Screen on either disc of the Interactive Chemistry program.

ActivChemistry™ Button

Notes
For those running the CD-ROM's programs on monitors set in 8-bit (256-color) mode, a palette shift may occur when going from the Interactive Chemistry program to the ActivChemistry program. If this happens:

- change your monitor settings to 16-bit (thousands of colors) resolution or higher and re-launch ActivChemistry or
- quit the Interactive Chemistry program and launch ActivChemistry directly from the top level of the CD-ROM.

Notes Button

The Notes utility allows the user to record, save, and print notes while using the Interactive Chemistry program. The utility uses the standard Notepad application provided with Apple Macintosh or Microsoft Windows operating systems.

Notes
Instructions for use of the Notes utility are available in the documentation for the Notepad function in the operating system user's guide.

If clicking on the Notes button does nothing, verify that the Notepad utility is installed on the system.

On Macintosh computers, the contents of the Notepad file are saved automatically. The file can be saved and renamed as a separate document, if desired.

In Windows, the first time you access the Notepad you will be prompted to create a file called "Chemnote.txt." Choose "Yes." To keep your notes from session to session, you must save them.

ActivChemistry™

The Saunders Interactive General Chemistry CD-ROM, Version 2.5, contains ActivChemistry™ simulation software, a unique interactive multimedia package.

1 Introduction

The ActivChemistry™ simulation software provides a sophisticated desktop laboratory used to simulate chemical processes and experiments. You construct an experiment by placing its simulated components on a Workbench and linking them together. These simulated components include elements, compounds, lab equipment, meters, calculators, and more. When you run a simulation, you can observe the processes, reactions, and outcomes that demonstrate the principles of chemistry –sometimes with explosive results!

ActivChemistry also provides ActivChemistry Lessons designed to reinforce key concepts from the main Interactive General Chemistry presentation. These chemistry lessons guide you through simulated experiments and quiz you on the information presented, all to reinforce your understanding of chcmical processes.

Working with ActivChemistry

An interactive chemistry session incorporates the following main features of ActivChemistry:

Components

These are ready-made software modules that simulate different components of a chemistry experiment. You select them from a menu and use them to construct a complete simulation.

The Workbench

This is the main window where you set up experiments using components and where you run the simulations.

Lessons

This window provides interactive information and instructions for running simulations.

The Shelf

This window allows you to store useful compounds that you have created for later use.

The Periodic Tables

ActivChemistry provides large and abbreviated versions of the Periodic Table of the Elements. These provide information about the elements and also samples for use in simulations.

Studying, Teaching, and Developing New Tools

You can use ActivChemistry to:

- Work through the chemistry lessons provided in this package or lessons created by an instructor. Lessons can even be completed and turned in as homework
- Run simple experiments while reading about chemistry concepts
- Explore independently by modifying experiments, creating "what if" scenarios, and researching chemical properties and trends using the Periodic Tables of Elements provided in this package

Chemistry teachers can use ActivChemistry to:

- Easily create and display simulated experiments that demonstrate chemistry concepts
- Write lessons that include simulations of experiments, discussions, and can even include exams

All ActivChemistry users can take advantage of ActivChemistry's expandability, by creating and sharing new simulations, components, and database extensions on the World Wide Web.

2 Getting Started with ActivChemistry

There are many ways to learn to use ActivChemistry. Some people like to explore on their own and others prefer to have some guidance. This section provides step-by-step instructions for learning to use ActivChemistry features. Once you begin, however, you can explore in any direction you like, because you can always return to the basic tour.

ActivChemistry Home

When you launch ActivChemistry, the software displays a title screen, and then opens the Home window. The graphic items on the Home window are hotspots. That means you can click on them to access the main ActivChemistry features. Home window hotspots include:

ActivChemistry Lessons

Displays a menu where you can open ready-made interactive chemistry lessons prepared by chemistry professors.

Demo Lab

Displays a simulated laboratory where you can point and click to explore chemistry concepts and ways to use ActivChemistry.

Workbench

Opens a new empty Workbench window where you can begin to construct chemistry simulations.

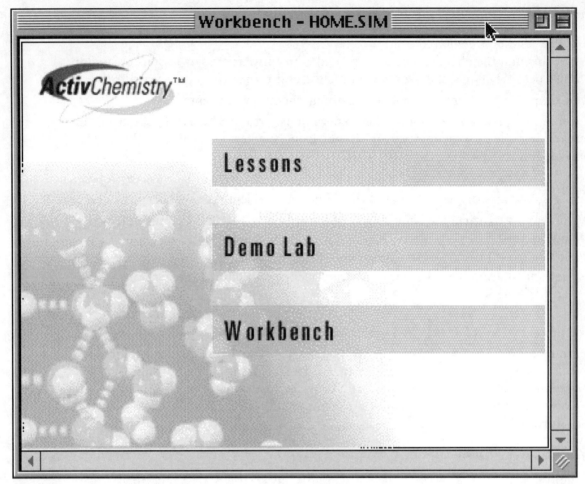

The Basic Software Tour begins at the Home window.

The Basic Software Tour

This tour provides a brief introduction to ActivChemistry. You can begin here for a basic introduction to the Workbench and Components, or with the Demo Lesson to learn about using the Lesson window, and then continue with the Demo Lab, to expand your capabilities. This tour assumes you are familiar with basic Macintosh or Windows operations such as selecting, clicking, dragging, and choosing menu items.

Creating a New Workbench

We'll begin at the Home window. On the Home window, click on the **Workbench** hotspot to open a new Workbench window and navigation bar.

Above the Workbench on your computer desktop, you will see a toolbar. For now we will focus on the Workbench tools, and simulation controls. Later we will show you how to find a complete description of toolbar functions in the Help menu.

Saving the Workbench

First, let's save the Workbench window. A Workbench window contains a single simulation.

Under the File menu, select **Save As**. In the file browser that appears, enter CHEMTOUR.sim as the name of the simulation and save it on your hard drive.

Working with Components and Tools

In ActivChemistry, you construct simulations by

- Placing components on the Workbench
- Adjusting their controls
- Adding compounds
- Linking components together
- Running the simulation to observe reactions and results

We'll build a small simulation to demonstrate this.

To place a simulation component on the Workbench, use the Place tool.

This window provides interactive information and instructions for running simulations.

This is automatically selected when you choose a component from the Components menu.

1. From the Components menu, select **Equipment**, and in the submenu that appears, select **Bottle**. When you move the cursor back to the Workbench, it changes to the Place tool.
2. Click once on the Workbench, and it places a Bottle component there.
3. Click three more times. Each time you click, a Bottle is placed on the Workbench, until you select another tool or component.

Now that you have four Bottles on the Workbench, you can use the Hand tool to arrange them.

1. Click on the **Hand** tool to select it.
2. In the Workbench window, click and drag a Bottle to select and move it.

3. Practice repositioning the bottles in any order.

But, now you have too many Bottles and need to remove some, using the Sponge tool.

1. Click on the **Sponge** tool to select it.
2. Then click once on a Bottle to remove it.

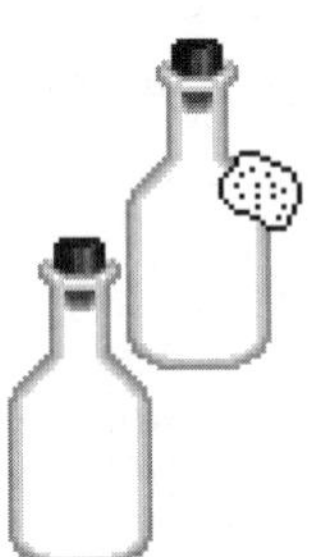

3. Remove two more bottles, so only one is left on the bench.

Building a Simulation

In this experiment, we want to observe the melting and boiling points of aluminum. We will do this by building a simulation. In addition to the Bottle, we will need three more items from the Components:

- A **Heat Pump** to heat the Bottle.
- Some **Aluminum** atoms to put in the Bottle.
- A **Thermometer** to measure the temperature at which aluminum melts and then boils.

Components in ActivChemistry can simulate most things found in a laboratory.

1. From the Components menu, in the Equipment submenu, select **Heat Pump**.
2. Then click on the Workbench to place it.
3. From the Components menu, go to the Measurement Devices submenu, then select **Thermometer**, and click on the Workbench to place it.
4. Select the Hand tool and use it to arrange the items as shown at the right.

Now we have everything we need except the aluminum. We can get this from the Tiny Periodic Table of Elements.

1. From the Windows menu, select **Tiny Periodic Table** to display it.
2. Click on the square marked **Al**, for aluminum.
3. Move the cursor back to the Workbench, and it changes to the Place tool.
4. Click once on the Workbench, and it places an Aluminum atom.

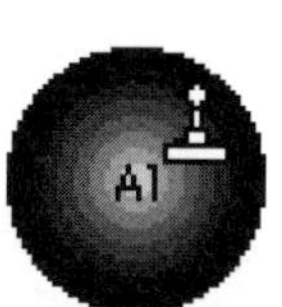

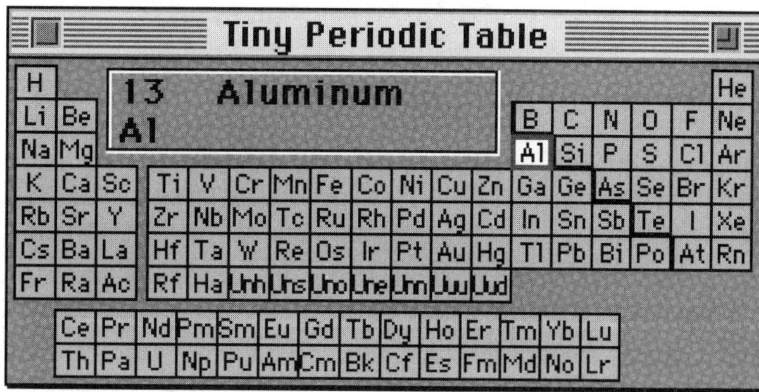

5. Close the Tiny Periodic Table.

Using Component Hotspots, Setup Windows, and Displays

Now that you have your components set up on the Workbench, let's take a look at the different component features.

You can display different component properties by selecting them from the Display menu. Try selecting different properties now. Not all components have all of the same properties.

Hotspots

Some components have hotspots that you can click on to change component settings. These can include buttons, arrows, corks, switches, etc. Most hotspots are easy to find. To highlight them as shown below, select **Help** from the Component Properties submenu of the Display menu. This also displays Help icons that you can click on to display Help dialog boxes.

Setup

Some components have Setup windows for entering precise settings. To display a Setup window, double-click on a non-hotspot area of a component, or click once on its Setup icon.

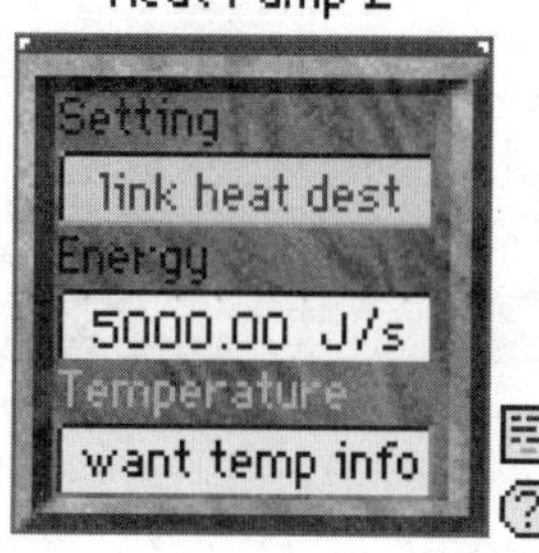

Now let's get back to building the simulation.

We can put some aluminum in the Bottle, using the Eyedropper tool.

1. Click on the **Eyedropper** tool to select it. When you move back to the Workbench, the cursor changes to an eyedropper.
2. Next, click the Eyedropper on the **aluminum** atom to sample some.
3. Move the Eyedropper over the Bottle.
4. Click on the **Bottle** and the Eyedropper drops some aluminum in it.

We didn't even have to take the cork off first! Let's do that now:

To uncork a Bottle, click on the top of the Bottle using the Hand tool.

It's time to save your work.

Select **Save Simulation** from the File menu.

We need to adjust the controls on the Heat Pump. Since we do not yet know the melting point of aluminum, set the temperature reasonably high.

1. Double-click on the **Heat Pump** to display its Setup window.
2. In the Desired Temperature field, type 4000 (Kelvins).
3. In the Energy Output Rate field, type 5000 (Joules per second).
4. Then, click **Done**. The Heat Pump displays those settings when all links are made.

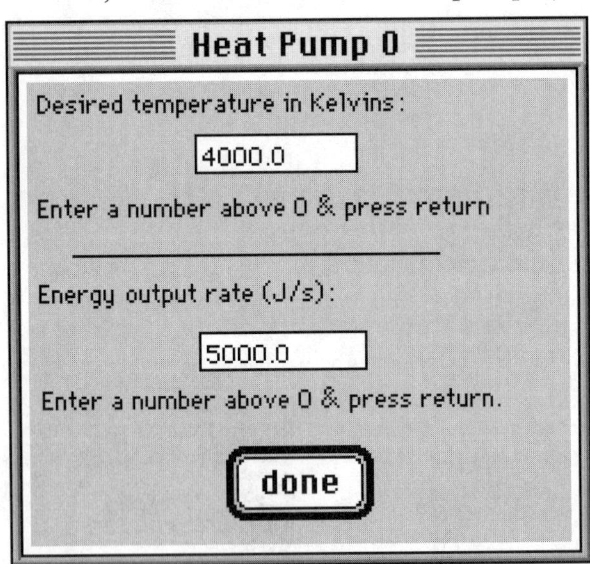

Now we can link the components together—the last step before running the simulation. To make a link, you select the Link tool,

and then click on the center of one component and drag to another component. ActivChemistry automatically implements the right kind of link, such as temperature, or material flow (solid, liquid, or gas).

1. Click on the **Link** tool to select it.
2. Click on and drag from the center of the **Heat Pump** to the center of the **Bottle** to make a heat link.
3. Click on and drag from the center of the **Bottle** to the center of the **Thermometer** to make a temperature link.
4. Click on and drag from the center of the **Thermometer** to the center of the **Heat Pump** to send a temperature reading back to the Heat Pump.

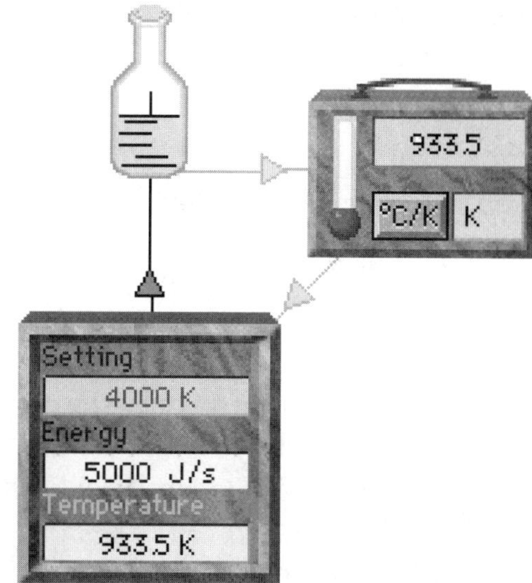

Now that the experiment is set up, you can run the simulation and observe the results. First, save your work.

Select **Save Simulation** from the File menu.

Running the Simulation

To run the simulation, you use the Simulation buttons on the toolbar.

1. Press the **Play/Stop** button.
2. Check the Thermometer component. If it is displaying °C (centigrade), click on the hotspot button using the Hand tool to change it to **K** (Kelvin).
3. Observe the experiment.

Even if nothing appears to happen right away, you can tell that the simulation is running by looking at the toolbar displays.

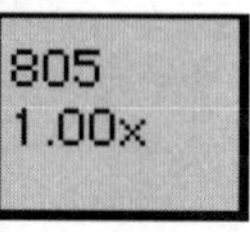

The Tick number increases, and the Planetary Atom icon orbits.

As time passes, the contents of the Bottle become hotter and hotter. When the temperature reaches 933.5 K, it plateaus briefly, and shifting lines appear in the aluminum in the Bottle. This means the aluminum has reached its melting point and is liquefied.

Then the temperature continues to rise. When the temperature reaches 2792 K, it plateaus again, and bubbles appear in the aluminum in the Bottle. This means the aluminum has reached its boiling point and is vaporizing.

Now you can stop the simulation.

Press the Play/Stop button,

again, to stop the simulation.

You have built a simulation and run it. Before we leave the simulation, let's store the components on a shelf for future use.

Creating a Shelf

A Shelf is a convenient place to store components and containers with compounds that you will use again.

1. From the File menu, select **New Shelf.** A new untitled Shelf window appears.
2. Click on the title bar of the Shelf window and drag it to the right of the Workbench window.
3. Click on the **Bottle** containing aluminum, drag it over to the shelf, and release it when the cursor looks like this.
4. Do the same with the **Thermometer** and the **Heat Pump.**

5. From the File menu, select **Save Shelf.**
6. In the file browser that appears, enter BOILAL.shf as the name of the shelf and save on your hard drive (you may wish to create an ActivChemistry folder to hold your ActivChemistry files).

You can open this shelf again when you want to construct a similar experiment.

Now you have completed the Basic Software Tour!

If you want to take a break, select **Quit** from the File menu. You will be asked to save your work.

Running the Demo Lesson

This interactive lesson demonstrates how to operate the Lesson window and run a more complicated experiment.

1. Launch the ActivChemistry software, if necessary.
2. From the Help menu, select **Demo Lesson**. This opens the Lesson window, which appears to the right of the Workbench window.
3. Begin reading the text, which will tell you how to proceed.
4. The lesson takes about 15 minutes to complete. When you are finished with the lesson, press the **Quit Lesson** button on the toolbar.

Opening Help Lessons

Now that you are familiar with lessons, you can use the lessons that appear under the Help menu to learn about ActivChemistry features in detail. You operate Help lessons just like regular lessons, but these focus on how to use ActivChemistry.

From the Help menu, select any one of the following: **Toolbar**, **Windows**, **Lessons**, **Menus**, or **Components**. (Components Help launches a simulation where you can place components on the Workbench and click on their Help buttons to access more information.)

Exploring the Demo Lab

The Demo Lab is a simulation that shows you many new ways to use ActivChemistry.

1. From the Help menu, select **Demo Lab**. Or, press the **Demo Lab** button on the Home window. This opens a simulated laboratory in the Workbench window, shown below.
2. Click on different areas of the illustration to launch simulations about different chemistry topics.
3. To highlight the active areas, select **Component Properties** in the Display menu, and in the submenu that appears, select **Help**. This outlines hotspots using green boxes.
4. A Lesson window also opens. Look there for information on how to proceed.

There is much to explore in the Demo Lab. As you work, you will become more familiar with ActivChemistry tools, components, and options. If you need to take a break, save your work and quit the program so you can return to it later.

Opening ActivChemistry Lessons

ActivChemistry Lessons are keyed to the chapters of the Saunders Interactive General Chemistry program. They illustrate key concepts in chemistry.

1. From the Help menu, select **Go Home**. This takes you back to the Home window.
2. Click the **ActivChemistry Lessons** hotspot. This displays a menu that lists the ActivChemistry Lessons included with this version of the software.
3. Click on a lesson in the list. This launches a Lesson window with steps to follow.

Note: If you launch ActivChemistry from within a chapter of the Saunders Interactive General Chemistry program, you will go to a menu of lessons specific to that chapter.

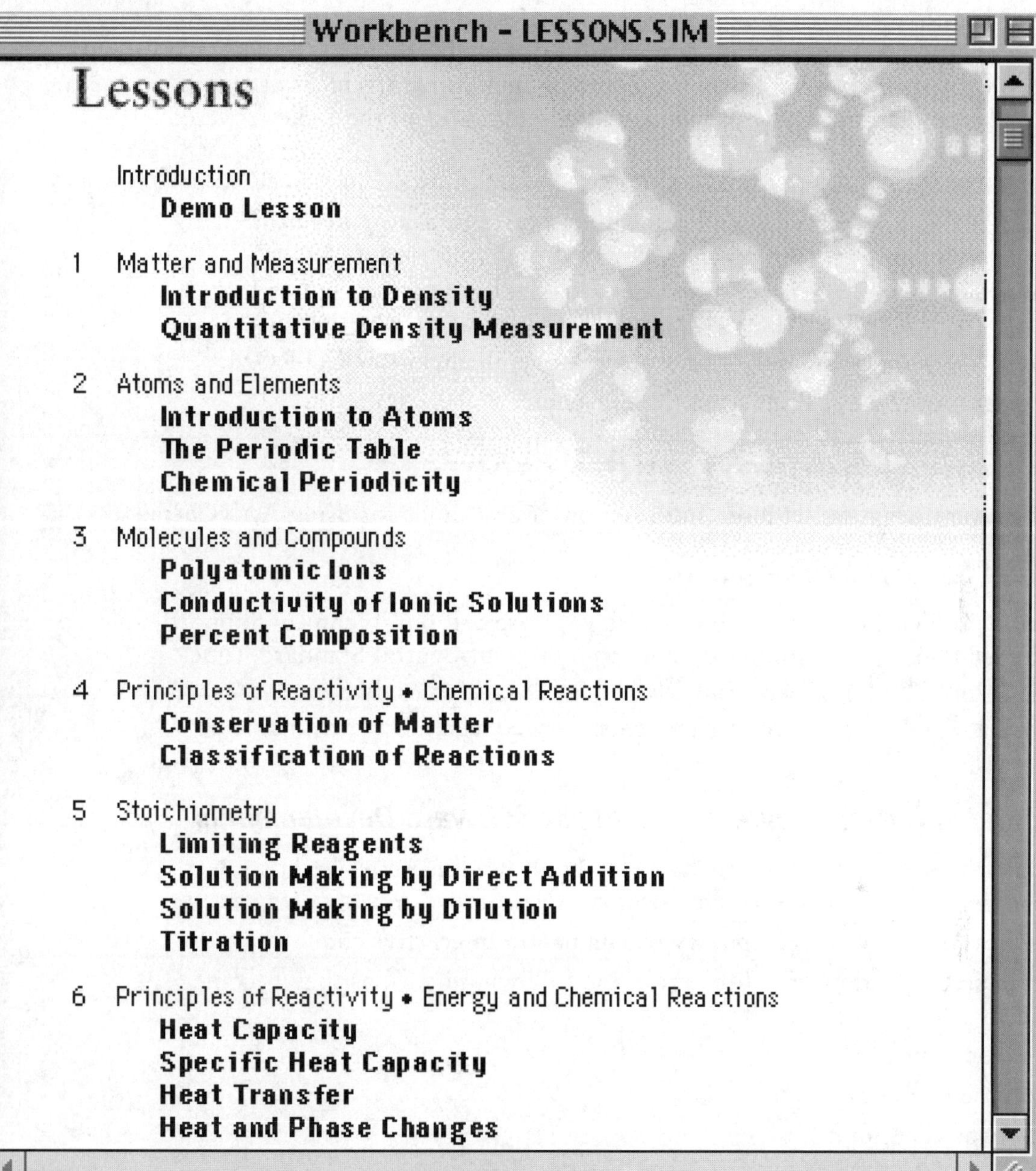

3 Viewing and Printing the User's Guide

One more resource you'll want to investigate is the *User's Guide*. It contains much more information about how to use ActivChemistry.

The ActivChemistry *User's Guide* is stored on the Saunders Interactive General Chemistry CD-ROM. You can view it and print it directly from the CD-ROM, or you can copy it to your hard drive and store it there for convenient access. It is a good idea to read the Table of Contents, to get an idea of the topics covered in the *User's Guide*.

The *User's Guide* is stored in an Acrobat page description file called **AC_GUIDE.PDF**. You can view and print it, and do special searches using the Acrobat software provided on the CD-ROM, but you cannot change the document.

- To open and view the *User's Guide*, double-click on **AC_GUIDE.PDF** in the DOCS folder on the Saunders Interactive General Chemistry CD-ROM.
- To print it out, select **Print** from the File menu.

4 Technical Support and Resources

There are several ways to get more information or assistance for using ActivChemistry

Technical Support by Telephone

If you have problems running ActivChemistry, please call our Technical Support Center at 1-800-447-9457 from 7:00 a.m. to 6:00 p.m. Central Standard Time, Monday through Friday, or visit our Web site at http://www.hbtechsupport.com.

Content-related information concerning chemistry concepts is not available at this number.

Contacting Salamander Interactive about Software Developments

The simulation engine, components, and databases used in ActivChemistry were developed by Salamander Interactive, located in San Jose, CA.

Their address on the Web is **http://www.SalamanderInteractive.com**

You can expect to find information about the following items, as they become available.

- New components
- Database upgrades
- A software development kit
- Other available products

CONTENTS

CHAPTER 1
Matter and Measurement

Screen 1.2 Matter and Measurement

1. In the photo using mercury and water as an example of differences in density, how do you know which beaker contains mercury and which contains water?

   ____

2. How could you distinguish gold from silver? A ruby from a diamond?

   ____

Screen 1.3 States of Matter

The melting point of a substance—the temperature at which the substance changes from the solid state to the liquid state—is a physical property. In the photo of solid bromine some liquid is evident. The photo was taken by freezing bromine in the freezing compartment of an ordinary home refrigerator and then moving the sample to the studio as quickly as possible. What does this tell you about the freezing point of bromine? Within what temperature range might bromine melt?

____

Screen 1.4 The Macroscopic and Microscopic Scales

Examine the scanning microscope picture of copper on the screen marked "50 million" magnification. Each "bump" represents a copper atom on the surface of a piece of solid copper. Describe how the atoms are arranged in the solid state.

____

____

Screen 1.5 Elements and Atoms

1. How many elements are presently known? __________
2. How many elements exist in nature? __________
3. Five different elements are pictured on this screen. Under "normal" conditions, which ones are solid? Which ones are liquid? Which ones are gases?

   ____

4. Examine the periodic table that is a "sidebar" to this screen. (This same *Periodic Table* database can be accessed from the Tools icon.)
   a) How many elements are metals? __________
   b) How many are nonmetals? __________

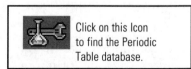

Click on this Icon to find the Periodic Table database.

c) Some elements have the properties of both metals and nonmetals, and we call them metalloids. How many elements have we placed in this category? ___________

5. Give the name or symbol in the table below, as appropriate. List the element as a metal, nonmetal, or metalloid and indicate whether it is a solid, liquid, or gas at 25 °C. To answer this question, use the *Periodic Table* database.

Name	Symbol	Element Type	State
Carbon	C	Nonmetal	Solid
	Al		
Titanium			
	K		
Silicon			
	Br		
Copper			
	F		
Xenon			
	Au		
Iron			

EXERCISE 1.1 *The Chemical Elements*

Use the *Periodic Table* database on the Tools icon to find the names of the elements Na, Cl, and Cr. Find the symbols for the elements zinc, nickel, and potassium.

Screen 1.6 Compounds and Molecules

Bring up the structure of the caffeine molecule by clicking on the red arrow next to the photo of solid caffeine. Each "ball" in this "ball and stick" model of the molecule represents an atom. (The "sticks" are chemical bonds, the "glue" that holds the molecule together.) The carbon atoms are gray, the hydrogen atoms are white, the oxygen atoms are red, and the nitrogen atoms are blue.

a) How many total atoms are there in the caffeine molecule?

____

b) How many carbon atoms? ___________ How many hydrogen atoms? ___________ How many oxygen atoms? ___________ And how many nitrogen atoms? ___________

c) Describe any features of this molecule that you think are interesting and/or important.

____

MOLECULAR MODELS ON THE CD-ROM

The caffeine molecule you see here was generated by a computer program from CAChe Scientific/Oxford Molecular. Many of the molecules seen in this interactive presentation (and many more) are also available in a folder marked *Models* within the CAChe folder, which you can access by first leaving this presentation. These models can be manipulated so you can determine much more about their structure. See the instructions for its use in Appendix A of this Workbook and in the Read Me file in the CAChe folder.

Screen 1.7 The Kinetic Molecular Theory

Describe at least two differences between solids, liquids, and gases.

Screen 1.8 Density

1. Explain why there is a difference in the density of the brick and the Styrofoam block shown on this screen, considering their structure on the molecular scale.

   ---

2. In your experience, which is denser—milk or rock? (Imagine a cup of milk and a cup of pebbles.)

   ---

3. The densities of the elements at 25 °C are listed in the *Periodic Table* database. Select (click on) an element in the table to view a summary of its properties. Answer the following questions:

   Click on this Icon to find the Periodic Table.

   a) What is the density of aluminum at 25 °C? ________ g/cm^3
   b) What is the density of uranium at 25 °C? ________ g/cm^3
   c) Look up the densities of lead, platinum, and mercury at 25 °C. Which is the densest? ________________
      Which is the least dense? ________________

4. The densities of common compounds are given in the *Database of Compounds* (CMPB_DB.PDF), located in the Tools folder on this disc, which can be accessed from the desktop. Answer the following questions:
   a) Acetone is a common solvent. Is it a liquid or solid at room temperature (25 °C)? ____________________________ (Its density at 25 °C is 0.78 g/cm^3.)
   b) Which is more dense at 25 °C, acetone or liquid mercury? ________________

EXERCISE 1.2 *Using Density*

The density of air is 1.12×10^{-3} g/cm^3. What is the volume of air, in cubic centimeters, of exactly 100 g of air?

Screen 1.9 Density on the Submicroscopic Scale

1. The models of gold and silver show how the atoms are arranged in the solid. (The lines do not depict "chemical bonds." They are there to better define the arrangement in space of the atoms.) This model shows the simplest repeating unit of atoms. Give a verbal description of this arrangement.

   ---

2. Look up potassium and chromium in the *Periodic Table* database.
   a) How are their atoms arranged in the solid?

   ____

   b) Which is denser?

   ____

   c) Which element has the heavier atoms? How do you know?

   ____

   ____

Screen 1.10 Temperature

1. You have a beaker of water at 5 °C and another beaker containing an equal amount of water at 95 °C.
   a) In which beaker are the water molecules moving faster? ____________________
   b) Describe what happens when the water in the two beakers is mixed.

   ____

   c) What do you suppose the temperature of the new water sample might be?

   ____

2. a) Which temperature is higher, 72 °F or 65 °C? __________
   b) Which temperature is higher, 72 °F or 310 K? __________
   c) Which temperature is lower, -25 °C or 200 K? __________
3. Would you feel hot or cold if the temperature was 35 °C? ____________________

EXERCISE 1.3 *Temperature*

Liquefied nitrogen boils at 77 K. What is this temperature in Celsius degrees?

> You should be aware that the animation of the reaction of hydrogen and oxygen is highly simplified. It is meant to represent only the fact that H_2 and O_2 molecules have combined to form water and that the number of H and O atoms is the same before and after the reaction.

Screen 1.11 Chemical Change

The video on this screen shows the reaction of hydrogen gas (in the balloon) with oxygen.

1. What is the source of the oxygen for this reaction?

   ____

2. Why is the hydrogen-filled balloon floating in air?

   ____

3. The narrator states that energy is released. What evidence is there for this energy release? In what form is this energy?

   ____

Screen 1.12 Chemical Changes on the Molecular Scale

1. Examine the video of the reaction of elemental phosphorus and chlorine and the animation of this reaction. Note that the animation is not meant to illustrate *how* the reaction occurs, but only that the number of P and Cl atoms is the same before and after reaction.
   a) How many P_4 molecules are there in the beginning? ____________
   b) How many Cl_2 molecules are there in the beginning? ____________
   c) How many PCl_3 molecules are formed? ____________
2. Is the fact that there are the same number of P atoms and Cl atoms before the reaction as after the reaction unique to this reaction or will this be observed generally in all chemical reactions?

   ____

Screen 1.13 Mixtures and Pure Substances *and* Screen 1.14 Separation of Mixtures

Review Screens 1.13 and 1.14 and then examine the test tubes on Screen 1.13. Describe how you might separate cobalt(II) hydroxide from the solution in the middle tube and how you could separate dissolved potassium chromate from water in the test tube on the right.

____

____

____

Screen 1.15 Units of Measurement

The speedometer in your car is probably marked off in two scales (such as the one on this screen). One scale is in units of miles per hour (white numerals on this screen) and the other is kilometers per hour (yellow numerals on this screen).

1. Does either scale use SI units? How should the speedometer be marked if SI units were used?

   ____
2. Use the speedometer to arrive at the approximate relation between kilometers and miles. That is, how many kilometers are there, approximately, per mile?

   ____

Screen 1.16 The Metric System *and* Screen 1.17 Using Numerical Information

1. On Screen 1.15 you arrived at an approximate relation between kilometers and miles. Can you confirm that relation in the table on Screen 1.16?

   ____
2. On Screen 1.15 you learned that a toilet uses 1.0 gallon per flush or 3.8 L per flush. Use the SI units for volume (and the fact that there are 4 quarts in a gallon) to confirm that 1.0 gallon is indeed equivalent to 3.8 L.

   ____

EXERCISE 1.4 *Using Numerical Information*

a) A platinum sheet is 2.50 cm on a side and has a mass of 1.656 g. The density of platinum is 21.45 g/cm^3. What is the thickness of the platinum sheet in millimeters?

b) A standard bottle has a volume of 750 mL. How many liters does this represent?

Screen 1.18 Return to the Puzzler

How is metallic iron separated from breakfast cereal? Can you think of an application of this technique in the recycling of household waste?

Additional Questions

1. A man can run a mile in 4 minutes. Can he run a kilometer in 4 minutes? Support your answer with a calculation.

2. In the supermarket, a pound of hamburger costs $2.50. If you saw a kilogram of hamburger meat for $2.50, would you buy it? Support your answer with a calculation.

Thinking Beyond

1. Examine the pictures of copper on Screen 1.4. In the photo of the magnification of the sample by 50 million, the scale along the side of the photo is in nanometers, where 1 nanometer is equivalent to 1×10^{-9} meters. Give a very rough approximation of the diameter of a copper atom in nanometers and meters. Compare your estimate with the radius of a copper atom in the *Periodic Table* database.

2. Heat capacity is a physical property. Specific heat capacity is the amount of heat energy required to raise the temperature of 1 gram of a substance by one degree. Which has the greater heat capacity, 1 gram of water or 1 gram of aluminum?

3. Compare the physical and chemical properties of water and gasoline.

____

____

____

STUDY QUESTIONS

1. Give the name of each of the following elements:

   (a) C (c) Cl (e) Mg

   (b) Na (d) P (f) Ca

2. Give the symbol for each of the following elements:

   (a) lithium (d) silicon

   (b) titanium (e) cobalt

   (c) iron (f) zinc

3. Ethylene glycol, $C_2H_6O_2$, is a liquid that is the base of the antifreeze you use in the radiator of your car. It has a density of 1.1135 g/cm^3 at 20 °C. If you need 500. mL of this liquid, what mass of the compound, in grams, is required?
4. Water has a density at 25 °C of 0.997 g/cm^3. If you have 500. mL of water, what is its mass in grams? In kilograms?
5. The "cup" is a volume widely used by cooks in the United States. One cup is equivalent to 225 mL. If 1 cup of olive oil has a mass of 205 g, what is the density of the oil (in grams per cubic centimeter)?
6. Many laboratories use 25 °C as a standard temperature. What is this temperature in kelvins?
7. Make the following temperature conversions:

°C	K
(a) 16	______
(b) ______	370
(c) -40	______

8. Solid gallium has a melting point of 29.8 °C. If you hold this metal in your hand, what is its physical state? That is, is it a solid or a liquid? Explain briefly.
9. The average lead pencil, new and unused, is 19 cm long. What is its length in millimeters? In meters?
10. A standard U.S. postage stamp is 2.5 cm long and 2.1 cm wide. What is the area of the stamp in square centimeters? In square meters?
11. A typical laboratory beaker has a volume of 800. mL. What is its volume in cubic centimeters? In liters? In cubic meters?
12. A new U.S. quarter has a mass of 5.63 g. What is its mass in kilograms? In milligrams?

13. Complete the following table of masses.

Milligrams	*Grams*	*Kilograms*
______	0.693	______
156	______	______
______	______	2.23

14. A standard sheet of notebook paper is 8 1/2 × 11 in. What are these dimensions in centimeters? What is the area of the paper in square centimeters?
15. The smallest repeating unit of a crystal of common salt is a cube with an edge length of 0.563 nm. What is the volume of this cube in cubic nanometers? In cubic centimeters?
16. The platinum-containing cancer drug cisplatin contains 65.0% platinum. If you have 1.53 g of the compound, how many grams of platinum can be recovered from this sample?
17. You can identify a metal by carefully determining its density. An unknown piece of metal, with a mass of 29.454 g, is 2.35 cm long, 1.34 cm wide, and 1.05 cm thick. Is the element nickel, titanium, zinc, or tin?
18. The density of pure water is given below at various temperatures.

t, °C	*d* (g/cm^3)
5	0.99999
15	0.99913
25	0.99707
35	0.99406

Use these data, and the plotting program on this disc, to predict the density of water at 40 °C.

CHAPTER 2
Atoms and Elements

Screen 2.2 Introduction to Atoms

1. Name an important force that is involved in chemistry.

2. A sulfur atom has 16 protons and 16 neutrons in the nucleus and 16 electrons around the nucleus. Would a sulfur atom be larger or smaller than an oxygen atom?

   Briefly discuss your reasoning.

Screen 2.3 Origins of Atomic Theory

1. Where does the word "atom" come from?

2. What is a difference between the "planetary" model of the atom and the more modern version?

Screen 2.5 The Dalton Atomic Theory

1. What laws must theories of matter satisfy?

2. View the photos of magnesium reacting with oxygen in the air.
   a) What do you observe as this chemical reaction proceeds? Include changes you see in the nature of the magnesium, the nature of the product, and evidence for energy evolved in the process.

   b) It is noted that it is "seemingly strange" that the magnesium gains weight as it burns. Why might this seem strange?

c) Can you name a process where mass is lost by a substance as it burns?

d) How does Dalton's theory explain the "seemingly strange" result that the magnesium gains weight as it burns?

Screen 2.6 Electricity and Electric Charge

1. What is an important principle of operation of the electroscope?

2. How does the attraction and repulsion of electric charges apply to the structure of the atom? Go back to Screen 2.2 and think again about the structure of the atom.

Screen 2.7 Evidence of Subatomic Particles

1. What elements were discovered by Madame Curie?
2. What types of radiation are emitted by radioactive elements, and what are their distinguishing characteristics?

Screen 2.8 Electrons

1. How could you demonstrate that an electron is a negatively charged particle?

2. What property of the electron was measured in Thomson's experiment?

3. How are electrons related to the beta rays emitted by radioactive elements?

Screen 2.9 Mass of the Electron

1. What is the connection between Thomson's experiments with electrons and Millikan's experiment.

Screen 2.10 Protons

1. How are postively charged particles generated in the "canal-ray" experiment?

2. Why does hydrogen have the largest charge-to-mass ratio of all the elements studied?

Screen 2.11 The Nucleus of the Atom

1. Why are so few particles deflected as they pass through the foil?

2. Why are even fewer particles deflected back in the direction from which they came?

Screen 2.12 Neutrons

1. How does the mass of a neutron compare with the masses of the proton and the electron?

Screen 2.13 Isotopes *and* Screen 2.14 Summary of Atomic Composition

1. Atomic masses are given in "atomic mass units," amu. On what is this scale based?

2. Define the atomic number of an atom or element.

3. What determines the mass number of an atom?

4. Give the symbol for each of the following:
   a) an iron atom with 28 neutrons ____________
   b) a copper atom with 36 neutrons ____________
5. How many electrons, protons, and neutrons are there in an atom of bromine-79, ^{79}Br?

EXERCISE 2.1 *The Structure of the Atom*

a) What is the mass number of a copper atom with 34 neutrons? ____________
b) How many protons, neutrons, and electrons are there in a ^{59}Ni atom? ____________
c) Silicon nas three isotopes with 14, 15, and 16 neutrons, respectively. What are the symbols for these isotopes?

Screen 2.15 Atomic Mass

1. What two factors determine the atomic mass of an element?

____

____

2. a) The atomic mass of lithium is __________
   b) What are the common isotopes of lithium? (Hint: Go the the *Periodic Table* database located in the Toolbox. Next, look up lithium and then look for its stable isotopes.) Lithium has two stable isotopes: __________
   c) Which is more abundant? __________ Explain your choice.

   ____

3. Magnesium has three stable isotopes, ^{24}Mg, ^{25}Mg, and ^{26}Mg.
   a) What is the average atomic mass of magnesium? __________
   b) One of these isotopes has an abundance or 78.99% and another is 10.00% abundant. What is the abundance of the third? __________
   c) What is the mass number of the most abundant isotope of the three? __________

EXERCISE 2.2 *Isotopes*

Verify that the atomic mass of chlorine is 35.45 amu, given the following information:

^{35}Cl mass = 34.96885 amu, percent abundance = 75.77

^{37}Cl mass = 36.96590 amu, percent abundance = 24.23

Screen 2.16 The Periodic Table

As described by Professor DiSalvo on Screen 1.6, the periodic table is one of the most important tools of chemistry. It is important to know the terminology of the table. The periodic table on this screen outlines that terminology. To find the name of each element, and other information about an element, open the *Periodic Table* database in the Toolbox.

1. Name an element in Group 2A. __________
2. Name an element in the third period. __________
3. What element is in the second period in Group 4A? __________
4. What element is in the third period in Group 6A? __________
5. What halogen is in the fifth period? __________
6. What alkaline earth element is in the third period? __________
7. What noble gas element is in the fourth period? __________
8. Name the nonmetal in Group 6A and the fourth period. __________
9. Name the elements in Group 1B. __________

   Are these elements metals, metalloid, or nonmetals? __________
10. What metalloid is in Group 3A? __________

Screen 2.17 Chemical Periodicity

1. After viewing the videos of the reactions of lithium, sodium, and potassium with water, how would you describe the relative speeds of the reactions?

2. a) Give a detailed description of the reaction of sodium with water.

   b) Speculate on the reason that sodium forms a little ball when it reacts with water.

   c) Why do you think the sodium ball scoots around on top of the water?

3. Like the alkali metals, most of the alkaline earth metals react with water to give hydrogen gas (and the metal hydroxide). Magnesium reacts with water, but only if the water is heated. Based on the trend in reactivity of the alkali metals, what do you suppose will be the relative reactivity of calcium?

Screen 2.18 The Mole

Suppose that 56 popcorn kernels occupy a volume of 10.0 cm^3. If you spread 1.00 mol of popcorn kernels over the state of Iowa, how deep would the pile of kernels be? Calculate the depth in units of meters and kilometers. (Iowa covers 55,875 square miles or 1.45×10^5 km^2.)

Screen 2.19 Moles and Molar Masses of the Elements

1. Which has more mass, 2.00 mol of sulfur or 2.00 mol of copper? (Click on the element in the plot at the upper right to see the masses of three different mole quantities.)

   ____

2. Examine the plot at the upper right of the screen and think about the relative slopes of the lines on the plot. (The slope of a line in this plot is the change in number of moles divided by the change in mass. For example, the slope for copper is 0.015 mol/g.)

   a) What is the slope of the line for sulfur? ______________

   b) What is the slope of the line for lead? ______________

   c) Why is the slope of the line for lead less than that of sulfur? ______________

   d) What would be the slope of the line for silver (Ag) in a similar plot? ______________

3. Based on the different volumes occupied by 1.00 mol of sulfur and 1.00 mol of copper, which do you suppose is denser?

   a) ______________________________

   Use the *Periodic Table* database tool (click on the Tools icon to see this tool) to look up the densities of these elements to check your answer.

   b) density (sulfur) = ________ g/cm^3 c) density (copper) = ________ g/cm^3

4. You are given a copper cylinder that has a radius of 0.70 cm and is 4.2 cm long. The density of copper is 8.96 g/cm^3. If the copper atoms in the bar occupy 74% of the available space in the bar (that is, 26% of the bar is empty space between the copper atoms), what is the radius of a copper atom? (Volume of a cylinder = $\pi r^2 l$, where l = length and r = radius. Volume of a sphere = $4\backslash 3\pi r^3$.)

EXERCISE 2.3 *Mass and Moles*

a) What is the mass, in grams, of 2.5 mol of aluminum?

b) How many moles are represented by 454 g of sulfur?

c) What is the volume of a piece of platinum that contains 1.0×10^{24} atoms?

Thinking Beyond

1. Using the *Periodic Table* database, note the trend in the sizes of atoms on moving from element number 3, lithium, to element number 10, neon. How can a neon atom be smaller, even though it contains about three times as much matter as a lithium atom?
2. Use the *Periodic Table* database to examine the number of isotopes of each element. Do you notice any trend?
3. Notice that isotopes of the first 20 elements have roughly the same number of neutrons as protons in their nucleus, but much larger atoms (such as ^{238}U) have many more neutrons than protons. Propose an explanation.

STUDY QUESTIONS

1. Give the mass number of each of the following atoms: (a) beryllium with 5 neutrons, (b) titanium with 26 neutrons, and (c) gallium with 39 neutrons.
2. Give the complete symbol ($^{A}_{Z}X$) for each of the following atoms: (a) sodium with 12 neutrons; (b) argon with 21 neutrons; and (c)gallium with 39 neutrons.
3. How many electrons, protons, and neutrons are there in an atom of (a) calcium-40, ^{40}Ca; (b) tin-119, ^{119}Sn; and (c) plutonium-244, ^{244}Pu?
4. Fill in the blanks in the table (one column per element).

Symbol	^{45}Sc	^{33}S	_____	_____
Number of protons	_____	_____	8	_____
Number of neutrons	_____	_____	9	31
Number of electrons in the neutral atom	_____	_____	_____	25

5. Radioactive americium-241 is used in household smoke detectors and in bone mineral analysis. Give the number of electrons, protons, and neutrons in an atom of americium-241.
6. Verify that the atomic mass of lithium is 6.94 amu, given the following information:
   ^{6}Li, exact mass = 6.015121 amu, percent abundance = 7.50%
   ^{7}Li, exact mass = 7.016003 amu, percent abundance = 92.50%
7. Gallium has two naturally occurring isotopes, ^{69}Ga and ^{71}Ga, with masses of 68.9257 amu and 70.9249 amu, respectively. Calculate the percent abundances of these isotopes of gallium.
8. Calculate the number of grams in
   (a) 2.5 mol of boron
   (b) 0.015 mol of oxygen
   (c) 1.25×10^{-3} mol of iron.
   (d) 653 mol of helium.
9. Calculate the number of moles represented by each of the following:
   (a) 127.08 g of Cu
   (b) 20.0 g of calcium
   (c) 16.75 g of Al
   (d) 0.012 g of potassium
   (e) 5.0 mg of americium

10. A chunk of sodium metal, Na, if thrown into a bucket of water, produces a dangerously violent explosion. If 50.4 g of sodium is used, how many moles of sodium does that represent? How many atoms? (Caution: Sodium is very reactive with water. The metal should be handled only by a knowledgeable chemist.)
11. If you have a 35.67-g piece of chromium metal on your car, how many atoms of chromium are in the piece?
12. What is the average mass of one copper atom?
13. What is the average mass of one silver atom?
14. A piece of copper wire is 7.6 m long (about 25 ft) and has a diameter of 2.0 mm. If copper has a density of 8.92 g/cm^3, how many moles of copper and how many atoms of copper are in the piece of wire?

CHAPTER 3
Molecules and Compounds

Screen 3.2 Elements that Exist as Molecules

1. Of the common elements, how many exist as diatomic molecules? ________________

2. a) What are the allotropes of carbon? ________________

   b) What are the structural characteristics of the carbon allotropes? Are there structural similarities or differences among these allotropes? (As you examine the structures on this screen recall that each "ball" in these "ball and stick" models represents an atom. Models of these allotropes are also available in the *Models* folder on this disc.)

3. What are the similarities and differences in the elemental forms of C, N, O, and F?

Screen 3.3 Molecular Compounds

1. When examining the molecular models, recall that carbon atoms are gray, the hydrogen atoms are white, the oxygen atoms are red, and the nitrogen atoms are blue.

   a) How many total atoms are there in the ethanol molecule? ________________

   b) Is the ammonia molecule flat? ________________ How would you describe its structure?

2. Find the following molecules in the *Models* folder or directory, and describe the structure of each one.

   a) BF_3 (boron trifluoride)

   b) C_6H_6 (benzene)

   c) C_2F_4 (tetrafluoroethylene, the building block of Teflon polymers)

Screen 3.4 Representing Compounds

After examining all of the ways of representing ethanol (click on the "More" button on the initial screen), look up acetone (C_3H_6O) in the folder sequence CAChe/ORGANIC/ALDE_KET. (Acetone is a common organic compound and is widely used as a solvent.)

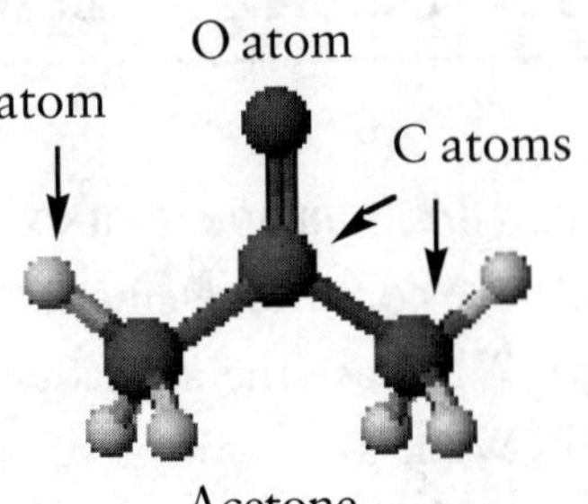

Acetone

1. Draw an expanded structural formula for the molecule.

2. Examine the molecular model (both as ball-and-stick and space-filling models) and describe what you see. Is any fragment of the acetone molecule similar to a part or parts of the ethanol molecule?

Screen 3.5 Binary Compounds of the Nonmetals

1. There are a large number of compounds that contain just two kinds of elements. In particular, there are a large number of compounds—the hydrocarbons—that contain just carbon and hydrogen. The models are contained in the *Models* folder within the CAChe folder on the top level of the CD-ROM. Look up the following:

   a) Ethane (ORGANIC/ALKANES sub-folders) Molecular Formula = ________________
   Expanded structural formula

   b) Propene (ORGANIC/ALKENES sub-folders) Molecular Formula = ________________
   Expanded structural formula

   c) Acetylene (INFRARED sub-folders) Molecular Formula = ________________
   Expanded structural formula

d) Describe the similarities or differences among these structures.

2. Many compounds, such as those just above, have historical names. On the screen depicting some of these compounds are two compounds with N and O that have such names. One can, however, name them systematically. Name each of the compounds below:

Formula	Name
N_2O	
NO	
NO_2	
N_2O_4	

Screen 3.6 Alkanes: A Class of Compounds

Like some of the compounds you explored on Screen 3.5, the alkanes are hydrocarbons. That is, they are binary compounds of C and H.

1. Examine the structures of methane, propane, and pentane, and recall the structure of ethane from your work with Screen 3.5. Based on these structures, draw the expanded structural formula for each of the following (both of which are in the *Models* folder):

   a) butane

   b) hexane

2. Each of the alkanes (except methane) is a chain of carbon atoms with H atoms attached. Are these carbon-atom chains straight or bent? How would you describe them?

3. How do the C atoms at the end of the chain differ from the C atoms in the "interior" of the chain?

Screen 3.7 Ions—Cations and Anions

1. When a magnesium atom forms an ion, are electrons gained or lost? ______________

   How many? ______________ What happens to the size of the atom when it forms an ion?

   ____

2. When a fluorine atom forms an ion, are electrons gained or lost? ______________

   How many? ______________ What happens to the size of the atom when it forms an ion? ____

Screen 3.7SB A Closer Look—Charges on Ions

After examining the "Partial Periodic Table" showing typical ion charges, answer the following questions:

1. Do any of the elements in this table form both a cation and an anion? Which element(s)?

   ____

2. In Group 3A the only element listed is aluminum. The other elements in the group (except B, a metalloid) have a similar chemical behavior, however. Does gallium form an anion or cation? ______________ What is the charge on this ion? ______________ How many protons does the ion have? ______________ How many electrons?

   ____

3. Common metals and nonmetals of the main groups of the periodic table gain or lose electrons to achieve the same number of electrons as the nearest succeeding or preceding noble gas. Complete the following table:

Element	Ion Formed	Number of Electrons Gained or Lost	Total Number of Electrons in the Ion	Noble Gas with the Same Number of Electrons
Li	Li^+	-1	2	He
Ca				
		+3		Ar
S				
		-2	36	

4. a) The transition metals form cations. In the process they generally lose at least two electrons, but may lose three or more, to form a cation.

Element	Ion Formed	Number of Electrons Lost	Total Number of Electrons in Cation	Element with the same Number of Electrons
Ti	Ti^{2+}	2	20	Ca
Fe	Fe^{3+}			
Cr		3		
Co			24	
Cu				Co
		2		V

b) Are these elements like the main group elements in that they form an ion with the same number of electrons as the preceding or succeeding noble gas?

EXERCISE 3.1 *Charges on Ions*

Predict possible charges for ions formed from (1) K, (2) Se, (3) Be, (4) V, (5) Co, and (6) Cs.

Screen 3.8 Polyatomic Ions

It is very important for you to know the names and formulas of common polyatomic ions. Complete the table below.

Formula	Name
NO_3^-	nitrate ion
	phosphate ion
CO_3^{2-}	
	acetate ion
$H_2PO_4^-$	
	sulfate
ClO^-	
	perchlorate ion
HCO_3^-	

Screen 3.9 Coulomb's Law

Use the tool on this screen to explore Coulomb's law, which controls many of the atom-atom and molecule-molecule interactions important to us in chemistry.

Place the movable sphere (an ion) about 2 cm to the left of the stationary sphere (another ion). Give the left sphere a charge of +1 and the right sphere a charge of -1. (Use the red arrows to raise or lower the ion charge as appropriate.) The gray arrows are pointing at one another indicating that the ions are attracted to one another. The size of the arrow indicates the magnitude of the force of attraction.

1. Give one of the spheres a charge of 0. What do you observe?

2. Raise the charge on one ion by one (e.g., increase the cation charge from +1 to +2). Now what do you observe about the magnitude of the interaction between the ions?

3. a) Raise the ion charge by yet another unit. (The cation is now +3 and the anion is -1, for example.) What do you observe about the magnitude of the attraction?

   b) What is the relation between ion charge and the magnitude of the attraction between ions?

4. Now give one ion a charge of +2 and the other ion a charge of -2. How does the force of attraction compare with the force of attraction when one ion is +1 and the other is -1?

5. Give one ion a charge of +2 and the other a charge of -2, and place them about 2 cm apart. Move the left ion until the ions are about 4 cm apart (this is the limit of movement of the left ion). Compare the magnitude of the force of attraction at the two positions.

   Move the ions so they are about 1 cm apart. Compare the magnitude of forces with the previous positions of the ion.

   What can you conclude about the relation between the ion positions and the magnitude of their attraction?

6. Finally, give each ion the same charge, say +2. What do you observe?

7. Summarize the relation between ion charge, the distance between ions, and the forces of attraction and repulsion.

Screen 3.10 Ionic Compounds

1. Examine the video of the reaction of sodium with chlorine.

   a) The sodium metal for the reaction is removed from a beaker, where chips of the metal are clearly seen under a clear, colorless liquid. Is this liquid water? Explain. (Refer to Screen 2.17.)

   ____

   b) What happens to the sodium in the course of the reaction? Does it gain or lose one or more electrons? ____________ If so, how many electrons? ____________ What is the final form of the sodium in sodium chloride? ____________

   c) Is energy involved in this reaction? Is there evidence for the evolution of energy? What is that evidence?

   ____

2. Complete the following table:

Cation	Anion	Formula of Ionic Compound
Na^+	Cl^-	NaCl
Ba^{2+}	Cl^-	
Ca^{2+}	SO_4^{2-}	
		$Mg_3(PO_4)_2$
K^+	CO_3^{2-}	
Ni^{2+}	Br^-	
		$FeCl_3$
		$Fe(OH)_2$
Al^{3+}	NO_3^-	
Na^+	HPO_4^{2-}	
		$Ba(CH_3CO_2)_2$

EXERCISE 3.2 *Formulas of Ionic Compounds*

a) Identify the constituent ions, and give the number of each, in each of the following ionic compounds: (a) NaF, (b) $Cu(NO_3)_2$, and (c) $NaCH_3CO_2$.

b) Iron is a transition metal and so can form ions with at least two different charges. Write formulas for the compounds formed between iron and chlorine.

c) Write formulas for all of the neutral ionic compounds that can be formed by combining the cations Na^+ or Ba^{2+} with the anions S^{2-} or PO_4^{3-}.

Screen 3.10SB A Closer Look—The Ionic Crystal Lattice

What is the relationship between the shape of the macroscopic crystal of KBr (potassium bromide) and the submicroscopic arrangement of its K^+ and Br^- ions?

Screen 3.11 Properties of Ionic Compounds

1. Why is the melting temperature of MgO (2826 °C) so much higher than the melting temperature of $MgCl_2$ (714 °C) or NaCl (801 °C)?

2. Speculate on the reason that a sharp blow to the crystal of KBr in the video led to a very clean break. Think about the submicroscopic structure of the solid and what would happen if a layer of positive and negative ions is displaced relative to a neighboring layer.

3. Glass shatters into irregular pieces when struck, whereas the salt crystal broke cleanly. What can this tell us about the submicroscopic structure of glass relative to a salt like NaCl or KBr?

Screen 3.12 Solubility of Ionic Compounds

The video on this screen shows that potassium permanganate, $KMnO_4$, dissolves readily in water to give K^+ and MnO_4^- ions. Examine the data in the table below and speculate on the relation between the solubility of ionic compounds and the types of ions in the compound. (Hint: Think back to your experiments with Coulomb's law on Screen 3.9.)

Ionic Compound	Cation	Anion	Solubility in Water (g compound/100 mL) at 25 °C
NaCl	Na^+	Cl^-	39.1
$NaNO_3$			180
$MgCl_2$			72.7
MgO			0.08
$AlCl_3$			69.9
Al_2O_3			Insoluble

Screen 3.13 Naming Ionic Compounds

Use the naming "tool" on this screen to complete the table below. (To use the "tool," first click on the "More" button. A list of cations and anions will appear. Choose one from each column to learn the name and formula of the resulting compound.)

Names of Common Ionic Compounds

Cation	Anion	Name	Formula
NH_4^+	Cl^-	Ammonium chloride	NH_4Cl
NH_4^+	CO_3^{2-}		
Ba^{2+}	Cl^-		
Ag^+	NO_3^-		
Ag^+	Cl^-		
NH_4^+	S^{2-}		
Pb^{2+}	S^{2-}		
Fe^{3+}	O^{2-}		
Na^+	O^{2-}		

Use the table of anions and cations to complete the following table:

Names of Common Ionic Compounds

Cation	Anion	Name	Formula
		Ammonium bromide	
Ba^{2+}			BaS
	Cl^-	Iron(II) chloride	
	F^-		PbF_2
Al^{3+}	CO_3^{2-}		
		Iron(III) oxide	
			$LiClO_4$
		Aluminum phosphate	
	Br^-	Lithium bromide	
			$Ba(NO_3)_2$
Al^{3+}		Aluminum oxide	
		Iron(III) carbonate	

EXERCISE 3.3 *Formulas and Names of Ionic Compounds*

1. Give the formula for each of the following ionic compounds.
   a) Ammonium nitrate ____________
   b) Cobalt(II) sulfate ____________
   c) Nickel(II) cyanide ____________
   d) Vanadium(III) oxide ____________
   e) Barium acetate ____________
   f) Calcium hypochlorite ____________
2. Name each of the following compounds:
   a) $MgBr_2$ ____________
   b) Li_2CO_3 ____________
   c) $KHSO_3$ ____________
   d) $KMnO_4$ ____________
   d) $(NH_4)_2S$ ____________
   f) $CuCl$ and $CuCl_2$ ____________

Screen 3.14 Hydrated Compounds

What are some uses of hydrated compounds?

____

Screen 3.15 Compounds, Molecules, and the Mole *and* Screen 3.16 Using Molar Mass

MOLAR MASS CALCULATOR

The molar mass of any compound can be calculated using the "Molarity Calculator" located in the Tools area.

1. Examine the graphs of molar mass versus mass on Screen 3.16.
   a) What are the units for the slope of the line: g/mol or mol/g? ____________
   b) Why is the slope of the plot for water so much greater than that for K_2CrO_4, potassium chromate?

      ____
   c) How would the slope for PBr_3 (Screen 3.15) compare with that for K_2CrO_4?

      ____
   d) Compare the slope of the line for CO_2 with that for H_2O and that for K_2CrO_4.

      ____
2. Calculate the molar mass of the compounds pictured on Screen 3.15. Note that the water in hydrated compounds is included in the molar mass.

Name	Formula	Molar Mass
Sodium chloride		
Potassium dichromate		
Nickel(II) chloride dihydrate	$NiCl_2 \cdot 6H_2O$	237.7 g/mol
Copper(II) sulfate pentahydrate		
Cobalt(II) chloride hexahydrate		

EXERCISE 3.4 *Molar Mass*

a) Calculate the molar mass of limestone, $CaCO_3$, and caffeine, $C_8H_{10}N_4O_2$.

b) If you have 454 g of $CaCO_3$, how many moles does this represent?

c) To have 2.50×10^{-3} mol of caffeine, how many grams must you have?

Screen 3.17 Percent Composition

1. Which has the greater percentage of N in 25.0 g of the compound: NO or NO_2?

____

2. Which has the greatest percentage of H in 12 g of compound: CH_4, C_2H_6, or C_5H_{12}?

____

3. a) You analyze a sample of an organic compound found in wine and determine it is 32.0% carbon. Is the compound succinic acid ($C_4H_6O_4$), malic acid ($C_4H_6O_5$), or tartaric acid ($C_4H_6O_6$)?

____

____

b) Which of the above organic compounds has the greatest percentage by mass of carbon? ____________ Which has the lowest percentage? ____________

EXERCISE 3.5 *Percent Composition*

Express the composition of each compound below in terms of the mass of each element in 1.00 mol of compound and the weight percent of each element:

Compound	Mass in 1.00 mol	Weight Percent
NaCl, sodium chloride		
C_8H_{18}, octane		
$(NH_4)_2SO_4$, ammonium sulfate		

Screen 3.18 Determining Empirical Formulas *and* Screen 3.19 Determining Molecular Formulas

Complete the table below.

Name	Molecular Formula	Molar Mass	Empirical Formula
Ethane	C_2H_6	30.1 g/mol	CH_3
	N_2O_4		
Benzene		78.1 g/mol	CH
Naphthalene		128.2 g/mol	C_5H_4
Tartaric acid	$C_4H_6O_6$		

EXERCISE 3.6 *Empirical and Molecular Formulas*

a) A boron hydride consists of 78.14% B and 21.86% H; its molar mass is 27.7 g/mol. What are the empirical and molecular formulas of this compound?

b) Chlorine gas is combined with 0.532 g of titanium metal, and 2.108 g of Ti_xCl_y is formed. What is the empirical formula of this compound?

Thinking Beyond

Here is a question that we'll answer in Chapter 8, but let's figure out what we can now. On Screen 3.7 there is an animation of the formation of the Mg^{2+} cation and the F^- anion.

1. Why do you suppose the magnesium atom decreases in size when it loses two electrons to become a cation?

2. Why do you suppose the fluorine atom increases in size when it gains an electron to become an anion?

STUDY QUESTIONS

1. Predict the charges on the ions formed by aluminum and selenium.
2. What charges are most commonly observed for ions of the following elements?
   - (a) Magnesium
   - (b) Zinc
   - (c) Iron
   - (d) Gallium
3. Give the symbol, including the correct charge, for each of the following ions:
   - (a) Strontium ion
   - (b) Aluminum ion
   - (c) Sulfide ion
   - (d) Cobalt(II) ion
   - (e) Titanium(IV) ion
   - (f) Hydrogen carbonate ion
   - (g) Perchlorate ion
   - (h) Ammonium ion

4. Cobalt is a transition metal and so can form ions with at least two different charges. Write the formulas for the compounds formed with oxide ions and each of two different cobalt ions.
5. Which of the following are correct formulas for compounds? For those that are not, give the correct formula.
   (a) AlCl (c) Ga_2O_3
   (b) NaF_2 (d) MgS
6. Solid magnesium oxide melts at 2826 °C. This property, combined with the fact that it is not an electric conductor, makes it an ideal heat insulator for electric wires in cooking ovens and toasters (see the photo). In contrast, solid NaCl melts at the relatively low temperature of 801 °C. What is the formula of magnesium oxide? Suggest a reason for its melting temperature being so much higher than that of NaCl.
7. Name each of the following ionic compounds:
   (a) K_2S (c) $(NH_4)_3PO_4$
   (b) $NiSO_4$ (d) $Ca(ClO)_2$
8. Give the formula for each of the following ionic compounds:
   (a) Ammonium carbonate
   (b) Calcium iodide
   (c) Copper(II) bromide
   (d) Aluminum phosphate
   (e) Silver(I) acetate
9. Write the formulas for all the compounds that can be made by combining each of the cations with each of the anions listed here. Name each compound formed.

Cations	*Anions*
K^+	CO_3^{2-}
Ba^{2+}	Br^-
NH_4^+	NO_3^-

10. Calculate the molar mass of each of the following compounds:
    (a) Fe_2O_3, iron(III) oxide
    (b) BF_3, boron trifluoride
    (c) N_2O, dinitrogen monoxide (laughing gas)
    (d) $MnCl_2 \cdot 4H_2O$, manganese(II) chloride tetrahydrate
    (e) $C_6H_8O_6$, ascorbic acid or vitamin C
11. How many moles are represented by 1.00 g of each of the following compounds?
    (a) CH_3OH, methanol
    (b) Cl_2CO, phosgene, a poisonous gas
    (c) NH_4NO_3, ammonium nitrate
    (d) $MgSO_4 \cdot 7H_2O$, magnesium sulfate heptahydrate (Epsom salt)

12. Acrylonitrile, C_2H_3CN, is used to make acrylic plastics. If you have 2.50 kg of acrylonitrile, how many moles of the compound are present?

13. An Alka-Seltzer tablet contains 324 mg of aspirin ($C_9H_8O_4$), 1904 mg of $NaHCO_3$, and 1000. mg of citric acid ($C_6H_8O_7$). (The last two compounds react with each other to provide the "fizz," bubbles of CO_2, when the tablet is put into water.)

    (a) Calculate the number of moles of each substance in the tablet.

    (b) If you take one tablet, how many molecules of aspirin are you consuming?

14. Sulfur trioxide, SO_3, is made in enormous quantities by combining oxygen and sulfur dioxide, SO_2. The trioxide is not usually isolated but is converted to sulfuric acid. If you have 1.00 kg of sulfur trioxide, how many moles does this represent? How many molecules? How many sulfur atoms? How many oxygen atoms?

15. Calculate the mass percent of each element in the following compounds.

    (a) PbS, lead(II) sulfide, galena

    (b) C_3H_8, propane, a hydrocarbon fuel

    (c) $CoCl_2 \cdot 6H_2O$, a beautiful red compound

16. Vinyl chloride, CH_2CHCl, is the basis of many important plastics (PVC) and fibers.

    (a) Calculate the molar mass.

    (b) Calculate the mass percent of each element in the compound.

    (c) How many grams of carbon are there in 454 g of vinyl chloride?

17. The empirical formula of succinic acid is $C_2H_3O_2$. Its molar mass is 118.1 g/mol. What is its molecular formula?

18. Acetylene is a colorless gas that is used as a fuel in welding torches, among other things. It is 92.26% C and 7.74% H. Its molar mass is 26.02 g/mol. Calculate the empirical and molecular formulas.

19. Nitrogen and oxygen form an extensive series of oxides with the general formula N_xO_y. One of them is a blue solid that comes apart, reversibly, in the gas phase. It contains 36.84% N. What is the empirical formula of this oxide?

20. Mandelic acid is an organic acid composed of carbon (63.15%), hydrogen (5.30%), and oxygen (31.55%). Its molar mass is 152.14 g/mol. Determine the empirical and molecular formulas of the acid.

21. Cacodyl, a compound containing arsenic, was reported in 1842 by the German chemist Robert Wilhelm Bunsen. It has an almost intolerable garlic-like odor. Its molar mass is 210 g/mol, and it is 22.88% C, 5.76% H, and 71.36% As. Determine its empirical and molecular formulas.

22. Elemental sulfur (1.256 g) is combined with fluorine, F_2, to give a compound with the formula SF_x, a very stable, colorless gas. If you have isolated 5.722 g of SF_x, what is the value of x?

CHAPTER 4
Principles of Reactivity: Chemical Reactions

Screen 4.2 Chemical Equations

1. Observe the reaction of Al and Br_2.
   a) Is elemental bromine a solid, liquid, or gas? ______________

      What color is the element? ______________

   b) As the reaction proceeds, what do you see? How do you know a reaction has occurred? Is there any evidence for energy changes in the reaction?

      ____

      ____

Screen 4.3 The Law of Conservation of Matter

1. Observe the decomposition of mercury(II) oxide and the accompanying animation.
   a) What visual evidence is there for the decomposition of the oxide?

      ____

   b) What happens to the oxygen evolved in the decomposition?

      ____

   c) In the animation, how many molecules of oxygen are evolved? __________ How many atoms of mercury? __________ How many "molecules" of HgO must have decomposed? __________ Does this agree with the balanced chemical equation for the decomposition?

      ____

2. In the reaction of aluminum and bromine on Screen 4.2,

   $2\ Al(s) + 3\ Br_2(\ell) \rightarrow Al_2Br_6(s)$

   how many molecules of bromine do you need to react completely with 2000 atoms of Al? __________ How many molecules of Al_2Br_6 can be obtained? __________

EXERCISE 4.1 *Stoichiometric Coefficients*

Consider the reaction of iron and oxygen.

$4\ Fe(s) + 3\ O_2(g) \rightarrow 2\ Fe_2O_3(s)$

a) What are the stoichiometric coefficients in this equation?

   ____

b) If you were to use 8000 atoms of iron, how many molecules of O_2 are required to consume the iron completely?

   ____

Screen 4.4 Balancing Chemical Equations

1. Reactions that form oxides

   Observe the video of the reaction of elemental phosphorus. Here the phosphorus is removed from a beaker full of water and placed on a laboratory spoon in the air. (The phosphorus used here is called "white" phosphorus, even though its color is often closer to yellow.)

   a) Is elemental phosphorus a solid, liquid, or gas? ___________

   b) What can you conclude about the relative tendency of phosphorus to react with water and air? Why is the element stored under water?

   ____

2. Combustion reactions

   Observe the animation of the reaction of propane and oxygen. How does this animation illustrate the principle of the conservation of matter?

   ____

EXERCISE 4.2 *Writing and Balancing Chemical Equations*

a) Pentane burns in air to give carbon dioxide and water. Write the balanced equation for this reaction.

b) Tetraethyllead, $Pb(C_2H_5)_4$, burns in air to give PbO(s), CO_2(g), and H_2O(g). Write the balanced equation for this reaction.

Screen 4.5 Compounds in Aqueous Solution

1. Observe the video of the reaction of magnesium and aqueous HCl (hydrochloric acid).

   a) Is magnesium a metal, nonmetal, or metalloid? Why did you classify it as you did?

   ____

   b) When $MgCl_2$, magnesium chloride, is placed in water, it is a "strong electrolyte." Suggest an experiment you could do to prove that the compound readily dissociates into ions in solution.

   ____

   ____

   c) Hydrogen gas is evolved in this reaction. Suggest an experiment you could do to prove that the gas is H_2 and not O_2 or CO_2.

   ____

   ____

2. Describe the chief difference between strong and weak electrolytes and between electrolytes and nonelectrolytes.

   ____

3. List some strong electrolytes other than aqueous HCl.

   ____

Screen 4.6 Solubility

This screen allows you to combine 10 different cations with 10 different anions to "prepare" as many as 100 different compounds. For each compound "prepared," you can learn its solubility in water, its name, and its formula. Let us use this "tool" to develop some general guidelines that will aid in predicting the water solubility of many other ionic compounds.

1. Explore the following pairs of ions:

Water Solubility of Common Ionic Compounds

Cation	Anion	Solubility	Name	Formula
NH_4^+	Cl^-	Soluble	Ammonium chloride	NH_4Cl
NH_4^+	CO_3^{2-}			
Ba^{2+}	Cl^-			
Ba^{2+}	CO_3^{2-}			
Ag^+	NO_3^-			
Ag^+	Cl^-			
NH_4^+	S^{2-}			
Pb^{2+}	S^{2-}			
Fe^{3+}	O^{2-}			
Na^+	O^{2-}			

2. Based on the results in the table above (and other experiments if you have the time), what general statements concerning the aqueous solubility of ionic compounds can be made?

____

____

3. Do your general statements agree with the "Summary" to Screen 4.6?

____

____

____

4. Complete the following table using the information on Screen 4.6 as a guide:

Water Solubility of Common Ionic Compounds

Cation	Anion	Solubility	Name	Formula
			Ammonium bromide	
Ba^{2+}		Not soluble		
	Cl^-	Soluble		
	Cl^-	Not soluble		
	CO_3^{2-}	Soluble		
			Iron(II) oxide	
				$LiClO_4$
			Ammonium phosphate	
	Br^-	Not soluble		
				$Ba(NO_3)_2$
Al^{3+}		Soluble		
Al^{3+}		Not soluble		
			Iron(III) carbonate	

EXERCISE 4.3 *Solubility of Ionic Compounds*

Predict whether each is soluble or not soluble in water.

a) KNO_3 ____________ c) CuO ____________

b) $CaCl_2$ ____________ d) $NaCH_3CO_2$ ____________

Screen 4.7 Acids

1. Describe the difference between a strong acid and a weak acid. Use HNO_3 and CH_3CO_2H (acetic acid) as examples. (In the case of acetic acid it is the H^+ attached to the O atom that is "lost" when the acid ionizes.)

____

____

2. Write a balanced chemical equation that describes the ionization of nitric acid.

____

Screen 4.8 Bases

1. Describe the difference between a strong base and a weak base. Use KOH and NH_3 (ammonia) as examples.

____

____

Screen 4.9 Less Obvious Acids and Bases

1. Write a balanced chemical equation that describes the behavior of MgO, magnesium oxide, in water. Is it an acidic oxide or a basic oxide?

____

____

EXERCISE 4.4 *Acids and Bases*

For each of the following, indicate whether you expect an acidic or basic solution when the compound is placed in water.

a) SeO_2 ______________

b) MgO ______________

c) P_4O_{10} ______________

Screen 4.10 Equations for Reactions in Aqueous Solution—Net Ionic Equations

1. Consider the following questions after watching the video of the addition of a colorless solution of lead(II) nitrate to a solution of yellow potassium chromate.
   a) What is the color of the precipitate in this reaction? ______________
   Give its name ______________ and formula ______________
   b) Explain why K^+ is considered a "spectator ion" and can be eliminated from the chemical equation described on this screen.

   ____

   c) Is there another spectator ion in this equation? If so, what is its identity?

   ____

2. The net ionic equation for the reaction on the screen is

   $CrO_4^{2-}(aq) + Pb^{2+}(aq) \rightarrow PbCrO_4(s)$

   a) Are there the same number of atoms of each kind on both sides of the equation?
   b) Is the equation balanced for electric charge? That is, is the total charge on the left the same as on the right? Why or why not?

   ____

3. Consider the equation for the reaction of magnesium carbonate with hydrochloric acid.

   $MgCO_3(s) + 2\ HCl(aq) \rightarrow MgCl_2(aq) + CO_2(g) + H_2O(\ell)$

   What is (are) the spectator ion(s) in this equation?

   ____

EXERCISE 4.5 *Net Ionic Equations*

Balance each of the following equations, and write net ionic equations:

a) $BaCl_2(aq) + Na_2SO_4(aq) \rightarrow BaSO_4(s) + NaCl(aq)$

b) Lead(II) nitrate reacts with potassium chloride to give lead(II) chloride and potassium nitrate.

Screen 4.11 Types of Reactions in Aqueous Solution

After reviewing each reaction on this screen, go back to Screen 4.1SB, the Chemical Puzzler. Identify each reaction below as one of the four types of reactions. In each case try to justify your choice.

1. $Pb(NO_3)_2(aq) + 2\ KI(aq) \rightarrow PbI_2(s) + 2\ KNO_3(aq)$

Reaction type ______________________________

Reason:

Net ionic equation:

2. $HCl(aq) + NaOH(aq) \rightarrow NaCl(aq) + H_2O(\ell)$

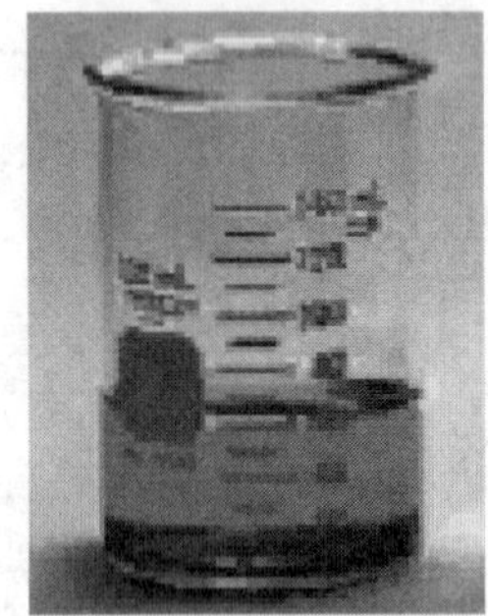

Reaction type ______________________________

Reason:

Net ionic equation:

3. $Zn(s) + 2\ HCl(aq) \rightarrow ZnCl_2(aq) + H_2(g)$

Reaction type ______________________________

Reason:

Net ionic equation:

4. $Cu(s) + AgNO_3(aq) \rightarrow Cu(NO_3)_2(aq) + 2\ Ag(s)$

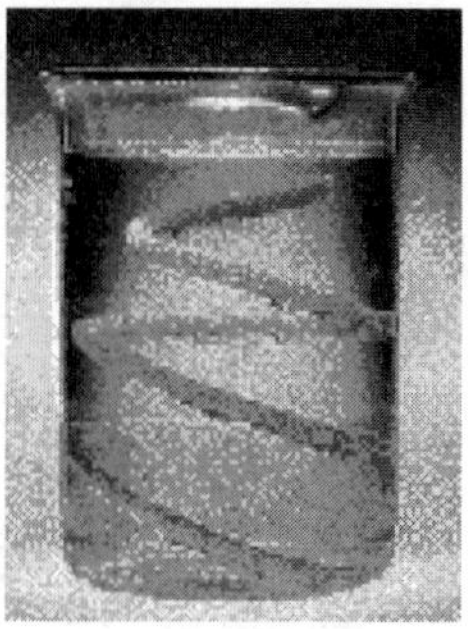

Reaction type ______________________________

Reason:

Net ionic equation:

Screen 4.12 Precipitation Reactions

1. One precipitation reaction described here has the net ionic equation:

   $Ag^+(aq) + Cl^-(aq) \rightarrow AgCl(s)$

   A few drops of a solution of a soluble silver compound are added to a solution of a soluble chloride-containing compound.

   a) What water-soluble silver compound might have been used? ____________________

   b) What water-soluble compound that contains chloride ion might have been used?

   ____

2. The other precipitation reaction described here has the net ionic equation:

   $Fe^{3+}(aq) + 3\ OH^-(aq) \rightarrow Fe(OH)_3(s)$

   A few drops of a solution of a soluble compound containing the hydroxide ion are added to a solution of a soluble iron(III) compound.

   a) What water-soluble iron(III) compound might have been used? ____________________

   b) What is its color? ____________________

   c) What water-soluble compound that contains hydroxide ion might have been used?

   ____________________ (Hint: Refer to the screens on acids and bases.)

3. After studying the example on Screen 4.12PR, decide what reaction you could use to prepare each of the following:

   a) $BaSO_4$

   ____

   b) Ag_3PO_4

   ____

   c) CdS

   ____

EXERCISE 4.6 *Precipitation Reactions*

When aqueous silver nitrate is mixed with an aqueous solution of potassium chromate, K_2CrO_4, does a precipitate form? If so, write the balanced equation for the reaction and then write the net ionic equation.

Screen 4.13 Acid-Base Reactions

1. Examine the bottom video on Screen 4.13SB.
   a) A probe from the pH meter is inserted into one of the solutions. Is this a solution of an acid or base? ______________________ What is in the other beaker, a solution of an acid or base? ______________________ Describe how you arrived at your answer.

   ____

   b) How can you tell when sufficient acid (or base) has been added to consume the base (or acid) completely?

   ____

EXERCISE 4.7 *Acid-Base Reactions*

Write the balanced, overall equation and the net ionic equation for the reaction of magnesium hydroxide with hydrochloric acid.

Screen 4.14 Gas-Forming Reactions

1. a) Name four important gases that may be produced in gas-forming reactions:

   ____

   b) Can you think of another common gas (a component of air) that could also be produced in a gas-forming reaction? ______________________ (This is the gas produced in an air bag in an automobile. See Chapter 12.)
2. Carbon dioxide can be produced by decomposing (by heating) some metal carbonates. Write a balanced chemical equation for the decomposition of nickel(II) carbonate.

   ____

3. Hydrogen can be produced in a variety of reactions. Balance each of the following chemical equations:
   a) ___Na(s) + ___HCl(aq) $\rightarrow$ ___NaCl(aq) + ___$H_2(g)$
   b) ___K(s) + ___$H_2O(\ell)$ $\rightarrow$ ___KOH(aq) + ___$H_2(g)$
4. Observe the video of the reaction of copper and nitric acid. How do you know that NO_2 gas has been generated?

   ____

EXERCISE 4.8 *A Gas-Forming Reaction*

Cerussite, $PbCO_3$, is a a lead-containing mineral. Write a balanced equation to show what happens when cerussite is treated with nitric acid. Give the name of each of the reaction products.

Screen 4.15 Oxidation-Reduction Reactions

1. Observe the photos of the reaction of magnesium with oxygen.
   a) What evidence is there that a reaction has occurred?

   ____

b) What is the name of the product of this reaction? ____________ What is its color? ____________ Its physical state? ____________

2. Iron reacts with oxygen (it rusts in air) to give iron(III) oxide.

   a) Write a balanced chemical equation for this reaction.

   ____

   b) What is the oxidizing agent in this reaction? ____________ What has been oxidized? ____________ What has been reduced? ____________

Screen 4.16 Redox Reactions and Electron Transfer

1. Observe the series of photos on this screen. Cite two observations that prove that reaction has occurred.

   ____

   ____

2. What is the nature of elemental silver? ____________

3. Describe one characteristic of copper(II) ions dissolved in water.

   ____

4. Consider the reaction of zinc metal with copper ions.

   $Zn(s) + Cu^{2+}(aq) \rightarrow Zn^{2+}(aq) + Cu(s)$

   a) Which species (Zn or Cu^{2+} ions) has been reduced? ____________ Which has been oxidized? ____________

   b) Has Zn transferred electrons to Cu^{2+} or has Cu^{2+} transferred electrons to Zn? That is, what is the direction of electron transfer?

   ____

   ____

Screen 4.17 Oxidation Numbers

1. Observe the reaction of NO gas with air to give NO_2 gas. What evidence is there that a reaction has occurred? (See the video of the reaction of copper with nitric acid on Screen 4.14.)

   ____

2. Sulfur dioxide reacts with oxygen to give sulfur trioxide. Give the oxidation number for each atom in the compounds involved in the reaction.

   $2\ SO_2(g) + O_2(g) \rightarrow 2\ SO_3(g)$

Compound	SO_2	O_2	SO_3
Oxidation Number of Sulfur Atom			
Oxidation Number of Oxygen Atom			

What has been oxidized? ____________ What has been reduced? ____________

EXERCISE 4.9 *Oxidation Numbers*

Assign an oxidation number to the underlined atom in each of the following molecules or ions.

a) $\underline{Fe}_2O_3$ b) $H_2\underline{S}O_4$ c) $\underline{C}O_3^{2-}$ d) $\underline{C}_6H_6$

Screen 4.18 Recognizing Oxidation-Reduction Reactions

1. In the experiment described in the video on this screen, hot iron wool is inserted into a flask containing chlorine gas.
   a) What is the color of the chlorine gas? ______________
   b) What is the "smoke" that is formed in the flask in the course of the reaction? ______________
   c) What is the name of the product of this reaction? ______________
   d) What is the oxidizing agent in the reaction? ______________ What is the reducing agent? ______________
2. Suppose you place copper wool in chlorine gas. The product expected is copper(II) chloride.
   a) Write a balanced chemical equation for this reaction.

   ____

   b) What is oxidized in this reaction? ______________ What is reduced? ______________
   c) What is the oxidizing agent? ______________ What is the reducing agent? ______________
3. In the video on another part of this screen, hot iron wool is sprayed with a stream of pure oxygen, O_2.
   a) What evidence is there that a reaction has occurred?

   ____

   b) What is the oxidizing agent in this reaction? ______________ What is the reducing agent? ______________

EXERCISE 4.10 *Oxidation-Reduction Reactions*

The following reaction is used to test for alcohol on the breath of a person.

$$3\ C_2H_5OH(aq) + 2\ Cr_2O_7^{2-}(aq) + 16\ H^+(aq) \rightarrow 3\ CH_3CO_2H(aq) + 4\ Cr^{3+}(aq) + 11\ H_2O(\ell)$$

Identify the oxidizing and reducing agents and the substance oxidized and the substance reduced. ______________________________

Screen 4.19 Return to the Puzzler

(Refer to the original Puzzler on Screen 4.1 and then to your answers on Screen 4.11.)

All four of the reactions illustrated and described here involve the exchange of some species. In precipitation, acid-base, and gas-forming reactions these are usually ions. In redox reactions, it is electrons.

1. Precipitation—How are the negative ions exchanged in the reaction depicted in the video?

   ____

2. Acid-base—Show that the reaction of hydrochloric acid and sodium hydroxide can be considered an exchange reaction.

3. Gas-forming reactions—A common gas-forming reaction involves the reaction of an acid with a metal carbonate.

$$CaCO_3(s) + 2\ HCl(aq) \rightarrow CaCl_2(aq) + H_2CO_3(aq)$$

$$H_2CO_3(aq) \rightarrow CO_2(g) + H_2O(\ell)$$

a) Describe the reaction of $CaCO_3$ and HCl as an exchange reaction.

b) The video shows the reaction of Zn with HCl. What is exchanged in this reaction?

4. Several of the gas-forming reactions we have described can also be described by another of our four reaction types: precipitation, acid-base, gas-forming, or redox. See the reaction of Zn with HCl and the reaction of Cu with HNO_3 on Screen 4.11 and then classify these in another of the four classes.

5. Why can oxidation-reduction reactions also be considered exchange reactions?

Thinking Beyond

In Chapter 12 we describe how automobile air bags work. They contain sodium azide, NaN_3. When rapid deceleration is detected, as in a crash, the compound is detonated according to the first, unbalanced chemical equation below. (The sodium produced in the initial explosion is quite reactive; one possibility is that it reacts with water vapor in the air to give sodium hydroxide and hydrogen.)

$$2\ NaN_3(s) \rightarrow 2\ Na(s) + 3\ N_2(g)$$

$$2\ Na(s) + 2\ H_2O(\ell) \rightarrow 2\ NaOH(s) + H_2(g)$$

1. Describe each of these reactions as one or more of the different reaction classes. If you have described them as redox reactions, what are the oxidation numbers of each species? What is oxidized and what is reduced? What is the oxidizing agent and what is the reducing agent in each reaction?

2. Why does the first reaction work to operate the air bag in a car?

STUDY QUESTIONS

1. Balance the following equations:
   (a) $Cr(s) + O_2(g) \rightarrow Cr_2O_3(s)$
   (b) $Cu_2S(s) + O_2(g) \rightarrow Cu(s) + SO_2(g)$
   (c) $C_6H_5CH_3(\ell) + O_2(g) \rightarrow H_2O(\ell) + CO_2(g)$
2. Balance the following equations and name the reaction products:
   (a) $MgO(s) + Fe(s) \rightarrow Fe_2O_3(s) + Mg(s)$
   (b) $AlCl_3(s) + H_2O(\ell) \rightarrow Al(OH)_3(s) + HCl(aq)$
   (c) $NaNO_3(s) + H_2SO_4(\ell) \rightarrow Na_2SO_4(s) + HNO_3(g)$
3. Balance the following equations:
   (a) The synthesis of urea, a common fertilizer
   $CO_2(g) + NH_3(g) \rightarrow CO(NH_2)_2(s) + H_2O(\ell)$
   (b) Reactions used to make uranium(VI) fluoride for the enrichment of natural uranium
   $UO_2(s) + HF(aq) \rightarrow UF_4(s) + H_2O(aq)$
   $UF_4(s) + F_2(g) \rightarrow UF_6(s)$
   (c) Reaction to make titanium(IV) chloride, which is then converted to titanium metal
   $TiO_2(s) + Cl_2(g) + C(s) \rightarrow TiCl_4(\ell) + CO$
   $TiCl_4(\ell) + Mg(s) \rightarrow Ti(s) + MgCl_2(s)$
4. Which compound or compounds in each of the following groups is (are) expected to be soluble in water?
   (a) FeO, $FeCl_2$, and $FeCO_3$
   (b) AgI, Ag_3PO_4, and $AgNO_3$
   (c) $NaCl$, Li_2CO_3, and $KMnO_4$
5. Give the formula for
   (a) A soluble compound containing the acetate ion
   (b) An insoluble sulfide
   (c) A soluble hydroxide
   (d) An insoluble chloride
6. Each compound below is water-soluble. What ions are produced in water?
   (a) KI
   (b) $Mg(CH_3CO_2)_2$
   (c) $KHSO_4$
   (d) KCN
7. Write a balanced equation for the ionization of nitric acid in water.
8. Oxalic acid, which is found in certain plants, can provide two hydrogen ions in water. Write balanced equations to show how oxalic acid, $H_2C_2O_4$, can supply one and then a second H^+ ion.
9. Write a balanced equation for the reaction of sulfur trioxide with water.
10. Balance each of the following equations, and then write the net ionic equation:

(a) $Zn(s) + HCl(aq) \rightarrow H_2(g) + ZnCl_2(aq)$

(b) $Mg(OH)_2(s) + HCl(aq) \rightarrow MgCl_2(aq) + H_2O(\ell)$

(c) $HNO_3(aq) + CaCO_3(s) \rightarrow Ca(NO_3)_2(aq) + H_2O(\ell) + CO_2(g)$

11. Balance each of the following equations, and then write the net ionic equation. Show states for all reactants and products (s, ℓ, g, aq).

(a) $Ba(OH)_2 + HNO_3 \rightarrow Ba(NO_3)_2 + H_2O$

(b) $BaCl_2 + Na_2CO_3 \rightarrow BaCO_3 + NaCl$

(c) $Na_3PO_4 + Ni(NO_3)_2 \rightarrow Ni_3(PO_4)_2 + NaNO_3$

12. Balance the equation for the following precipitation reaction, and then write the net ionic equation. Indicate the state of each species (s, ℓ, aq, or g).

$$CdCl_2 + NaOH \rightarrow Cd(OH)_2 + NaCl$$

13. Predict the products of each precipitation reaction, and then balance the completed equation.

(a) $NiCl_2(aq) + (NH_4)_2S(aq) \rightarrow$

(b) $Mn(NO_3)_2(aq) + Na_3PO_4(aq) \rightarrow$

14. Write an overall, balanced equation for the precipitation reaction that occurs when aqueous lead(II) nitrate is mixed with an aqueous solution of potassium hydroxide. Name each product.

15. Complete and balance the following acid-base reactions. Name the reactants and products.

(a) $CH_3CO_2H(aq) + Mg(OH)_2(s) \rightarrow$

(b) $HClO_4(aq) + NH_3(aq) \rightarrow$

16. Write a balanced equation for the reaction of barium hydroxide with nitric acid to give barium nitrate, a compound used in pyrotechnics such as green flares.

17. The beautiful mineral rhodochrosite is manganese(II) carbonate. Write an overall, balanced equation for the reaction of the mineral with hydrochloric acid. Name each reactant and product.

18. Determine the oxidation number of each element in the following ions or compounds:

(a) BrO_3^-

(b) $C_2O_4^{2-}$

(c) F_2

(d) CaH_2

(e) H_4SiO_4

(f) SO_4^{2-}

19. The following reaction can be used to prepare iodine in the laboratory. Determine the oxidation number of each atom in the following equation:

$$2\ NaI(s) + 2\ H_2SO_4(aq) + MnO_2(s) \rightarrow$$
$$Na_2SO_4(aq) + MnSO_4(aq) + I_2(s) + 2\ H_2O(\ell)$$

20. Which of the following reactions is (are) oxidation-reduction reactions? Explain your answer briefly. Classify the remaining reactions.

(a) $CdCl_2(aq) + Na_2S(aq) \rightarrow CdS(s) + 2\ NaCl(aq)$

(b) $2\ Ca(s) + O_2(g) \rightarrow 2\ CaO(s)$

(c) $Ca(OH)_2(s) + 2\ HCl(aq) \rightarrow CaCl_2(aq) + 2\ H_2O(\ell)$

21. In each of the following reactions, tell which reactant is oxidized and which is reduced. Designate the oxidizing agent and reducing agent.

(a) $2\ Mg(s) + O_2(g) \rightarrow 2\ MgO(s)$

(b) $C_2H_4(g) + 3\ O_2(g) \rightarrow 2\ CO_2(g) + 2\ H_2O(g)$

(c) $Si(s) + 2\ Cl_2(g) \rightarrow SiCl_4(\ell)$

22. The mineral dolomite contains magnesium carbonate. Name the spectator ions in the reaction of magnesium carbonate and hydrochloric acid and write the net ionic equation.

$$MgCO_3(s) + 2\ H^+(aq) + 2\ Cl^-(aq) \rightarrow CO_2(g) + Mg^{2+}(aq) + 2\ Cl^-(aq) + H_2O(\ell)$$

What type of reaction is this?

23. The compound $(NH_4)_2S$ reacts with $Hg(NO_3)_2$ to give HgS and NH_4NO_3.

(a) Write the overall balanced equation for the reaction. Indicate the state (s or aq) for each compound.

(b) Name each compound.

(c) What type of reaction is this?

24. Classify each of the reactions as an acid-base reaction, a precipitation, or a gas-forming reaction. Show states for the products (s, ℓ, g, aq), and then balance the completed equation. Write the net ionic equation.

(a) $MnCl_2(aq) + Na_2S(aq) \rightarrow MnS + NaCl$

(b) $K_2CO_3(aq) + ZnCl_2(aq) \rightarrow ZnCO_3 + KCl$

(c) $K_2CO_3(aq) + HClO_4(aq) \rightarrow KClO_4 + CO_2 + H_2O$

25. Vitamin C is the simple compound $C_6H_8O_6$. One method for determining the amount of vitamin C in a sample is to react it with a solution of bromine, Br_2.

$$C_6H_8O_6(aq) + Br_2(aq) \rightarrow 2\ HBr(aq) + C_6H_6O_6(aq)$$

What is oxidized and what is reduced in this reaction? Which substance is the oxidizing agent and which is the reducing agent?

26. The types of reactions described in this chapter can be used to prepare compounds. For example, insoluble barium chromate can be made by a precipitation reaction involving the soluble compounds $BaCl_2$ and K_2CrO_4.

$$BaCl_2(aq) + K_2CrO_4(aq) \rightarrow BaCrO_4(s) + 2\ KCl(aq)$$

The product, $BaCrO_4$, can be separated from water-soluble $BaCl_2$ by filtering the product mixture. The insoluble $BaCrO_4$ is trapped on the filter paper, and aqueous KCl passes through the paper. Suggest a precipitation reaction and a gas-forming reaction by which barium sulfate can be made.

CHAPTER 5
Stoichiometry

Screen 5.2 Weight Relations in Chemical Reactions

1. Show that 0.010 mol of P_4 is equivalent to 1.24 g.

____

____

2. Explain why 0.010 mol of P_4 requires 0.060 mol of Cl_2 for complete reaction and that they produce 0.040 mol of PCl_3.

____

____

3. Observe the reaction of P_4 and Cl_2.
   a) Is elemental chlorine a solid, liquid, or gas? _______________
   b) What color is P_4? _______________ Cl_2? _______________
   c) As the reaction proceeds, what do you see? How do you know a reaction has occurred? Is there any evidence for energy changes in the reaction?

   ____

   d) How would you describe, in your own words, the structure of the P_4 molecule? Of the PCl_3 molecule? (To have an even better picture of structures, find them in the *Models* folder and examine their structures.)

   P_4

   ____

   PCl_3

   ____

   e) In the animation at the bottom of the screen, how many Cl atoms are there before reaction? _______________ After reaction? _______________

The animation of the reaction of phosphorus and chlorine is highly simplified. It is meant to show only that the number of atoms of each kind is preserved upon reaction. It does not show *how* the reaction occurs.

Screen 5.3 Calculations in Stoichiometry

1. Observe the series of photos of magnesium reacting with air.
   a) How do you know a reaction has occurred?

   ____

   b) Describe the physical states of the reactants and products. How could you tell them apart?

   ____

2. Describe the way you would calculate the number of grams of PCl_3 produced from the reaction of 25.0 g of P_4 and excess Cl_2 according to the equation

$P_4(s) + 6\ Cl_2(g) \rightarrow 4\ PCl_3(\ell)$

EXERCISE 5.1 *Weight Relations in Chemical Reactions*

What mass of carbon, in grams, can be consumed by 454 g of O_2 in a combustion to give carbon monoxide? What mass of CO can be produced?

$2\ C(g) + O_2(g) \rightarrow 2\ CO(g)$

Screen 5.3 A Closer Look: Simple Stoichiometry

Use the plot of "Grams Hg" versus "Grams HgS."

1. Explain why the slope of the line is not 1. Why is it less than 1?

2. Use the graph to decide how many grams of Hg are expected if you begin with 7 g of HgS.

3. How does the slope of the line relate to the weight percent of Hg in HgS?

4. Calculate the mass of Hg expected if you begin a reaction with 10. g of HgS and excess O_2.

5. Gaseous NO reacts readily with O_2 to give NO_2, a common air pollutant.
   a) What is the balanced equation for the reaction?

   b) Calculate the mass of NO_2 expected if you begin with 1.0 g, 5.0 g, and 10. g of NO and excess O_2. Use the *Plotting Tool* on the disc to plot these data as given on this screen. Use the plot to decide how many grams of NO_2 are formed if you begin with 4.0 g of NO and excess O_2.

Plotting Tool

The Least Squares plotting tool is found in the Tools folder on this disc.

Screen 5.4 Reactions Controlled by the Supply of One Reactant

1. Explain how the video of methanol combustion illustrates the fact that methanol is the limiting reagent in this reaction.

2. If 2.5 moles of methanol burn in oxygen, how many moles of O_2 are required?

3. Examine the problem associated with this screen (Screen 5.4PR).
   a) Which of the diagrams at the right is correct? Why?

   b) Examine the diagram above the one for the correct answer. Why is the former not correct? Do the same for the diagram below the correct answer.

Screen 5.5 Limiting Reactants

1. Examine the series of photos of the reaction between varying amounts of Zn metal and a constant amount of hydrochloric acid.
   a) What gas inflates the balloons in the demonstration?

   b) Why are the volumes of gas in the first two balloons greater than the volume in the balloon on the right?

   c) In which flask is the Zn metal the limiting reactant and in which flask is the HCl(aq) the limiting reactant? Explain.

2. Review the reaction of methanol and oxygen on the previous screen. If you combine 5.0 mol of methanol with 12.5 mol of O_2, which is the limiting reactant?

EXERCISE 5.2 *Limiting Reactants*

You have 20.0 g of elemental sulfur, S_8, and 160. g of O_2. Which is the limiting reactant in the combustion of S_8 in oxygen to give SO_2? What amount of which reactant (in moles) is left after complete reaction? What mass of SO_2, in grams, is formed in the complete reaction?

Screen 5.6 Percent Yield

Diborane, B_2H_6, can be prepared by the following reaction (in a solvent other than water):

$$3\ NaBH_4 + 4\ BF_3 \rightarrow 3\ NaBF_4 + 2\ B_2H_6$$

If you begin with 18.9 g of $NaBH_4$ (and excess BF_3), and you isolate 7.50 g of B_2H_6, what is the percent yield of B_2H_6?

Screen 5.7SB Current Issues in Chemistry

Examine the map of the United States. Why are there areas of low pH in upstate New York and in Pennsylvania, West Virginia, and Virginia?

Screen 5.7PR Using Stoichiometry (1)

You have 2.357 g of a mixture of $BaCl_2$ and $BaCl_2 \cdot 2\ H_2O$. If experiment shows that the mixture has a mass of only 2.108 g after heating to drive off all of the water of hydration in $BaCl_2 \cdot 2\ H_2O$, what is the weight percent of $BaCl_2 \cdot 2\ H_2O$ in the original mixture?

Screen 5.8 Using Stoichiometry (2)

1. Assume you burn the white solid naphthalene, $C_{10}H_8$, in excess oxygen.
   a) Write the balanced equation for the combustion reaction.

   ____

   b) If you burn 0.100 g of naphthalene, what masses of H_2O and CO_2 are formed?

2. What is the empirical formula of quinone, an organic compound that contains only C, H, and O, if 0.105 g of the compound produces 0.257 g of CO_2 and 0.0350 g of H_2O when burned in oxygen?

EXERCISE 5.3 *Determining the Empirical and Molecular Formulas of a Hydrocarbon*

A 0.523-g sample of a hydrocarbon was burned in air to produce 1.612 g of CO_2 and 0.7425 g of H_2O. A separate experiment gave a molar mass of 114 g/mol for the unknown compound. Determine the empirical and molecular formulas for the hydrocarbon.

EXERCISE 5.4 *Formula Determination by Combustion Analysis*

A molecule of a new compound is composed of only C, H, and Cr. When 0.178 g of the compound is burned in air, the products are CO_2 (0.452 g), H_2O (0.0924 g), and Cr_2O_3. What is the empirical formula for the compound?

Screen 5.9 Solutions

1. Describe what happens as $KMnO_4$ is added to water. Is the mixture homogeneous or heterogeneous?

2. What is the name of the compound $KMnO_4$?

3. In the animation, the K^+ ions are shown as yellow spheres, whereas the MnO_4^- ions are a collection of blue and smaller red spheres.

   a) Describe what you see as the animation proceeds. How does an ionic solid dissolve in water? What forces are at work that allow the solid to dissolve?

This animation of the dissolution of $KMnO_4$ is highly simplified. The MnO_4^- ion is tetrahedral, and each ion is certainly surrounded by more than four water molecules.

   b) Water molecules are attached to both positive and negative ions. What does this tell you about the nature of water?

   c) How does the orientation of the water molecules differ when they encounter a K^+ ion as compared with a MnO_4^- ion? What does this tell you about the nature of the water molecule? That is, describe how electric charge is distributed in the water molecule.

Screen 5.10 Solution Concentration

Some $K_2Cr_2O_7$, 2.335 g, is dissolved in enough water to make exactly 500 mL of solution. What is the molarity of the potassium dichromate solution?

EXERCISE 5.5 *Solution Molarity*

You dissolve 26.3 g of $NaHCO_3$ in enough water to make exactly 200 mL of solution. What is the concentration of $NaHCO_3$?

Screen 5.10 Molarity Calculator (Tool)

An icon on the right side of this screen will take you to the *Molarity Calculator*. (You can also gain access to this tool by clicking on the Tool button in the navigation bar.) To gain practice using this tool, answer the following questions.

1. What mass of NaCl is required to make 50. mL of a 0.10 M solution?

Click on this icon to find the Molarity Calculator.

2. What mass of $KMnO_4$ is required to make 250. mL of a 0.0552 M solution?

3. On the previous page you calculated the concentration of 500 mL of a solution that contains 2.335 g of $K_2Cr_2O_7$. Verify your calculation by putting in the molarity you calculated and a volume of 500 mL. Does the required mass equal 2.335 g?

4. What mass of $CuSO_4 \cdot 5H_2O$ must be used to make 250 mL of a 0.16 M solution of the compound?

5. Use the tool to estimate the mass required to make 200 mL of a 0.050 M solution of Na_2CO_3.

6. Use the tool to estimate the mass required to make 75 mL of a 0.015 M solution of K_2CrO_4.

Screen 5.10SB A Closer Look—Ion Concentration in Solution

1. In the animation the green spheres represent negative ions (say Br^-), and the gray spheres represent positive ions (say Mg^{2+}).
   a) What is the ratio of positive to negative ions? What is the formula of the salt?

   b) Refer to your notes about Screen 5.9 and the nature of water molecules. If you had not been told the green spheres were negative ions, how could you have figured this out?

Screen 5.11 Preparing Solutions of Known Concentration (1)

1. Sufficient water was added to hydrated nickel(II) chloride to make exactly 250 mL of a 0.140 M solution. What is the concentration of Ni^{2+} ion? ________ Of the Cl^- ion? ________

2. Why is it important to shake the volumetric flask thoroughly before using the solution you have made?

3. Examine the glassware on Screen 5.11SB. Which piece of glassware is least accurate?

EXERCISE 5.6 *Preparing Solutions of Known Concentration*

An experiment requires 500. mL of a 0.0200-M solution of $KMnO_4$. You have some solid $KMnO_4$, and a 500. mL volumetric flask. Describe how you would make the required solution.

Screen 5.12 Preparing Solutions of Known Concentration (2)

1. Why is the color of the diluted solution K_2CrO_4 less intense than the color of the original, stock solution?

2. Explain why the equation $c_{stock}V_{stock} = c_{dilute}V_{dilute}$ can be used to calculate, for example, the concentration of a solution after it has been diluted. Does the volume need to be expressed in liters or can you use solution volumes in milliliters?

EXERCISE 5.7 *Preparing a Solution by Dilution*

1. You need exactly 250 mL of 1.00 M NaOH for an experiment, but you only have a large bottle of 2.00 M NaOH. Describe how to make the 1.00 M NaOH in the desired volume.
2. A solution of $CuSO_4$ has a concentration of 0.15 M. If you mix 6.0 mL of this solution with enough water to make 10.0 mL of solution, what is the concentration of $CuSO_4$ in the new solution?

Screen 5.13 Stoichiometry of Reactions in Solution

Observe the video of the reaction of an iron(II) salt with potassium permanganate.

1. Which beaker originally contains the iron(II) salt and which one the potassium permanganate? How can you tell?

2. What observations can you make as the reaction proceeds? That is, what is the evidence that a reaction has occurred?

EXERCISE 5.8 *Solution Stoichiometry*

Metal carbonates react with acids to give carbon dioxide, water, and the metal salt.

$$Na_2CO_3(aq) + 2\ HCl(aq) \rightarrow 2\ NaCl(aq) + H_2O(l) + CO_2(g)$$

If you combine 50.0 mL of 0.450 M HCl and an excess of Na_2CO_3, what mass of NaCl (in grams) is produced?

Screen 5.14 Titrations

1. Observe the video of a titration of oxalic acid with aqueous NaOH.
   a) What is the approximate volume of NaOH solution used?

   ---

   b) If the NaOH solution had a concentration of 0.10 M, what mass of oxalic acid must have been present in the flask?

   ---

2. You have 0.954 g of an unknown acid, H_2A, which reacts with NaOH according to the balanced equation

   $$H_2A(aq) + 2\ NaOH(aq) \rightarrow Na_2A(aq) + 2\ H_2O(\ell)$$

   If 36.04 mL of 0.509 M NaOH is required to titrate the acid to the equivalence point, what is the molar mass of the acid?

XERCISE 5.9 *Acid-Base Titration*

25.0-mL sample of vinegar requires 28.33 mL of a 0.953-M solution of NaOH for titration the equivalence point. What mass (in grams) of acetic acid (CH_3CO_2H) is in the vinegar ample, and what is the concentration of the acid in the sample?

$$CH_3CO_2H(aq) + NaOH(aq) \rightarrow NaCH_3CO_2(aq) + H_2O(\ell)$$

Screen 5.15 Titration Simulation

The experiment on this screen simulates the technique of titration. You can add aqueous sodium hydroxide from the buret in small increments to a flask containing 25.0 mL of aqueous acetic acid (CH_3CO_2H). The indicator, an organic dye, should turn color just at the point at which the acid has been completely consumed by the base.

$$CH_3CO_2H(aq) + NaOH(aq) \rightarrow NaCH_3CO_2(aq) + H_2O(\ell)$$

Run the titration, and determine the volume of NaOH added. If the concentration of the NaOH solution is 0.0886 M,

Using the Titration Tool Box

This tool allows you to simulate the titration of 25.0 mL of aqueous acetic acid solution of unknown concentration (in the flask) with aqueous sodium hydroxide (in the buret). Each time you return to the screen, a different acid concentration is chosen at random. The concentration of NaOH also varies.

1. How many moles of NaOH were used in the titration?

2. How many moles of CH_3CO_2H were in the flask?

3. What was the concentration of the aqueous acetic acid solution? When you have finished your calculations, use the summary screen to check your work (remember that you are working with a different concentration of NaOH).

4. What volume of 0.0886 M NaOH would have been required if the acid concentration had been 0.075 M?

Screen 5.16 Return to the Puzzler

Explain why you eventually see no reaction after adding a large amount of sodium bicarbonate to vinegar.

Thinking Beyond

1. List four reactions in the environment where one reactant limits the extent to which the reaction proceeds.

2. Are there reactions where some factor other than amounts of reactants limits the extent to which the reaction proceeds. Some examples might include thermal or light energy, time, or pressure.

STUDY QUESTIONS

1. Aluminum reacts with oxygen to give aluminum oxide.

   $4\ Al(s) + 3\ O_2(g) \rightarrow 2\ Al_2O_3(s)$

   If you have 6.0 mol of Al, how many moles of O_2 are needed for complete reaction? What mass of Al_2O_3, in grams, can be produced?

2. Cobalt metal reacts with hydrochloric acid according to the following *unbalanced* equation:

   $Co(s) + HCl(aq) \rightarrow CoCl_2(aq) + H_2(g)$

   If you begin with 2.56 g of cobalt metal and excess hydrochloric acid, what mass of cobalt(II) chloride can be obtained? What mass of hydrogen gas can be obtained?

3. Nitrogen monoxide is oxidized in air to give brown nitrogen dioxide.

   $2\ NO(g) + O_2(g) \rightarrow 2\ NO_2(g)$

   Starting with 2.2 moles of NO, how many moles and how many grams of O_2 are required for complete reaction? What mass of NO_2, in grams, is produced?

4. The equation for one of the reactions in the process of reducing iron ore to the metal is

   $Fe_2O_3(s) + 3\ CO(g) \rightarrow 2\ Fe(s) + 3\ CO_2(g)$

   (a) What is the maximum mass of iron, in grams, that can be obtained from 454 g (1.00 lb) of iron(III) oxide?

   (b) What mass of CO is required to reduce the iron(III) oxide to iron metal?

5. Aluminum chloride, $AlCl_3$, is an inexpensive reagent used in many industrial processes. It is made by treating scrap aluminum with chlorine according to the following balanced equation:

$$2\ Al(s) + 3\ Cl_2(g) \rightarrow 2\ AlCl_3(s)$$

(a) Which reactant is limiting if 2.70 g of Al and 4.05 g of Cl_2 are mixed?

(b) What mass of $AlCl_3$ can be produced?

(c) What mass of the excess reactant remains when the reaction is completed?

6. Methanol, CH_3OH, is a clean-burning, easily handled fuel. It can be made by the direct reaction of CO and H_2 (obtained from heating coal with steam).

$$CO(g) + 2\ H_2(g) \rightarrow CH_3OH(\ell)$$

Starting with a mixture of 12.0 g of H_2 and 74.5 g of CO, which is the limiting reactant? What mass of the excess reactant (in grams) remains after reaction is complete? What is the theoretical yield of methanol?

7. Ammonia gas can be prepared by the reaction of a basic oxide like calcium oxide with ammonium chloride, an acidic salt.

$$CaO(s) + 2\ NH_4Cl(s) \rightarrow 2\ NH_3(g) + H_2O(g) + CaCl_2(s)$$

If 112 g of CaO and 224 g of NH_4Cl are mixed, what is the maximum possible yield of NH_3? What mass of the excess reactant remains after the maximum amount of ammonia has been formed?

8. Ammonia gas can be prepared by the following reaction:

$$CaO(s) + 2\ NH_4Cl(s) \rightarrow 2\ NH_3(g) + H_2O(g) + CaCl_2(s)$$

If 100. g of ammonia is obtained, but the theoretical yield is 136 g, what is the percent yield of this gas?

9. The reaction of zinc and chlorine has been used as the basis of a car battery.

$$Zn(s) + Cl_2(g) \rightarrow ZnCl_2(s)$$

What is the theoretical yield of $ZnCl_2$ if 35.5 g of zinc is allowed to react with excess chlorine? If only 65.2 g of zinc chloride is obtained, what is the percent yield of the compound?

10. A mixture of $CuSO_4$ and $CuSO_4 \cdot 5\ H_2O$ has a mass of 1.245 g, but, after heating to drive off all the water, the mass is only 0.832 g. What is the weight percent of $CuSO_4 \cdot 5\ H_2O$ in the mixture?

11. A 1.25-g sample contains some of the very reactive compound $Al(C_6H_5)_3$. On treating the compound with aqueous HCl, 0.951 g of C_6H_6 is obtained.

$$Al(C_6H_5)_3(s) + 3\ HCl(aq) \rightarrow AlCl_3(aq) + 3\ C_6H_6(\ell)$$

Assuming that $Al(C_6H_5)_3$ was converted completely to products, what is the weight percent of $Al(C_6H_5)_3$ in the original 1.25-g sample?

12. Styrene, the building block of polystyrene, is a hydrocarbon, a compound consisting only of C and H. If 0.438 g of styrene is burned in oxygen and produces 1.481 g of CO_2 and 0.303 g of H_2O, what is the empirical formula of styrene?

13. Propanoic acid, an organic acid, contains only C, H, and O. If 0.236 g of the acid burns completely in O_2, and gives 0.421 g of CO_2 and 0.172 g of H_2O, what is the empirical formula of the acid?

14. Silicon and hydrogen form a series of compounds with the general formula Si_xH_y. To find the formula of one of them, a 6.22-g sample of the compound is burned in oxygen. On doing so, all of the Si is converted to 11.64 g of SiO_2 and all of the H to 6.980 g of H_2O. What is the empirical formula of the silicon compound?

15. Assume 6.73 g of Na_2CO_3 is dissolved in enough water to make 250. mL of solution. What is the molarity of the sodium carbonate? What are the molar concentrations of the Na^+ and CO_3^{2-} ions?

16. What is the mass, in grams, of solute in 250. mL of a 0.0125 M solution of $KMnO_4$?

17. What volume of 0.123 M NaOH, in milliliters, contains 25.0 g of NaOH?

18. If 4.00 mL of 0.0250 M $CuSO_4$ is diluted to 10.0 mL with pure water, what is the molarity of copper(II) sulfate in the diluted solution?

19. If you need 1.00 L of 0.125 M H_2SO_4, which method below do you use to prepare this solution? Calculate the concentration of the sulfuric acid in each of the other cases as well.

    (a) Dilute 36.0 mL of 1.25 M H_2SO_4 to a volume of 1.00 L.

    (b) Dilute 20.8 mL of 6.00 M H_2SO_4 to a volume of 1.00 L.

    (c) Add 950. mL of water to 50.0 mL of 3.00 M H_2SO_4.

    (d) Add 500. mL of water to 500. mL of 0.500 M H_2SO_4.

20. For each solution, tell what ions exist in aqueous solution, and give the concentration of each.

    (a) 0.12 M $BaCl_2$

    (b) 0.0125 M $CuSO_4$

    (c) 0.146 M $AlCl_3$

    (d) 0.500 M $K_2Cr_2O_7$

21. How many grams of Na_2CO_3 are required for complete reaction with 25.0 mL of 0.155 M HNO_3?

    $Na_2CO_3(aq) + 2\ HNO_3(aq) \rightarrow 2\ NaNO_3(aq) + CO_2(g) + H_2O(\ell)$

22. One of the most important industrial processes in our economy is the electrolysis of brine solutions (aqueous solutions of NaCl). When an electric current is passed through an aqueous solution of salt, the NaCl and water produce $H_2(g)$, $Cl_2(g)$, and NaOH—all valuable industrial chemicals.

    $2\ NaCl(aq) + 2\ H_2O(\ell) \rightarrow H_2(g) + Cl_2(g) + 2\ NaOH(aq)$

    What mass of NaOH can be formed from 10.0 L of 0.15 M NaCl? What mass of chlorine can be obtained?

23. In the photographic developing process silver bromide is dissolved by adding sodium thiosulfate.

    $AgBr(s) + 2\ Na_2S_2O_3(aq) \rightarrow Na_3Ag(S_2O_3)_2(aq) + NaBr(aq)$

    If you want to dissolve 0.250 g of AgBr, how many milliliters of 0.0138 M $Na_2S_2O_3$ should you add?

24. How many milliliters of 0.812 M HCl are required to titrate 1.33 g of NaOH to the equivalence point?

$NaOH(aq) + HCl(aq) \rightarrow NaCl(aq) + H_2O(\ell)$

25. How many milliliters of 0.955 M HCl are needed to titrate 2.152 g of Na_2CO_3 to the equivalence point?

$Na_2CO_3(aq) + 2\ HCl(aq) \rightarrow 2\ NaCl(aq) + CO_2(g) + H_2O(\ell)$

26. A noncarbonated soft drink contains an unknown amount of citric acid, $H_3C_6H_5O_7$. If 100. mL of the soft drink requires 33.51 mL of 0.0102 M NaOH to neutralize the citric acid completely, how many grams of citric acid does the soft drink contain per 100. mL? The reaction of citric acid and NaOH is

$H_3C_6H_5O_7(aq) + 3\ NaOH(aq) \rightarrow Na_3C_6H_5O_7(aq) + 3\ H_2O(\ell)$

CHAPTER 6
Principles of Reactivity: Energy and Chemical Reactions

Screen 6.2 Product-Favored Systems

1. Reactions that produce appreciable quantities of products are considered product-favored systems. Based on your experience, predict if each system below could be designated as product- or reactant-favored. Examine your prediction by going to the next screen.
   a) Rusting of iron ______
   b) Reaction of a diamond with O_2 at room temperature ______
   c) Combustion of gasoline ______
   d) The decomposition of sand, SiO_2, to elemental Si and O_2 ______
2. What are the products of the oxidation of the Gummi Bear?

   ______

3. What makes us believe this reaction is product-favored?

   ______

4. What do you suppose would happen if you put a cookie in the molten potassium chlorate?

   ______

   What about a piece of celery?

   ______

Screen 6.3 Thermodynamics and Kinetics

1. Chemists say that substances are thermodynamically stable or kinetically stable or both. Describe the difference between the kinetic stability of a substance and its thermodynamic stability. Give examples.

   ______

   ______

   ______

Screen 6.4 *and* Screen 6.5 Energy

1. Describe the type of energy exhibited by each of the following:
   a) A book resting on the edge of a desk ______
   b) A Frisbee sailing through the air ______
   c) Two cations very near each other ______
   d) Water droplets at the top of a waterfall ______

2. Energy conversion. Answer the following questions that explore some of the many possible energy conversions.
   a) To what forms of energy can one convert mechanical energy?

   ____

   b) To what forms of energy can one convert chemical energy?

   ____

   c) Give two examples of the conversion of chemical energy into radiant energy.

   ____

3. A Closer Look: Heat, Hotness, and Thermometers
   a) What is the difference between "heat" and "temperature"?

   ____

   b) How does the video illustrate heat transfer?

   ____

   How do you know that the heated copper bar is "hotter" than the water?

   ____

   ____

Screen 6.6 Energy Units

1. A nonsugar sweetener provides 16 kJ of nutritional energy. What is this energy in kilocalories?

   ____

   In dietary Calories? ______________________________

2. What was James Joule's contribution to the progress of science? How does the video illustrate this contribution?

   ____

   ____

Screen 6.7 Heat Capacity

Predict which might have the larger heat capacity, 2000 g (2 kg) of copper or the mug of water shown on this screen? Explain your reasoning and then go on to Screen 6.8.

____

____

____

Screen 6.8 Heat Capacity of Pure Substances

This screen and the next one (6.9) take up the concept of "specific heat capacity," the heat capacity per gram of a pure substance. As a substance with a given mass undergoes a temperature change, ΔT, the heat transferred to or from the substance, q, can be calculated from the equation q = (specific heat capacity, J/g · K)(mass, g)(change in T, K)

where ΔT = final temperature of substance - initial temperature of substance.

1. The "tool" on Screen 6.8 allows you to do an experiment. Give your answers to the three questions posed on the screen. Note that heat is added to each substance at the same rate, 50. joules per second.

   a) What effect does heating the blocks for a longer time have on the final temperature?

   ______

   b) What is the effect of the mass of the blocks on the observed temperature change?

   ______

   c) Use this experiment to calculate the specific heat capacities of wood, copper, and glass.

   Wood

   ______

   Copper

   ______

   Glass

   ______

   Do your answers agree with the "official" specific heat capacities for these substances? (For the values, click on the "Summary" button.)

   ______

   d) How does this "experiment" verify the relationship between heat transferred and the mass, specific heat, and temperature change of a substance? Explain fully.

   ______

   ______

   ______

2. Examine the specific heat capacity values in the table.

   a) Which would require more heat to be warmed from room temperature to 35 °C, 100 g of water or 100 g of aluminum?

   ______

   b) Which substance would evolve the most heat on cooling from 100 °C to 25 °C, 1 kg of glass or 1 kg of water?

   ______

   c) If water had been included in the "experiment" you did above, what would its final temperature be if you had heated 5.0 g for 5.0 seconds? Is this temperature lower or higher than the temperatures you observed for the wood, copper, or glass under these conditions? Explain.

   ______

   ______

   ______

Screen 6.9 Calculating Heat Transfer

Use this screen in conjunction with Screen 6.8.

1. In this question we want to calculate the quantity of heat required to heat 50.0 g of water from 25 °C to 37 °C, the temperature of your body.
   a) Will this quantity be a positive or negative value? ______________________
   b) What is the actual calculated value? ______________________
   c) What quantity of aluminum would you need to absorb this same amount of heat?
   ____

EXERCISE 6.1 *Using Specific Heat*

If 24.1 kJ is used to warm a piece of aluminum with a mass of 250. g, what is its final temperature if its initial temperature is 5.0 °C?

EXERCISE 6.2 *Using Specific Heat*

A 15.5-g piece of chromium, heated to 100.0 °C, is dropped into 55.5 g of water at 16.5 °C. The final temperature of the water and metal is 18.9 °C. What is the specific heat of chromium?

Screen 6.10 Heat Transfer Between Substances

You should be aware that the animation of heat transfer between substances is highly simplified.

1. When a hotter object comes into contact with a cooler one, explain what happens in terms of molecular motions.
   ____
   ____
2. What does it mean when we say that two objects have come to thermal equilibrium?
   ____

EXERCISE 6.3 *Heat Transfer Between Substances*

A piece of iron (400. g) is heated in a flame and then dropped into a beaker containing 1000. g of water. The original temperature of the water was 20.0 °C, but it is 32.8 °C after the iron bar has been dropped into the water and both have come to thermal equilibrium. What was the original temperature of the iron bar?

Screen 6.11 Heat Associated with Phase Changes

1. Describe what has happened after 1000 joules of heat energy have been added to 10.0 g of ice at 0 °C and to 10.0 g of iron at 0 °C.
   ____
   ____
2. Describe what would have happened to the 10.0 g of ice if 3500 J had been added?
   ____
   ____

What would the temperature of the system be at this point?

EXERCISE 6.4 *Changes of State*

What quantity of heat must be absorbed to warm 25.0 g of liquid methanol, CH_3OH, from 25.0 °C to its boiling point (64.6 °C) and then to evaporate the methanol completely at that temperature? The heat capacity of liquid methanol is 2.53 J/g · K. The heat of vaporization of the compound is 2.00×10^3 J/g.

EXERCISE 6.5 *Using a Change of State to Determine Heat Capacity*

An "ice calorimeter" can be used to determine the heat capacity of a metal. A piece of hot metal is dropped into a weighed quantity of ice. The quantity of heat given up by the metal can be determined from the amount of ice melted. Suppose a piece of metal with a mass of 9.85 g is heated to 100.0 °C and then dropped onto ice. When the metal's temperature has dropped to 0.0 °C, it is found that 1.32 g of ice has been melted to water at 0.0 °C. What is the heat capacity of the metal?

Screen 6.12 Energy Changes in Chemical Processes

1. A Gummi Bear has been placed in molten potassium chlorate (Screen 6.2). How might you define the changes in the system and the surroundings in this case?

2. Is the observed chemical change exothermic or endothermic?

Screen 6.13 The First Law of Thermodynamics

How does the demonstration on this screen illustrate the first law of thermodynamics?

Screen 6.14 Enthalpy Change and ΔH

1. Is the process observed upon adding heat energy to solid CO_2 an endothermic or exothermic process? Why is it endo- or exothermic?

2. Is there an increase in chemical potential energy (positive enthalpy change, +ΔH) or a decrease (negative enthalpy change, -ΔH) when gaseous CO_2 condenses to form a solid?

EXERCISE 6.6 *Changes of State and ΔH*

The enthalpy change for the sublimation of solid iodine is 62.4 kJ/mol.

$I_2(s) \rightarrow I_2(g)$ $\Delta H = 62.4$ kJ

a) What quantity of heat energy is required to sublime 10.0 g of the solid?

b) If 3.45 g of iodine vapor condenses to solid iodine, what quantity of energy is involved? Is the process exo- or endothermic?

Screen 6.15 Enthalpy Changes for Chemical Reactions

The energies of the various changes in state or in the nature of the substances involved on this screen can be found by moving the cursor over the energy level diagram at the right.

1. a) What is the enthalpy change, when 1 mol of water evaporates at 25 °C?

   b) What is the enthalpy change for the condensation of water vapor at 25 °C to water at 25 °C?

   c) How does this compare with ΔH for the evaporation of water at 25 °C? Are they the same or different? If so, in what way?

2. Is the conversion of water to its elements at 25 °C exothermic or endothermic? Is heat energy evolved or absorbed by the system in this process?

3. Methane and propane are used as fuels.

   $CH_4(g) + 2\ O_2(g) \rightarrow CO_2(g) + 2\ H_2O(\ell)$ $\Delta H = -890.3$ kJ

   $C_3H_8(g) + 5\ O_2(g) \rightarrow 3\ CO_2(g) + 4\ H_2O(\ell)$ $\Delta H = -2220$ kJ

   a) Which fuel provides more heat energy per mole?

   b) Which fuel provides more heat energy per gram?

   c) If you needed to burn propane to obtain 1.0×10^4 kJ of heat energy, how many grams of propane would you burn?

EXERCISE 6.7 *Heat Energy Calculation*

How much heat energy is required to decompose 12.6 g of liquid water to the elements?

Screen 6.16 Hess's Law

1. What is the most exothermic step in the formation of sulfuric acid from its elements?

___

2. Using the reactions below, with their respective enthalpy changes, calculate the enthalpy change for the combustion of methane.

$C(s) + O_2(g) \rightarrow CO_2(g)$ $\Delta H = -393.5$ kJ

$C(s) + 2\ H_2(g) \rightarrow CH_4(g)$ $\Delta H = -74.8$ kJ

$2\ H_2(g) + O_2(g) \rightarrow 2\ H_2O(\ell)$ $\Delta H = -571.6$ kJ

The balanced equation for the combustion of methane is

$CH_4(g) + 2\ O_2(g) \rightarrow CO_2(g) + 2\ H_2O(\ell)$

and the enthalpy change for the reaction is

How does this calculation illustrate Hess's law?

___

___

EXERCISE 6.11 *Using Hess's Law*

To obtain the metallic lead, lead(II) sulfide (PbS) is first roasted in air to form lead(II) oxide (PbO).

$PbS(s) + 3/2\ O_2(g) \rightarrow PbO(s) + SO_2(g)$ $\Delta H = -413.7$ kJ

and then lead(II) oxide is reduced with carbon to the metal.

$PbO(s) + C(s) \rightarrow Pb(s) + CO(g)$ $\Delta H = +106.8$ kJ

a) What is the enthalpy change for the following reaction?

$PbS(s) + C(s) + 3/2\ O_2(g) \rightarrow Pb(s) + CO(g) + SO_2(g)$ $\Delta H = ?$

Is the reaction exothermic or endothermic?

b) How much energy, in kilojoules, is required (or evolved) when 454 g (1.00 pound) of PbS is converted to lead?

Screen 6.17 Standard Enthalpy of Formation

1. Circle any of the following enthalpy changes that can be designated as a standard enthalpy of formation.

   a) $C(s) + O_2(g) \rightarrow CO_2(g)$ $\Delta H° = -393.5$ kJ

   b) $CH_4(g) \rightarrow C(s) + 2\ H_2(g)$ $\Delta H° = +74.8$ kJ

   c) $NO(g) + 1/2\ O_2(g) \rightarrow NO_2(g)$ $\Delta H° = -57.07$ kJ

   d) $Mg(s) + O_2(g) + H_2(g) \rightarrow Mg(OH)_2(s)$ $\Delta H° = -924.54$ kJ

2. List the values for the standard molar enthalpy of formation of the hydrocarbons methane, ethane, propane, and butane. These data are found in the *Enthalpies of Formation* associated with this screen.

Compound	*Enthalpy of Formation, kJ/mol*
Methane, CH_4	______________
Ethane, C_2H_6	______________
Propane, C_3H_8	______________
Butane, C_4H_{10}	______________

   a) What trend do you see in these values?

   ____

   b) Use the *Plotting Tool* on this disc to plot the ΔH°_f values for these compounds versus the number of carbon atoms in the molecule. Use this plot to "predict" the value for pentane, C_5H_{12}.

   ____

   ____

3. What is meant by the following symbols?

   a) $\Delta H°_{rxn}$ ______________

   b) $\Delta H°_{vap}$ ______________

4. Define the standard state or standard conditions in terms of the state of an element.

   ____

5. Circle any of the following that do not represent the standard state of an element.

   a) $C(g)$

   b) $Sn(s)$

   c) $Mg(\ell)$

   d) $H_2(s)$

   e) $Cl_2(g)$

6. Use the table of standard enthalpies of formation on this screen to calculate the enthalpy change for the combustion of 1.00 mol of butane, C_4H_{10}.

$$2\ C_4H_{10}(g) + 13\ O_2(g) \rightarrow 8\ CO_2(g) + 20\ H_2O(\ell)$$

   a) $\Delta H^\circ_{combustion}$ = ______________

   b) What quantity of heat would be evolved if only 15.0 g of butane burned in excess O_2?

EXERCISE 6.9 *Using Enthalpies of Formation*

Benzene, C_6H_6, is an important hydrocarbon. Calculate its enthalpy of combustion; that is, find the value of $\Delta H°$ for the following reaction.

$$C_6H_6(\ell) + 15/2\ O_2(g) \rightarrow 6\ CO_2(g) + 3\ H_2O(\ell)$$

The enthalpy of formation of benzene, $\Delta H°_f\ [C_6H_6(\ell)]$, is +49.0 kJ/mol.

Screen 6.18 Calorimetry

1. Use the calorimeter "tool" to determine the temperature change and total heat transferred for three different masses of benzoic acid. What is the relationship between the mass of compound and these two outcomes? Use this to estimate ΔT and total energy if 100 mg of benzoic acid is used in the calorimeter.

Mass of Compound	ΔT	*Total energy transferred*
50	____	____________ kJ
200	____	____________ kJ
500	____	____________ kJ
100	____	____________ kJ (estimated)

2. Octane, C_8H_{18}, a primary constituent of gasoline, burns in air.

$$C_8H_{18}(\ell) + 25/2\ O_2(g) \rightarrow 8\ CO_2(g) + 9\ H_2O(\ell)$$

Suppose that a 1.00-g sample of octane is burned in a calorimeter that contains 1.20 kg of water. The temperature of the water and the calorimeter rises from 25.00 °C (298.15 K) to 33.20 °C (306.35 K). If the heat capacity of the calorimeter, $C_{calorimeter}$, is known to be 837 J/K, calculate the heat transferred in the combustion of the 1.00-g sample of C_8H_{18}.

Screen 6.19 Return to Puzzler

1. What quantity of energy is required to boil a cup of water?

2. How many burning peanuts are required to supply this heat energy?

Thinking Beyond

1. On Screen 6.15 you learned that the conversion of water to its elements is endothermic; the enthalpy change is positive. Thinking on the molecular level, why do you think this should be true?

2. Gaseous water is thermodynamically less stable than liquid water. Still, a puddle of water on the ground evaporates. Why?

3. When we eat food it reacts with oxygen in an exothermic process—just like a Gummi Bear reacts exothermically with the oxidizing agent potassium chlorate—to provide the energy to run our bodies. The formation of carbohydrates such as sugar from CO_2 and H_2O is an endothermic process. What is the source of energy for this process?

STUDY QUESTIONS

1. You pick up a six-pack of soft drinks from the floor, but it slips from your hand and smashes onto your foot. Comment on the work and energy involved in this sequence. What forms of energy are involved and at what stages of the process?
2. Criticize the following statements:
   (a) An enthalpy of formation refers to a reaction in which 1 mol of one or more reactants produces some quantity of product.
   (b) The standard enthalpy of formation of O_2 as a gas at 25 °C and a pressure of 1 bar is 15.0 kJ/mol.
   (c) The thermal energy transferred from 10.0 g of ice as it melts is $q = +6$ kJ.
3. A 2-in. piece of a two-layer chocolate cake with frosting provides 1670 kJ of energy. What is this in Calories?
4. A total of 74.8 J of heat is required to raise the temperature of 18.69 g of silver from 10.0 to 27.0 °C. What is the specific heat of silver?
5. Which gives up more heat on cooling from 50 °C to 10 °C, 50.0 g of water or 100. g of ethanol (specific heat = 2.46 J/g · K)?

6. How much heat energy in kilojoules is required to heat all the aluminum in a roll of aluminum foil (500. g) from room temperature (25 °C) to the temperature of a hot oven (255 °C)?
7. A 192-g piece of copper is heated to 100.0 °C in a boiling water bath and then dropped into a beaker containing 750. mL of water (density = 1.00 g/cm3) at 4.0 °C. What is the final temperature of the copper and water after they come to thermal equilibrium? (The specific heat of copper is 0.385 J/g · K)
8. A 150.0-g sample of a metal at 80.0 °C is placed in 150.0 g of water at 20.0 °C. The temperature of the final system (metal and water) is 23.3 °C. What is the specific heat of the metal?
9. Calculate the quantity of heat required to convert the water in five ice cubes (60.1 g) from $H_2O(s)$ at 0.0 °C to $H_2O(g)$ at 100.0 °C. The heat of fusion of ice at 0 °C is 333 J/g; the heat of vaporization of liquid water at 100 °C is 2260 J/g.
10. How much heat energy (in joules) is required to raise the temperature of 454 g of tin (1.00 lb) from room temperature (25.0 °C) to its melting point, 231.9 °C, and then melt the tin at that temperature? The specific heat of tin is 0.227 J/g · K, and the metal requires 59.2 J/g to convert the solid to a liquid.
11. Nitrogen monoxide has recently been found to be involved in a wide range of biological processes. The gas reacts with oxygen to give brown NO_2 gas.

    $2\ NO(g) + O_2(g) \rightarrow 2\ NO_2(g)$ $\quad\quad \Delta H°_{rxn} = -114.1\ kJ$

    Is the reaction endothermic or exothermic? If 1.25 g of NO is converted completely to NO_2, what quantity of heat is absorbed or evolved?
12. Isooctane (2,2,4-trimethylpentane) burns in air to give water and carbon dioxide.

    $2\ C_8H_{18}(\ell) + 25\ O_2(g) \rightarrow 16\ CO_2(g) + 18\ H_2O(\ell)$ $\quad\quad \Delta H°_{rxn} = -10{,}922\ kJ$

    Is the combustion exothermic or endothermic? If you burn 1.00 L of the hydrocarbon (density = 0.6878 g/mL), what quantity of heat is involved?
13. Methanol, CH_3OH, is a possible automobile fuel. The alcohol produces energy in a combustion reaction with O_2.

    $2\ CH_3OH(g) + 5\ O_2(g) \rightarrow 2\ CO_2(g) + 4\ H_2O(\ell)$

    A 0.115-g sample of methanol evolves 1110 J when burned at constant pressure. What is the enthalpy change, $\Delta H°_{rxn}$, for the reaction? What is the enthalpy change per mole of methanol (often called the molar heat of combustion)?
14. Using the reactions below, find the enthalpy change for the formation of PbO(s) from lead metal and oxygen gas.

    $Pb(s) + CO(g) \rightarrow PbO(s) + C(s)$ $\quad\quad \Delta H°_{rxn} = -106.8\ kJ$

    $2\ C(s) + O_2(g) \rightarrow 2\ CO(g)$ $\quad\quad \Delta H°_{rxn} = -221.0\ kJ$

    If 250 g of lead reacts with oxygen to form lead(II) oxide, what quantity of heat (in kilojoules) is absorbed or evolved?
15. The standard molar enthalpy of formation of solid chromium(III) oxide is -1139.7 kJ/mol. Write the balanced equation for which the enthalpy of reaction is -1139.7 kJ.

16. An important step in the production of sulfuric acid is

$$SO_2(g) + 1/2\ O_2(g) \rightarrow SO_3(g)$$

It is also a key reaction in the formation of acid rain, beginning with the air pollutant SO_2. Using the data in the compound database, calculate the enthalpy change for the reaction. Is the reaction exothermic or endothermic?

17. The first step in the production of nitric acid from ammonia involves the oxidation of NH_3.

$$4\ NH_3(g) + 5\ O_2(g) \rightarrow 4\ NO(g) + 6\ H_2O(g)$$

(a) Use the information in the compound database to find the enthalpy change for this reaction. Is the reaction exothermic or endothermic?

(b) What quantity of heat is evolved or absorbed if 10.0 g of NH_3 are oxidized?

18. How much heat energy (in kilojoules) is evolved by a reaction in a calorimeter in which the temperature of the calorimeter and water increases from 19.50 °C to 22.83 °C? The calorimeter has a heat capacity of 650. J/K and contains 320. g of water.

19. You can find the amount of heat evolved in the combustion of carbon by carrying out the reaction in a combustion calorimeter. Suppose you burn 0.300 g of C(graphite) in an excess of $O_2(g)$ to give $CO_2(g)$.

$$C(graphite) + O_2(g) \rightarrow CO_2(g)$$

The temperature of the calorimeter, which contains 775 g of water, increases from 25.00 °C to 27.38 °C. The heat capacity of the calorimeter is 893 J/K. What quantity of heat is evolved per mole of C?

20. The combustion of diborane, B_2H_6, proceeds according to the equation

$$B_2H_6(g) + 3\ O_2(g) \rightarrow B_2O_3(s) + 3\ H_2O(g)$$

and 1941 kJ of heat energy is liberated per mole of $B_2H_6(g)$ (at constant pressure). Calculate the molar enthalpy of formation of $B_2H_6(g)$ using this information, the data in the compound database, and the fact that ΔH_f° for $B_2O_3(s)$ is -1271.9 kJ/mol.

CHAPTER 7
Atomic Structure

Screen 7.3 Electromagnetic Radiation

1. Why do we refer to radiation as "electromagnetic?"

____

____

2. A line 10 cm long is draw below. Draw a standing wave with a wavelength of 2 cm on this line. Label the amplitude of your wave.

____

3. Which has the greater frequency, a wave with a wavelength of 780 nm or one with a wavelength of 550 nm?

____

4. If blue light has a wavelength of 550 nm, what is the frequency of the light?

____

Screen 7.4 The Electromagnetic Spectrum

1. Use the spectrum "tool" to answer the following questions:
   a) As you move the slider from the blue region to the red region, what happens to the wavelength of the light? To its frequency?

   ____

   ____

   b) Place the slider somewhere in the blue region of the spectrum. What is the approximate wavelength of this radiation? In nanometers? __________ nm In meters? __________ m
   c) Place the slider somewhere in the orange region of the spectrum. What is the approximate wavelength of this radiation? In nanometers? __________ nm In meters? __________ m
   d) Which color of light in the visible spectrum has the longest wavelength? __________
   The shortest wavelength? __________
2. Which color of light in the visible spectrum has the highest frequency? __________
3. Is the frequency of radiation used in a microwave oven higher or lower than that of your favorite FM radio station. (Assume the station broadcasts at a frequency of 91.7 MHz where MHz = 10^6 s^{-1}). __________
4. Is the wavelength of x-rays shorter or longer than that of ultraviolet light? __________

Screen 7.5 Planck's Equation

1. Place the following types of radiation in order of increasing frequency and in order of increasing energy: Red light; yellow light; microwaves; x-rays; and ultraviolet light.

   Increasing frequency: ______________________________

   Increasing energy: ______________________________

2. When a hot iron bar emits yellow light, what is the approximate energy of 1 photon of this light. What is the energy of 1.00 mol of photons of this light?

EXERCISE 7.1 *Photon Energies*

Compare the energy of a mole of photons of blue light with the energy of a mole of photons of microwave radiation having a frequency of 2.45 GHz (1 GHz = 10^9 s^{-1}). Which has the greater energy?

Screen 7.6 Atomic Line Spectra

1. Why is the spectrum of light in the animation on this screen called a "line" spectrum? Explain fully.

   ____

   ____

2. As the colors of the "lines" progress from red to green, blue, and violet, is the light more or less energetic?

   ____

3. Calculate the difference in wavelength of light between each pair of lines. What do you observe? Is there a trend in the differences? Are these differences becoming larger or smaller on going from long wavelength to short wavelength? (This observation will be useful on Screen 7.7.)

   ____

   ____

4. Examine the Balmer equation. As n increases from 3 to higher values, what happens to the wavelength of the radiation emitted by the excited atom? Does the wavelength become longer or shorter? How does this observation relate to the experimental observation in the question immediately above?

   ____

   ____

   ____

   ____

Screen 7.7 Bohr's Model of the Hydrogen Atom

1. How does Bohr's theory describe the energy of an electron in an atom? (Specify the parameters involved in the equation relating energy to fundamental properties of the atom.)

2. How can we detect the movement of an electron from a higher energy level in an atom to a lower energy level?

3. What does it mean when an electron is moved to an energy level with n = infinity?

4. What do we mean when we say that electrons occupy quantized energy levels?

5. When a hydrogen atom is in its ground state, in what quantum level is the electron located? That is, what is the value of n for the electron?

6. How could you detect the movement of an electron from an energy level, say $n = 2$, to a higher level, say $n = 4$? What experiment would you do?

7. Look at the video of the "glowing" pickle on Screen 7.7P. Why does the pickle "glow" when an electric current is applied and why is it thought that the light given off is yellow?

Screen 7.8 Wave Properties of the Electron

1. Use de Broglie's equation to answer the following questions:
   a) The wavelength of a particle (increases)(decreases) __________ when the mass of the particle decreases.
   b) What happens to the wavelength when the velocity of the particle increases?

2. How can one prove experimentally that electrons have wave properties?

Screen 7.9 Heisenberg's Uncertainty Principle

What is the importance of Heisenberg's uncertainty principle with respect to the structure of the atom?

____

____

Screen 7.10 Schrödinger's Equation and Wave Functions

1. What was Schrödinger's contribution to our understanding of the structure of the atom?

   ____

   ____

2. The value of the wave function relative to the electron's distance from the nucleus is plotted for an orbital on this screen. (Click on "Probability Density.") How do you interpret this?

   ____

   ____

3. How many quantum numbers must be used to describe an orbital?

   ____

   ____

Screen 7.11 Shells, Subshells, and Orbitals

1. An example of an orbital is $4p_z$. In what shell is the electron located? In what subshell? In what orbital within the subshell?

   ____

   ____

2. What is the relation between the value of the quantum number n and the number of subshells in a given shell?

   ____

   ____

3. a) What is the relation between the type of subshell and the number of orbitals in that subshell?

   ____

   b) How many orbitals in an s subshell? ____________________

   c) In a p subshell? ____________________

   d) In a d subshell? ____________________

   e) In an f subshell? ____________________

4. What subshell has 1 orbital? __________ 3 orbitals? __________ 5 orbitals? __________
5. Is there a subshell with 4 orbitals? ____________________

Screen 7.12 Quantum Numbers and Orbitals

1. What are the three quantum numbers that are used to designate an orbital?

____

2. What quantum number distinguishes an s orbital from a p orbital? __________
3. What is the value of ℓ for a d orbital? __________
4. Give a set of quantum numbers for a 3s orbital. How does this differ from a 3p orbital?

____

5. Give a set of three quantum numbers for a 4p orbital.

____

6. Give a set of three quantum numbers for a 5d orbital.

____

EXERCISE 7.2 *Using Quantum Numbers*

Complete the following statements:

1. When n = 2, the values of ℓ can be __________ and __________ .
2. When ℓ = 1, the values of m_ℓ can be __________ , __________ , and __________ , and the subshell has the letter label __________ .
3. When ℓ = 2, the subshell is called a __________ subshell.
4. When a subshell is labeled s, the value of ℓ is __________ , and m_ℓ has the value __________ .
5. When a subshell is labeled p, __________ orbitals occur within the subshell.
6. When a subshell is labeled f, there are __________ values of m_ℓ, and __________ orbitals occur within the subshell.

Screen 7.13 Shapes of Atomic Orbitals

1. Examine the 1s, 2s, and 3s orbitals. What is their general shape? How do these differ from one another? What are their similarities?

____

____

____

2. Examine the 2p and 3p orbitals. What is their general shape? How do these differ from one another? What are their similarities?

____

____

____

3. Examine the 3d orbitals. How would you describe their general shape?

____

4. What is the relation between ℓ and the number of planar nodes for an orbital?

____

EXERCISE 7.3 *Orbitals Shapes*

Give the n and ℓ values for each of the following orbitals:

6p n = ________ and ℓ = ________

4p n = ________ and ℓ = ________

5d n = ________ and ℓ = ________

4f n = ________ and ℓ = ________

Thinking Beyond

1. What do you suppose a 4f orbital will look like? How many planar nodes will it have? How many spherical nodes?

____

____

2. This chapter depicts the work of a number of scientists. Determine the age of each scientist below when they made the indicated discovery.

Scientist	*Discovery*	*Age at that time*
James Maxwell	Theory of electromagnetic radiation, 1864	________
Max Planck	Theory of energy quantization in 1900	________
Niels Bohr	Model of the hydrogen atom, 1913	________
Louis de Broglie	Theory of matter waves, 1924	________
Werner Heisenberg	Uncertainty principle, 1927	________

STUDY QUESTIONS

1. The colors of the visible spectrum, and the wavelengths corresponding to the colors, are given on Screen 7.4.
   (a) What colors of light involve less energy than green light?
   (b) Which color of light has photons of greater energy, yellow or blue?
   (c) Which color of light has the greater frequency, blue or green?
2. Green light has a wavelength of approximately 5.0×10^2 nm. What is the frequency of this light? What is the energy in joules of one photon of green light? What is the energy in joules of 1.0 mol of photons of green light?
3. The most prominent line in the line spectrum of aluminum is found at 396.15 nm. What is the frequency of this line? What is the energy of 1 photon with this wavelength? Of 1.00 mol of these photons?
4. The most prominent line in the spectrum of neon is found at 865.438 nm. Other lines are found at 837.761 nm, 878.062 nm, 878.375 nm, and 1885.387 nm.
   (a) Which of these lines represents the most energetic light?
   (b) What is the frequency of the most prominent line? What is the energy of one photon with this wavelength?

5. Place the following types of radiation in order of increasing energy per photon:
   (a) Radar signals
   (b) Radiation within a microwave oven
   (c) γ-rays from a nuclear reaction
   (d) Red light from a neon sign
   (e) Ultraviolet radiation from a sun lamp
6. The energy emitted when an electron moves from a higher energy state to one of lower energy in any atom can be observed as electromagnetic radiation.
   (a) Which involves the emission of less energy in the H atom, an electron moving from n = 4 to n = 2 or an electron moving from n = 3 to n = 2?
   (b) Which involves the emission of the greater energy in the H atom, an electron changing from n = 4 to n = 1 or an electron changing from n = 5 to n = 2?
7. An electron moves from the n = 5 to the n = 1 quantum level and emits a photon with an energy of 2.093×10^{-18} J. How much energy must the atom absorb to move an electron from n = 1 to n = 5?
8. An orbital is designated 2s, and another orbital is designated 4p. Which is the larger orbital? Which has more planar nodes?
9. Answer the following questions:
   (a) When n = 4, what are the possible values of ℓ?
   (b) When ℓ is 2, what are the possible values of m_ℓ?
   (c) For a 4s orbital, what are the possible values of n, ℓ, and m_ℓ?
   (d) For a 4f orbital, what are the possible values of n, ℓ, and m_ℓ?
10. A possible excited state of the H atom has the electron in a 4p orbital. List all possible sets of quantum numbers n, ℓ, and m_ℓ for this electron.
11. How many subshells occur in the electron shell with the principal quantum number n = 4?
12. What is the maximum number of orbitals that can be identified by each of the following sets of quantum numbers? When "none" is the correct answer, explain your reasoning.
    (a) n = 4, ℓ = 3
    (b) n = 5
    (c) n = 2, ℓ = 2
    (d) n = 3, ℓ = 1, m_ℓ = -1
13. Answer the following questions as a review of this chapter:
    (a) The quantum number n describes the __________ of an atomic orbital and the quantum number ℓ describes its __________ .
    (b) When n = 3, the possible values of ℓ are __________ .
    (c) What type of orbital corresponds to ℓ = 3? __________
    (d) For a 4d orbital, the value of n is __________ , the value of ℓ is __________ , and a possible value of m_ℓ is __________ .
    (e) An atomic orbital with 3 nodal planes is __________ .

(f) Which of the following orbitals cannot exist according to modern quantum theory: 2s, 3p, 2d, 4g, 5p, 6p? ____________________

(g) Which of the following is not a valid set of quantum numbers?

n	ℓ	m_ℓ
3	2	1
2	1	2
4	3	0

(h) What is the maximum number of orbitals that can be associated with each of the following sets of quantum numbers? (One possible answer is "none.")

(i) $n = 2$ and $\ell = 1$

(ii) $n = 3$

(iii) $n = 3$ and $\ell = 3$

(iv) $n = 2$, $\ell = 1$, and $m_\ell = 0$

14. In what way does Bohr's model of the atom violate the uncertainty principle?

15. An advertising sign gives off red light and green light.

(a) Which light has the higher energy photons?

(b) One of the colors has a wavelength of 680 nm and the other has a wavelength of 500 nm. Identify which color has which wavelength.

(c) Which light has the higher frequency?

16. Complete the following table:

Quantum Number	Atomic Property Determined by Quantum Number
____________	Orbital size
____________	Relative orbital orientation
____________	Orbital shape

17. Technetium is not found naturally on earth; it must be synthesized in the laboratory. Nonetheless, because it is radioactive it has valuable medical uses. For example, the element in the form of sodium pertechnetate ($NaTcO_4$) is used in imaging studies of the brain, thyroid, and salivary glands and in renal blood flow studies, among other things.

(a) In what group and period of the periodic table is the element found?

(b) The valence electrons of technetium are found in the 5s and 4d subshells. What is a set of quantum numbers (n, ℓ, and m_ℓ) for one of the electrons of the 5s subshell?

(c) Technetium emits a γ-ray with an energy of 0.141 MeV. (1 MeV = 1 million electron volts where 1 eV = 9.6485×10^4 J/mol.) What are the wavelength and frequency of a γ-ray photon with an energy of 0.141 MeV?

CHAPTER 8
Atomic Electron Configurations and Chemical Periodicity

Screen 8.2 *and* Screen 8.3 Spinning Electrons and Magnetism

1. Describe an experiment that would allow you to detect the presence of unpaired electrons in a molecule.

2. What is the difference between a substance that is diamagnetic, one that is paramagnetic, and one that is ferromagnetic? Give an example of a ferromagnetic substance.

Screen 8.4 The Pauli Exclusion Principle

1. State the Pauli principle:

2. Give the four quantum numbers for an electron in
   a) a 2s orbital
   b) a 2p orbital
3. What is the difference between a 1s orbital, a 2s orbital, and a 3s orbital?

4. How do the 2p orbitals differ from a 2s orbital?

5. If another electron is added to the 3s orbital of sodium, this would give the electron configuration for the element magnesium. Give a set of four quantum numbers for this electron. In what way does this set differ from the quantum numbers for the 3s electron of sodium (the set given on this screen)?

Screen 8.5 Atomic Subshell Energies

1. How do subshell energies differ on going from hydrogen to elements with more than one electron?

2. What are the two general rules regarding subshell energies?

3. Review the problem screen associated with Screen 8.5, and then answer the following:
   a) When two subshells have the same n value, which is filled first? Use 3s and 3p subshells of calcium as an example.

   b) Place the following subshells in order of energy: 3s, 4d, 1s, 2p, 2s, 4p, 4s, 3p.

EXERCISE 8.1 *Order of Subshell Assignments*

To which of the following subshells should an electron be assigned first?

1. 4s or 4p
2. 5d or 6s
3. 4f or 5s

Screen 8.6 Effective Nuclear Charge, Z*

The animation on this screen shows the effective nuclear charge experienced by the 2s electron of Li as it penetrates closer to the nucleus. (The distance is measured in units of a_o, Bohr radii, which is equivalent to 0.0529 nm.) Why is the effective nuclear charge felt by this electron only +1 beyond about $2a_o$, whereas it increases rapidly to +3 as the electron approaches the nucleus?

Screen 8.7 Atomic Electron Configurations

Click on this icon in the upper left of Screen 8.7 to go directly to the periodic table displaying electron configurations.

This screen has many parts. Most important is the "Electron Configuration" tool at the upper left of the screen. Click here for a periodic table from which the configuration of any element may be obtained in both spectroscopic and box notations. (Note that electron configurations are also found in the *Periodic Table* database, which you can access using the icon on the tool bar at the bottom of the screen.) Also, click on "More" for important definitions and other aspects of electron configurations.

1. Give the electron configuration for magnesium using
   a) Spectroscopic notation
   b) Noble gas notation
   c) Box notation
   d) In what "block" of the periodic table is magnesium found?

e) Show how the configuration of magnesium is related to the period and group within which the element is located.

____

____

2. Give the electron configuration for phosphorus using
   a) Spectroscopic notation ______________________
   b) Noble gas notation ______________________
   c) Box notation ______________________
   d) In what "block" of the periodic table is phosphorus found? ______________
   e) Show how the configuration of phosphorus is related to the period and group within which phosphorus is located.

____

____

3. Give the electron configuration for vanadium using
   a) Spectroscopic notation ______________________
   b) Noble gas notation ______________________
   c) Box notation ______________________
   d) In what "block" of the periodic table is vanadium found? ______________
   e) Show how the configuration of vanadium is related to the period and group within which vanadium is located.

____

____

4. Give the electron configuration for cerium, Ce, using
   a) Noble gas notation ______________________
   b) In what "block" of the periodic table is cerium found? ______________
   c) Show how the configuration of cerium is related to the period and group within which the element is located.

____

____

5. Examine the electron configurations of Fe, Ru, and Os (iron, ruthenium, and osmium). All three have configurations of $d^x s^y$. How is the quantum number n of the d subshell related to the position of the element in the periodic table? How is the quantum number n of the s subshell related to the position of the element in the periodic table?
   a) ____

____

____

b) Now do the same exercise for two elements in the f-block, say Sm and Pu (samarium and plutonium). How are the n quantum numbers of the f and s orbitals related to the position of the element in the periodic table?

6. Illustrate Hund's rule with the electron configurations of Al, Si, P, and S.

EXERCISE 8.2 *Electron Configurations*

1. What element has the configuration $1s^2 2s^2 2p^6 3s^2 3p^5$?
2. Using the spectroscopic notation and a box notation, depict the configuration for sulfur.
3. Use the spectroscopic notation (with the noble gas notation) to depict the configuration for each of the following:
   a) Ge, germanium
   b) Co, cobalt
   c) W, tungsten
   d) Hg, mercury
   e) U, uranium

Screen 8.8 Electron Configurations of Ions

1. Depict the electron configuration for each of the following (you may use the noble gas notation):
   a) K^+
   b) Ca^{2+}

   Compare these configurations. Are they similar? If not, how do they differ?

2. Depict the electron configuration for each of the following (you may use the noble gas notation)
   a) O^{2-}
   b) F^-

   Compare these configurations. Are they similar? If not, how do they differ?

EXERCISE 8.3 *Ion Configurations*

Depict electron configurations for V^{2+}, V^{3+}, and Co^{3+}. Use the noble gas notation and an orbital box diagram. Are any of these ions paramagnetic? If, so give the number of unpaired electrons.

Screen 8.9 Atomic Properties and Periodic Trends

1. Designate the valence electrons for each of the following elements:
   a) sodium ____________________
   b) calcium ____________________
   c) aluminum ____________________
   d) phosphorus ____________________
   e) chlorine ____________________

Screen 8.10 Atomic Properties and Periodic Trends: Size

1. Why does the size of elements in a group increase on moving down the group?

____

____

2. Why does the size of elements decrease on moving across a period?

____

____

3. How is the radius of a Cl atom defined? What type of experiment might you do to find this value?

____

____

4. If a C atom is attached or "bonded" to a Cl atom, the calculated distance between the atoms is the sum of their radii. Calculate the distance between the following pairs of atoms. Then use models in the *Models* folder or directory to examine the distance in the designated molecules. (See the directions for using these models in Appendix A of this manual. Look particularly for the method of measuring bond distances.) Note that the distances given on these models are in Ångstrom units, where 1 Ångstrom = 100 pm.

Molecule	Atom Distance	Calculated, pm	Measured, pm
BF_3	B—F		
PF_3	P—F		
C_2H_5OH, ethanol	C—H		
C_2H_5OH, ethanol	C—C		
C_2H_5OH, ethanol	C—O		

EXERCISE 8.4 *Periodic Trends in Atomic Radii*

Place the three elements Al, C, and Si in order of increasing radii.

Screen 8.11 Atomic Properties and Periodic Trends: Ion Sizes

1. In each pair of ions below, designate the larger ion.
   a) Na^+ and Cs^+ __________
   b) I^- and Cl^- __________
   c) Ca^{2+} and Cl^- __________
2. Consider the ions Mg^{2+} and F^-.
   a) What are their electron configurations?
      Mg^{2+} = __________ F^- = __________
   b) What are their radii? Mg^{2+} = __________ pm F^- = __________ pm
   c) How do you account for the fact that these ions have such different radii when their electron configurations are the same?
3. The distance between a cation and an anion in an ionic solid may be estimated using a model of the solid. Use the appropriate ionic radii to calculate the distance expected between Na^+ and Cl^- and then use the model of the NaCl crystal lattice in the *Models* folder to examine this distance. Choose a chloride ion (a green sphere) that is as close as possible to a sodium ion (a silver sphere). (See the directions for using these models in Appendix A of this manual. Look particularly for the method of measuring bond distances.)

Screen 8.12 Atomic Properties and Periodic Trends: Ionization Energy

1. Explain why ionization energy increases on moving across a given period.
2. Explain why ionization energy decreases on moving down a group.
3. Examine the ionization of magnesium to form its ions
   a) As successive electrons are removed from magnesium, why does the ionization energy increase?

b) Why is the ion depicted as smaller than the atom as an electron is removed?

c) The difference between the first and second ionization energies of magnesium is about 700 kJ/mol. Why is the difference between the second and third ionization energies over 6000 kJ? What implications does this have for the structure of atoms in general? What implications does this have for the chemistry of magnesium?

Screen 8.13 Atomic Properties and Periodic Trends: Electron Affinity

1. Explain why the electron affinity of atoms becomes more positive on moving across a given period.

2. Explain why the electron affinity of atoms generally becomes less positive on moving down a given group.

3. The affinity of a Group 7A atom for an electron is quite large. (The electron affinity of Cl, for example, is a large, positive value.) In contrast, the affinity of a Group 1A atom such as lithium for an electron is quite low. What implications does this have for the chemistry of Group 1A and Group 7A elements?

EXERCISE 8.5 *Periodic Trends*

Compare the three elements Al, C, and Si.

1. Place the three elements in order of increasing atomic radius.
2. Rank the elements in order of increasing ionization energy.
3. Which element, Al or Si, is expected to have the less positive electron affinity?

PERIODIC TABLE TOOL

Click on the "toolbox" and then on *Periodic Table*. Once there, click on "Periodic Trends."

1. What are the most dense elements in the periodic table?

2. Why are the transition elements in the sixth period so much more dense than the transition metals of the fourth and fifth periods? (See Screen 8.14 for information on this topic.)

3. Why do the Group 1A elements have second ionization energies so much larger than the Group 2A elements?

Screen 8.15 Chemical Reactions and Periodic Properties (1)

Examine the reactions of lithium, sodium, and potassium with water.

1. Write the balanced equation for the reaction of sodium and water.

2. Is there a correlation between the relative reactivities of these metals and any of their atomic properties?

Screen 8.16 Chemical Reactions and Periodic Properties (2)

We can imagine the formation of an ionic compound in the gas phase. For example, we could consider the processes

$$K^+(g) + Cl^-(g) \rightarrow KCl(g)$$

$$Mg^{2+}(g) + O^{2-}(g) \rightarrow MgO(g)$$

1. There are three factors that determine the enthalpy change for these processes ($\Delta H_{formation}$). Those factors are

2. Based on the factors that determine $\Delta H_{formation}$, which of the two reactions just above should have the more negative value for this enthalpy change? Explain.

Screen 8.17 Lattice Energy

1. What is the difference between $\Delta H_{formation}$ as defined on Screen 8.16 and the lattice energy of an ionic solid?

2. Why does the lattice energy decline (become less negative) in the series LiF > LiCl > LiBr > LiI?

3. Why does the lattice energy decline (become less negative) in the series LiF > NaF > KF > RbI > CsI?

4. Which compound might have the larger lattice energy (more negative value), KCl or MgO? Explain your choice.

Screen 8.18 Lattice Energies and Thermodynamic Cycles

Using the approach on this screen, calculate the molar enthalpy of formation of sodium fluoride, NaF. In addition to data already on this screen, other required values are

Electron affinity of F = -328.0 kJ/mol

Energy to dissociate F_2 molecules to F atoms in the gas phase = 78.99 kJ/mol of F atoms formed

Lattice energy for NaF = -926 kJ/mol

Screen 8.19 Return to the Puzzler

Why is it not reasonable to expect a compound such as $MgCl_3$ to exist?

Thinking Beyond

1. What would the periodic table look like if the rule governing possible values of the quantum number , were , = 0, 1, 2, 3, ...n (instead of the maximum value of , being n-1)?

---

2. On Screen 8.12 we explain that the ionization energy of oxygen is less than that of nitrogen because the addition of an electron to an already half-filled 2p orbital leads to greater electron-electron repulsions. If view of this observation, why is the ionization energy of beryllium not lower than that of lithium?

STUDY QUESTIONS

1. What are the electron configurations for Mg and Cl? Write these configurations using both the spectroscopic notation and orbital box diagrams.
2. Using the spectroscopic notation, give the electron configuration of vanadium, V.
3. Depict the electron configuration of a germanium atom using the spectroscopic and noble gas notations.
4. Using the spectroscopic and noble gas notations, write electron configurations for atoms of the following elements and then check your answers with the *Periodic Table* in the "toolbox."
   (a) Strontium, Sr (b) Zirconium, Zr.
   (c) Rhodium, Rh (d) Tin, Sn
5. Using orbital box diagrams, depict the electron configurations of the following ions: (a) Na^+, (b) Al^{3+}, and (c) Cl^-.
6. Using orbital box diagrams and the noble gas notation, depict the electron configurations of (a) Ti, (b) Ti^{2+}, and (c) Ti^{4+}. Are either of the ions paramagnetic?
7. Element 25 can be found at the bottom of the sea in the form of oxide "nodules."
   (a) Depict the electron configuration of this element using the noble gas notation and an orbital box diagram.
   (b) Using an orbital box diagram, show the electrons beyond those of the preceding noble gas for the 2+ ion.
   (c) Is the 2+ ion paramagnetic?

8. Depict the electron configuration for magnesium using the orbital box and noble gas notations. Give a complete set of four quantum numbers for each of the electrons beyond those of the preceding noble gas.
9. What is the maximum number of electrons that can be identified with each of the following sets of quantum numbers? In some cases, the answer may be "none." In such cases, explain why "none" is the correct answer.
   (a) $n = 2$ and $\ell = 1$
   (b) $n = 3$
   (c) $n = 3$ and $\ell = 3$
   (d) $n = 4$, , = 1, and $m_\ell = -1$, and $m_s = -1/2$
   (f) $n = 5$, , = 0, $m_\ell = +1$
10. Arrange the following elements in order of increasing size: Al, B, C, K, and Na.
11. Arrange the following atoms in the order of increasing ionization energy: Li, K, C, and N.
12. Compare the elements Li, K, C, and N.
   (a) Which has the largest atomic radius?
   (b) Which has the most negative electron affinity?
   (c) Place the elements in order of increasing ionization energy.
13. Periodic trends. Explain and answer briefly.
   (a) Place the following elements in order of increasing ionization energy: F, O, and S.
   (b) Which has the largest ionization energy: O, S, or Se?
   (c) Which has the most negative electron affinity: Se, Cl, or Br?
   (d) Which has the largest radius: O^{2-}, F^- or F?
14. Using a thermodynamic cycle (Screen 8.18), calculate the molar enthalpy of formation of lithium chloride. How well does your calculation agree with the known value for the compound (ΔH_f°[LiCl(s)] = -408.7 kJ/mol)? (In addition to information from Screen 8.18 you need to know that the enthalpy of formation of Li(g) is +159.4 kJ/mol, the ionization energy of Li is +520 kJ/mol, and the lattice energy of LiCl is -852 kJ/mol.)
15. Element 109, now named meitnerium, was produced in August 1982, by a team at Germany's Institute for Heavy Ion Research. Depict its electron configuration using the spectroscopic and noble gas notations. Name another element found in the same group as 109.
16. Name the element corresponding to each characteristic below:
   (a) The element with the electron configuration $1s^22s^22p^63s^23p^3$.
   (b) The element in the alkaline earth group that has the smallest atomic radius.
   (c) The element in Group 5A that has the largest ionization energy.
   (d) The element whose +2 ion has the configuration $[Kr]4d^5$.
   (e) The element with the most negative electron affinity in Group 7A.
   (f) The element whose electron configuration is $[Ar]3d^{10}4s^2$.

17. Answer the following questions about the elements with electron configurations below.

A= ...$3p^64s^2$ B = ... $3p^63d^{10}4s^24p^5$

(a) Is element A a metal, metalloid, or nonmetal?

(b) Is element B a metal, metalloid, or nonmetal?

(c) Which element is expected to have the larger ionization energy?

(d) Which element is the smaller of the two?

18. Answer each of the following questions:

(a) Of the elements O, S, and F, which has the largest atomic radius?

(b) Which is larger, Cl or Cl^-?

(c) Which should have the largest difference between the first and second ionization energy: Si, Na, P, or Mg?

(d) Which has the largest ionization energy: O, S, or Se?

(e) Which of the following has the largest radius: Xe, O^{2-}, N^{3-}, or F^-?

19. The configuration of an element is given here.

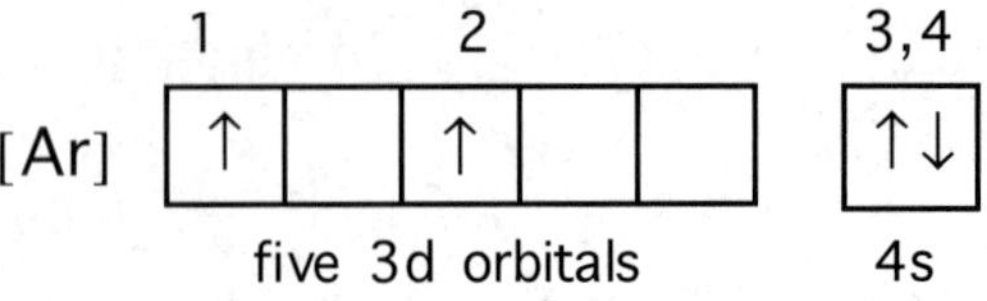

(a) What is the identity of the element?

(b) In what group and period is the element found?

(c) Is the element a nonmetal, a main group element, a transition element, a lanthanide element, or an actinide element?

(d) Is the element diamagnetic or paramagnetic? If paramagnetic, how many unpaired electrons are there?

(e) Write a complete set of quantum numbers for electrons 1, 2, and 4.

Electron	**n**	**ℓ**	**m_ℓ**	**m_s**
1	____	____	____	____
2	____	____	____	____
4	____	____	____	____

(f) If two electrons are removed to form the 2+ ion, what two electrons are removed? Is the ion diamagnetic or paramagnetic?

20. The ionization energies for the removal of the first electron in Si, P, S, and Cl are as listed below. Briefly rationalize this trend.

Element	**First Ionization Energy (kJ/mol)**
Si	780
P	1060
S	1005
Cl	1255

CHAPTER 9
Bonding and Molecular Structure: Fundamental Concepts

Screen 9.2 Valence Electrons

1. What is the relationship between the number of valence electrons and the position of an element in the periodic table?

   ____

   Give an example. __________________________________

2. How many valence electrons would iodine have? ____________________

Screen 9.3 Chemical Bond Formation

1. What are the major coulombic interactions between two atoms?

   ____

2. What are the three major types of chemical bonds?

   ____

3. What is the difference between an ionic bond and a covalent bond?

   ____

4. What is the difference in carbon-carbon bonding in ethane, ethene (ethylene), and acetylene?

   ____

Screen 9.4 Lewis Electron Dot Structures

1. How is the octet rule illustrated by the I atom in ICl?

   ____

2. What is indicated by the pair of lines between the C atoms in C_2H_4?

   ____

3. How is the octet rule illustrated by one of the C atoms in C_2H_4?

   ____

Screen 9.5 Drawing Lewis Electron Dot Structures

Criticize the dot structures below. Why do they *not* illustrate the octet rule?

a) :N̤̈—N=O̤:

b) :N̤̈—N≡O̤:

c) H—C̤(—:O:)—H

2. Use the examples on this screen as an aid in drawing Lewis electron dot structures of the following molecules or ions:

a) CH_2Cl_2 Number of valence electrons __

b) $NHCl_2$ Number of valence electrons __

c) CS_2 Number of valence electrons __

d) HCO_2^- Number of valence electrons __

e) NO_2^- Number of valence electrons __

f) SO_3^{2-} Number of valence electrons __

Screen 9.6 Resonance Structures

Draw resonance structures for the following ions:

a) HCO_2^- Number of valence electrons __

b) NCO^- Number of valence electrons __

c) NO_3^- Number of valence electrons __

EXERCISE 9.1 *Drawing Lewis Structures*

1. Draw Lewis electron dot structures for (a) NO_2^+ and (b) Cl_2CO.
2. Draw resonance structures for the carbonate ion, CO_3^{2-}.

Screen 9.7 Electron-Deficient Compounds

1. What is meant by an "electron-deficient" compound?

___

2. Is SO_3 an electron-deficient compound? Why or why not?

___

3. Is a reaction between BF_3 and H_2O possible? Explain briefly. (Hint: In what way is water similar or dissimilar to ammonia in terms of their Lewis dot structures?

___

___

Screen 9.8 Free Radicals

1. Are any of the following molecules free radicals? Explain briefly in each case.
   a) O_3

   ___

   b) HO

   ___

   c) CO

   ___

   d) CH_3

   ___

Screen 9.9 Bond Properties

1. What is the bond order for
   a) the C—S bond in CS_2? ___

   :S̈=C=S̈:

   b) the N—Cl bond in $NHCl_2$? ___
   c) the N—O bonds in NO_2^-? ___

   :Ö—N̈=Ö: ⊖ (lower O also carries a lone pair below)

   d) the S—O bonds in SO_3? ___

2. Fill in the following table. Click on the "table" icon on this screen for tables of data for bond length and bond energy.

Bond	Bond Length	Bond Energy	Bond Order
C—C in $H_3C—CH_3$			
C=C in $H_2C=CH_2$			
C≡C in HC≡CH			
C—O in $H_3C—OH$			
C=O in $H_2C=O$			

Is there a correlation between bond order and bond length? Between bond order and bond strength? Explain briefly.

3. Go to the *Models* folder and find the models for each of the molecules in the following table. Measure the indicated bond distance.

MOLECULAR MODELS

The models in the *Models* folder were created with software from CAChe/Oxford Molecular. They can be viewed with the "CAChe Visualizer for Education." Directions for using this software are given in Appendix A. Note that distances are given in Ångstrom (Å) units, where 1 Å = 10^{-10} m.

Bond, Compound Name, and Formula	Measured Bond Distance (Å)	Bond Order
C—C bond in ethane, C_2H_6		
C—C bond in butane, C_4H_{10}		
C=C bond in ethylene, C_2H_4		
C≡C bond in acetylene, C_2H_2		
CC bond in benzene, C_6H_6		
C=O bond in acetone, $(CH_3)_2CO$		
C=O bond in acetic acid, CH_3CO_2H		

a) Are all C—C single bonds the same length? All C=C double bonds? All C=O bonds? Comment.

b) Note that benzene, C_6H_6, has two resonance structures. What is the calculated carbon-carbon bond order for benzene?

H
C
HC CH
HC CH
C
H
↔
H
C
HC CH
HC CH
C
H

c) Is there a correlation between the bond order of bonds between carbon atoms and the measured bond length?

4. a) Which ion should have the longer N—O bonds, NO_2^- or NO_3^-? Explain your answer.

b) Which ion should have the stronger N—O bonds, NO_2^- or NO_3^-? Explain your answer.

c) Use the models of these ions in the *Models* folder to examine the bond order of the N—O bonds in these two ions. What is the calculated bond order? Is there a correlation between bond order and measured bond length?

5. a) Examine the animation of bond breaking on Screen 9.9. As two bonded atoms move apart, and the bond eventually breaks, what happens to the energy of the system?

b) In the animation of bond energy, the energy increases as the bond distance is less than 74 pm. Why does the energy increase?

Screen 9.10 Bond Energy and ΔH_{rxn}

Use this screen, and the problem on this screen, to calculate the enthalpy change for the following reaction. Use bond energies from the table on Screen 9.9.

$$CO(g) + Cl_2(g) \rightarrow Cl_2CO(g)$$

EXERCISE 9.2 *Using Bond Energies*

Using the bond energies in the table on Screen 9.9, calculate the ΔH°_{rxn} for the combustion of methane gas, CH_4, in oxygen to give carbon dioxide and water vapor.

Screen 9.11 Bond Polarity and Electronegativity

1. How is the electronegativity of an atom defined?

2. Examine the table of electronegativity values. Is there a relation between the position of an element in the periodic table and its electronegativity? (Trends are most readily apparent in the 3-D table.)

3. Which of the following bonds are polar? For each polar bond, indicate the negative and positive ends of the bond.

Bond	Polar?	Negative Atom	Positive Atom
C—Cl			
B—H			
S—O			
Si—O			
N—H			
C—C			

EXERCISE 9.3 *Bond Polarity*

Decide which bond in each of the following bond pairs is more polar. For each polar bond indicate the negative and positive atoms.

1. H—F and H—I
2. B—C and B—F
3. C—Si and C—S

Screen 9.12 Oxidation Numbers

1. a) Draw a dot structure for the ClO_2^- ion in the ionic limit.

   b) Indicate the oxidation numbers of each atom in this case. Cl = ________ O = ________

   c) How do these compare with the oxidation numbers of Cl and O in ClO_4^-?

2. Draw a dot structure for H_2CO, formaldehyde, in the ionic limit. What are the oxidation numbers of each atom in this case? Do they agree with the assumptions we made regarding oxidation numbers on Screen 4.17?

Screen 9.13 Formal Charge

1. Draw the possible resonance structures for the NCS^- ion. Assign formal charges to each atom in each resonance structure. Which resonance structure is the most important? (See Screen 9.13PR for help with this question.)

2. It is evident that the charges on atoms in molecules are usually not zero, nor are they the charges given by oxidation numbers. Instead, except in molecules such as H_2 or Cl_2 (called homonuclear diatomic molecules), there is usually a small positive or negative partial charge on each atom.

   Atom partial charges have been calculated for many of the molecules in the *Models* folder. These relative partial charges can be viewed for these molecules (or the calculated partial charges can be found in the "Internals" file for each molecule). (See instructions for using the "CAChe Visualizer for Education" in Appendix A.)

   For each molecule pictured below, indicate on each atom whether the atom is negative or positive. How do the calculated partial charges agree with the formal charges you calculate? (Be sure to take resonance into account.)

:F:
|
:F—B—F:

:O: ⊖
||
H—C—O:

:N=C=O: ⊖

:O—S=O:

Screen 9.14 Molecular Shape *and* Screen 9.15 Ideal Repulsion Shapes

What is the basis of the VSEPR model for determining molecular structure?

____

____

Screen 9.16 Determining Molecular Shape

Several hundred models of molecules of all types have been included on the CD-ROM. Let us use a few of these to explore shapes of molecules and see what general conclusions can be drawn. The models used in this question are found in the *Models* folder. (To learn how to measure bond angles, read the instructions for using the "CAChe Visualizer in Education" in Appendix A of this manual.)

Name and Formula	General Shape of Molecule or Ion	Bond Angle	Electron-Pair Geometry
Carbon dioxide, CO_2		O—C—O	
Boron trifluoride, BF_3		F—B—F	
Formaldehyde, H_2CO		O—C—H H—C—H	
Methane, CH_4		H—C—H	
Sulfite ion, SO_3^{2-}		O—S—O	
Water, H_2O		H—O—H	
Dinitrogen oxide, N_2O		N—N—O	

The same principles you have just used to determine the structure of simple molecules and ions can be applied to much more complex structures. Measure the bond angles for each of these molecules and give the geometry around each of the indicated atoms.

a) Acetic acid: The electron pair geometry around the CH_3 carbon atom is

```
  H :O:
  |  ||  ..
H-C--C--O-H
  |      ..
  H
```

Acetic acid

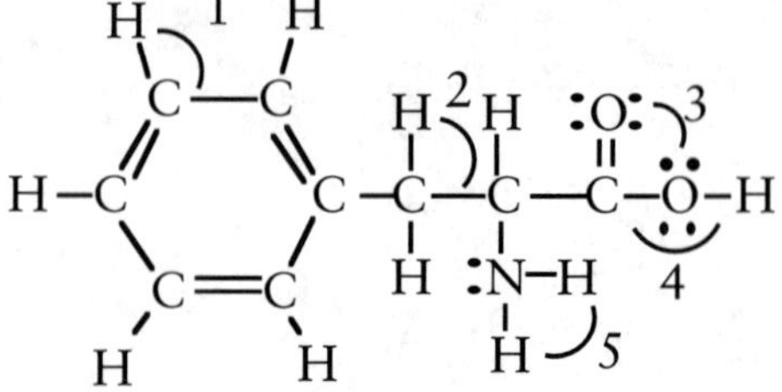

Phenylalanine, an amino acid

_______________________________ and that around the other C atom is _______________________________. The geometry around the O atom of the C—O—H link is _______________________________.

b) Phenylalanine: Angle 1 = _____ Angle 2 = _____ Angle 3 = _____ Angle 4 = _____ and Angle 5 = _____.

What is the electron pair geometry around the N atom? _______________________________

Around the C atoms of the C_6 ring? ____________________

EXERCISE 9.4 *Predicting Molecular Shapes*

1. What is the shape of the dichloromethane molecule, CH_2Cl_2? What is the predicted Cl—C—H bond angle?
2. What is the effect on the molecular geometry of adding an F^- ion to BF_3 to give BF_4^-?

Screen 9.17 Molecular Polarity

1. Use the *Molecular Polarity* tool on this screen to explore the polarity of molecules.
   a) Is BF_3 a polar molecule? Describe what happens as F is replaced by H on BF_3. Does the polarity change as F is replaced by H? What happens when two F atoms are replaced by H?

   ____

   ____

   b) Is $BeCl_2$ a polar molecule? Describe what happens when Cl is replaced by Br.

   ____

   ____

2. Examine the data in the table of dipole moments (click on the table icon for this data table).
   a) Why do the dipole moments of the hydrogen halides decline on going from HF to HI?

   ____

   b) Why is the dipole moment of ClF larger than that of BrF?

   ____

3. Locate the following molecules in the *Models* folder and complete the table.

Name and Formula	Is the Molecule Polar or Nonpolar?
Phosphorus trichloride, PCl_3	
Sulfur trioxide, SO_3	
Ethyl chloride, CH_3CH_2Cl	
Acetonitrile, CH_3CN	
Tetrafluoroethylene, C_2F_4	
Phosgene, Cl_2CO	
Methanol, CH_3OH	

EXERCISE 9.5 *Molecular Polarity*

For each of the following molecules, decide whether the molecule is polar and which side is positive and which is negative.

1. $BFCl_2$
2. NH_2Cl
3. SCl_2

Thinking Beyond

There are at least two reasons that it is important to know if an atom is negative or positive in a molecule: (i) This determines molecular polarity, which in turn affects the physical and chemical properties of the molecule. (ii) It allows us to predict in some cases the possible manner in which a molecule will react with another molecule. You may want to use the models in the *Models* folder to answer the following questions.

1. An H^+ ion will most readily attack a negatively charged atom in an ion or molecule. Examine the partial charges in the NCO^- ion and decide to which atom H^+ will attach itself in the NCO^- ion, the N or the O, to form hydrogen cyanate?

   ____

2. A number of molecules can act as acids in aqueous solution by donating an H^+ ion. The H atom most likely donated by a molecule will be the one bearing the most positive charge. Examine the partial charges in acetic acid, CH_3CO_2H, and decide which of the two types of H atoms—those attached to C or the one attached to O—is donated when the molecule acts as an acid.

   ____

   ____

   ____

3. Examine the structure of urea, $(NH_2)_2CO$. Its electron dot structure shows a lone pair of electrons on each N atom, which should lead to trigonal pyramidal geometry about the N atoms. Yet, each N atom is nearly trigonal planar. Speculate on the reason for this.

   ____

   ____

STUDY QUESTIONS

1. Give the periodic group number and number of valence electrons for each of the following atoms:
   (a) N (c) Na (e) F
   (b) B (d) Mg (f) S
2. Draw Lewis structures for the following molecules or ions:
   (a) NF_3 (c) HOBr
   (b) ClO_3^- (d) SO_3^{2-}
3. Draw Lewis structures for the following molecules:
   (a) $CHClF_2$, a chlorofluorocarbon (CFC).
   (b) Formic acid, HCO_2H.
   (c) Acetonitrile, H_3CCN
   (d) Methanol, H_3COH
4. Show all possible resonance structures for each molecule or ion.
   (a) SO_2 (b) SO_3 (c) SCN^-

5. Give the number of bonds for each of the following molecules. Tell the bond order for each bond.

   (a) H_2CO (b) SO_3^{2-} (c) NO_2^+

6. In each pair of bonds below, predict which is the shorter. If possible, check your prediction by consulting the table on Screen 9.9.

   (a) B—Cl or Ga—Cl

   (b) C—O or Sn—O

   (c) P—S or P—O

   (d) The C=C or the C=O bond in acrolein, H_2C=CH—C(H)=O. (A model of this molecule is found in the *Models* folder.)

7. Consider the carbon-oxygen bond in formaldehyde (H_2CO) and carbon monoxide (CO). In which molecule is the CO bond shorter? In which molecule is the CO bond stronger?

8. Compare the nitrogen-oxygen bond lengths in NO_2^+ and in NO_3^-. In which ion is the bond longer? Explain briefly.

9. Hydrogenation reactions, the addition of H_2 to a molecule, are widely used in industry to transform one compound into another. For example, the molecule called propene is converted to propane by addition of H_2.

```
    H   H H                     H   H H
    |   | |                     |   | |
  H—C═C—C—H  +  H2  ──►  H—C—C—C—H
          |                     |   | |
          H                     H   H H
```

   Use the bond energies on Screen 9.9 to estimate the enthalpy change for this hydrogenation reaction.

10. The compound oxygen difluoride is quite unstable, giving oxygen and HF on reaction with water.

$$OF_2(g) + H_2O(g) \rightarrow O{=}O(g) + 2\ HF(g) \qquad \Delta H^\circ_{rxn} = -318\ kJ$$

   Using bond energies, calculate the bond dissociation energy of the O-F bond in OF_2

11. Acrolein, H_2C=CH—C(H)=O, is the starting material for certain plastics. (A model of this molecule is found in the *Models* folder.)

   (a) Which bonds in the molecule are polar and which are nonpolar?

   (b) Which is the most polar bond in the molecule? Which atom is the negative end of the bond dipole?

12. Two resonance structures are possible for NO_2^-. Draw these structures and then find the formal charge on each atom in each resonance structure.

13. Draw the Lewis electron dot structure for each of the following molecules or ions. Describe the electron-pair geometry and the molecular geometry.

   (a) NH_2Cl (c) SCN-

   (b) Cl_2O (O is the central atom) (d) HOF

14. The following molecules or ions all have two oxygen atoms attached to a central atom. Draw the Lewis structure for each one and then describe the electron-pair geometry and the molecular geometry. Comment on similarities and differences in the series.

(a) CO_2 (c) O_3 (e) SO_2

(b) NO_2^- (d) ClO_2^-

15. Acetylacetone has the structure below. Estimate the values of the indicated angles. (A model of this molecule is found in the *Models* folder.)

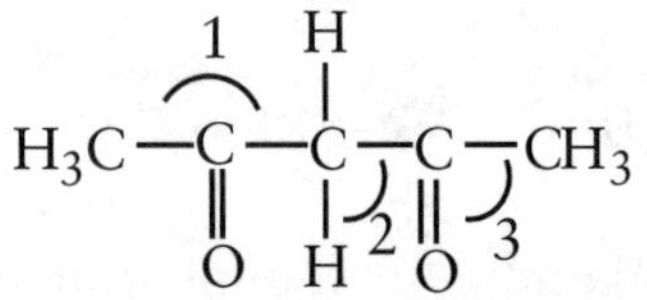

H_3C—C(=O)—C(H)(H)—C(=O)—CH_3 (angles 1, 2, 3 indicated)

16. Consider the following molecules:

(a) H_2O (b) NH_3 (c) CO_2 (d) ClF (e) CCl_4

(i) Which compound has the most polar bonds?

(ii) Which compounds in the list are not polar?

(iii) Which atom in ClF is more negatively charged?

17. Which of the following molecules is (are) polar? For each polar molecule indicate the direction of polarity, that is, which is the negative and which is the positive end of the molecule?

(a) $BeCl_2$ (c) CH_3Cl

(b) HBF_2 (d) SO_3

18. What are the orders of the N—O bonds in NO_2^- and NO_2^+? The nitrogen-oxygen bond length in one of these ions is 110 pm and in the other 124 pm. Which bond length corresponds to which ion? Explain briefly.

19. The molecule pictured below is acrylonitrile, the building block of the synthetic fiber Orlon. (A model of this molecule is in the *Models* folder.)

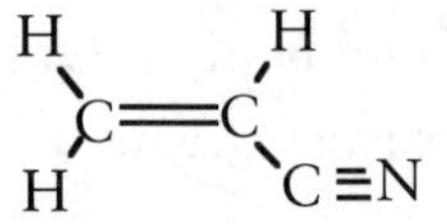

H_2C=C(H)—C≡N

(a) Give the approximate values of the HCH angle and the CCN angle.

(b) Which is the shorter carbon-carbon bond?

(c) Which is the stronger carbon-carbon bond?

(d) Which is the most polar bond, and what is the negative end of the bond dipole?

CHAPTER 10
Bonding and Molecular Structure: Orbital Hybridization, Molecular Orbitals, and Metallic Bonding

Screen 10.1 Chemical Puzzler

The video on this screen is the same as the one on Screen 8.3. What property of an atom, ion, or molecule causes it to be paramagnetic?

Screen 10.2 Models of Chemical Bonding

1. This screen points out the main tenants of two theories of bonding. List aspects of the theories that are similar.

2. List aspects of the theories that are dissimilar.

Screen 10.3 Valence Bond Theory

1. This screen describes attractive and repulsive forces that occur when two atoms come near each other. What must be true about the relative strengths of those attractive and repulsive forces if a covalent bond is to form?

2. It is stated that for a bond to form, orbitals on adjacent atoms must overlap and that each pair of overlapping orbitals will contain two electrons. Explain why neon does not form a diatomic molecule, Ne_2, whereas fluorine forms F_2.

3. When two atoms are widely separated, the energy of the system is zero. As they approach, the energy drops, reaches a minimum, and then increases as the atoms approach still more closely. Explain these observations.

Screen 10.4 Hybrid Orbitals

Hybrid orbitals are an extension of our treatment of orbital shapes in Chapters 7 and 8. What observation leads us to propose this extension to the original wave mechanics theory?

Screen 10.5 Sigma Bonding

The hybrid orbitals of each carbon in the molecule ethane, $H_3C—CH_3$, are similar to those used in methane. Draw a picture of the sigma bonding interactions between a C atom of ethane and one of the three H atoms to which it is bonded.

The C—C bond in ethane involves the overlap of two hybrid orbitals, one from each C atom. Sketch this interaction.

Screen 10.6 Determining Hybrid Orbitals

1. Examine the *Hybrid Orbitals* tool on this screen. Use this tool to systematically combine atomic orbitals to form hybrid atomic orbitals.
   a) What is the relationship between the number of hybrid orbitals produced and the number of atomic orbitals used to create them?

   b) Do hybrid atomic orbitals form between different p orbitals without involving s orbitals?

   c) What is the relationship between the energy of hybrid atomic orbitals and the atomic orbitals from which they are formed?

   d) Compare the shapes of the hybrid orbitals formed from an s orbital and a p_x orbital with the hybrid atomic orbitals formed from an s orbital and a p_z orbital.

   e) Compare the shape of the hybrid orbitals formed from s, p_x, and p_y orbitals with the hybrid atomic orbitals formed from s, p_x, and p_z orbitals.

2. What hybrid orbital set is used by the carbon atom in each of the ions or molecules in the table?

Name and Formula	Geometry of Molecule/Ion	Hybrid Orbitals Used by C Atom
Carbon dioxide, CO_2		
Formaldehyde, H_2CO		
Carbonate ion, CO_3^{2-}		
Carbon tetrachloride, CCl_4		

3. Decide on the hybrid atomic orbitals used by each of the indicated atoms in the molecule glycine.
The N in the NH_2 group ____________
The C in the CH_2 group ____________
The C in the CO_2 group ____________
The O in the OH group ____________

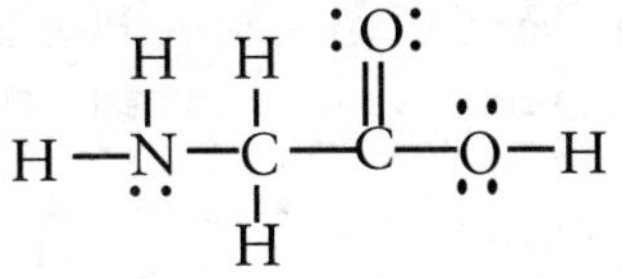

H—N(H)—C(H)(H)—C(=O)—O—H

EXERCISE 10.1 ***Hybrid Orbitals and Bonding***

Describe the bonding in SCl_2 using hybrid orbitals.

Screen 10.7 Multiple Bonding

1. Ethylene is a flat molecule. Why do the two planar triangles centered on the C atoms align in the same plane even though twisting would lead to lower electron-electron repulsive forces?

2. Carefully examine the π bonding in the benzene molecule. How does this picture relate to the fact that benzene has two resonance structures?

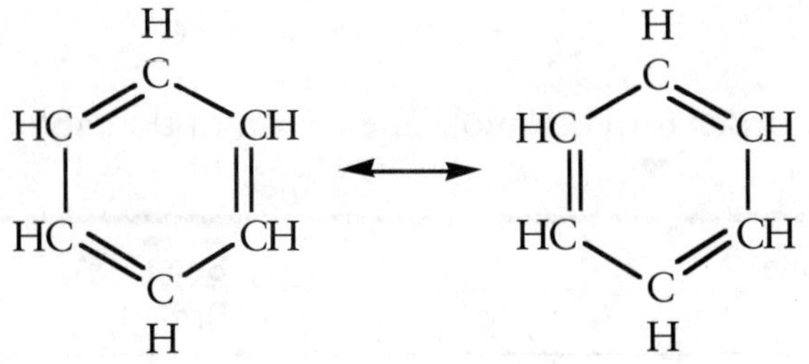

EXERCISE 10.2 ***Bonding and Hybridization***

Analyze the bonding in acetonitrile, $H_3C—C≡N$. Estimate values for the H—C—H, H—C—C, and C—C—N angles, and indicate the hybridization of both carbon atoms and the nitrogen atom. (See the *Models* folder for a model of this molecule.)

Screen 10.8 Molecular Fluxionality

1. Observe the animations of the rotations of trans-2-butene and butane about their carbon-carbon bonds. As one end of trans-2-butane rotates relative to the other, the energy increases greatly (from 27 kJ/mol to 233 kJ/mol) and then drops to 30 kJ/mol when the rotation has proceeded halfway (to cis-2-butene). In contrast, the rotation of the butane molecule requires much less energy for rotation (only about 65 kJ/mol). However, when butane has reached the halfway point in its rotation, the energy has reached a maximum. Why are the energy changes in the two systems almost the opposite of each other?

____

____

____

____

2. The structure of propene, CH_3—CH=CH_2, is pictured here. Circle the carbon-hydrogen group that can rotate freely with respect to the rest of the molecule.

```
    H   H   H
    |   |   |
H—C═C—C—H
            |
            H
```

3. Can the two CH_2 fragments of allene (see Screen 10.7SB) rotate with respect to each other? Briefly explain why or why not.

____

____

Screen 10.9 Molecular Orbital Theory

In a practical sense, what is meant by the term "antibonding?"

____

____

Screen 10.10 Molecular Electron Configurations

Give the molecular electron configuration and bond order for each molecule or ion in the table.

Name and Formula	Molecular Orbital Electron Configuration	Bond Order
Hydrogen molecule anion, H_2^-		
Hydrogen molecule cation, H_2^+		
HHe		
Dihelium cation, He_2^+		

Screen 10.11 Homonuclear, Diatomic Molecules

A Closer Look: Molecular Orbital Energies

A more sophisticated approach to molecular orbital theory takes into account the fact that σ_s and σ_p molecular orbitals interact. This interaction is most important, however, only for elements early in the 2nd period. This means that the MO diagrams for B_2, C_2, and N_2 are different from those of O_2 and F_2. In particular, the σ_{2p} molecular orbital is lower in energy than the π_{2p} orbital for O_2 and F_2.

1. What is the highest energy occupied molecular orbital (called the HOMO in the jargon of the theory) for the N_2 molecule? __________
2. What is the highest energy occupied molecular orbital for the O_2 molecule? __________
3. What is the lowest energy *un*occupied molecular orbital (called the LUMO in the jargon of the theory) for the O_2 molecule? __________
4. Write the molecular orbital electron configuration for the peroxide ion, O_2^{2-}.

   ____

5. Is the superoxide ion, O_2^-, expected to be more strongly or less strongly bound than neutral O_2? Explain briefly.

   ____

   ____

6. Explain why dineon, Ne_2, does not exist. Could the ion Ne^{2+} exist?

   ____

   ____

   ____

Screen 10.12 Early Return to the Puzzler: Paramagnetism

1. This screen states that molecular orbital theory can "predict" that O_2 has unpaired electrons. In fact, O_2 was known to be paramagnetic before molecular orbital theory was invented, so the theory is, in this case, just explaining the known facts. What experimental evidence (presented on Screen 10.12) supports the claim that O_2 molecules have unpaired electrons?

   ____

   ____

2. Use the molecular orbital diagram on this screen and those on Screen 10.11 to predict if the following ions will be paramagnetic or diamagnetic and give the bond order of each.

Formula	Magnetic Behavior	Bond Order
O_2^-		
F_2^{2+}		
N_2^{2+}		
C_2^{2-}		

EXERCISE 10.3 *Molecular Orbitals*

The cations O_2^+ and N_2^+ are important components of the earth's upper atmosphere. Write the electron configuration for O_2^+. Predict its bond order and magnetic behavior.

Screen 10.13 Molecular Orbitals and Vision

1. The key to understanding why light induces changes on the molecular scale and leads to vision is the breaking of a π bond in the molecule rhodopsin. Explain why the π bond breaks and why the breaking of the π bond allows rotation of the right-hand fragment of the molecule.

Screen 10.14 Metallic Bonding: Band Theory

1. What is meant by a *band* of orbitals?

2. The example shown in the animation is that of solid lithium, which forms from the overlap of 2s orbitals on each of many Li atoms. In this case, the band would be half full of electrons (the upper half of the band would contain vacant orbitals). Considering the electron configuration of lithium, explain this.

3. Why is band theory not used to explain the bonding in molecules such as gaseous ammonia, NH_3, or nitrogen, N_2?

Screen 10.15 Conductors and Insulators

1. What is the primary difference in the band structures of conductors and insulators?

2. Solid silicon has an electronic structure much like that of carbon in the diamond form, with a filled valence band and an empty conduction band at much higher energy. Is pure silicon an insulator or a conductor?

3. Solid germanium has a filled valence band and an empty conduction band as well, but the band gap is much smaller than in diamond or silicon. How do you expect germanium to act as a conductor in relation to diamond and to a metal such as lithium?

Screen 10.16 Semiconductors

1. Describe how semiconductors differ from conductors and from insulators.

2. The animation on this screen shows how atoms with unfilled orbitals can be inserted into pure silicon to form synthetic semiconductors. In this case, vacant orbitals now exist to which electrons in the valence band can jump. What types of atoms can be used for this process, those from periodic Group 3A or from Group 5A?

Thinking Beyond

1. While it would be nice to have a single theory that explains all of chemical bonding interactions, we actually use many: valence bond theory, molecular orbital theory, and band theory. Make a list of the types of structures or properties of molecules that each of these is best used to describe.

2. Hybrid orbital theory predicts that the geometry around each N atom of urea, $(H_2N)_2C{=}O$ would be trigonal pyramidal. In fact, the entire molecule is flat. What does the flat geometry around each N tell you about the hybrid orbitals it utilizes? Give an explanation for why this anomaly occurs.

3. In solid metals, atomic orbitals overlap to form bands of orbitals that are not completely occupied (see Screen 10.14). In Mg, each Mg atom has a filled 3s orbital and a band formed from those orbitals would be expected to be full, but solid Mg is still a good conductor of electricity. How can you explain this?

STUDY QUESTIONS

1. Draw the Lewis structure for OF_2. What are its electron-pair and molecular geometries? Describe the bonding in the molecule in terms of hybrid orbitals.

2. Give the hybrid orbital set used by sulfur in each of the following molecules or ions:

(a) SO_2 (c) SO_3^{2-}

(b) SO_3 (d) SO_4^{2-}

How does the hybrid orbital set change on adding O atoms to sulfur or on changing the charge on the species?

3. What is the hybridization of the nitrogen atom in the nitrate ion, NO_3^-? Describe the orbitals involved in the formation of an N=O bond.

4. Acrolein, a component of photochemical smog, has a pungent odor and irritates eyes and mucous membranes.

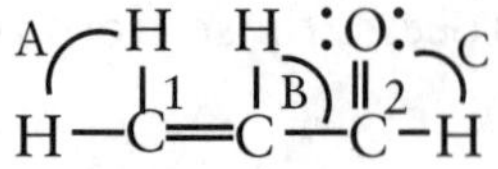

(a) What are the hybridizations of carbon atoms 1 and 2?

(b) What are the approximate values of angles A, B, and C?

Acrolein

5. Lactic acid is a natural compound found in sour milk.

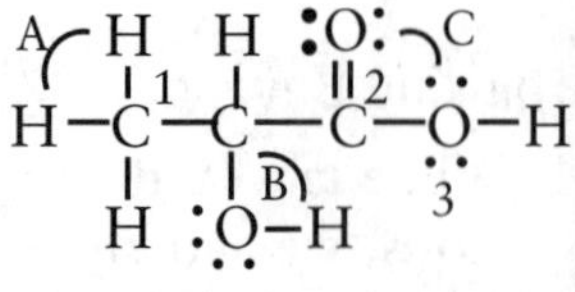

(a) How many π bonds are there in lactic acid? How many σ bonds?

(b) Give the hybridization of each atom 1 through 3.

(c) Which CO bond is the shortest in the molecule?

(d) Give the approximate values of the bond angles A through C.

Lactic acid

6. Hydrogen, H_2, can be ionized to give H_2^+. Write the electron configuration of the ion in molecular orbital terms. What is the bond order of the ion? Is the hydrogen-hydrogen bond stronger or weaker in H_2^+ than in H_2?

7. Calcium carbide, CaC_2, contains the acetylide ion, C_2^{2-}. Sketch the molecular orbital energy level diagram for the ion. How many net σ and π bonds does the ion have? What is the carbon-carbon bond order? How has the bond order changed on adding electrons to C_2 to obtain C_2^{2-}? Is the C_2^{2-} ion paramagnetic?

8. Assuming that we can apply the energy level diagram for homonuclear diatomic molecules to heteronuclear diatomics (Screen 10.11; use the diagram for O_2), write the electron configuration for carbon monoxide, CO. Is the molecule diamagnetic or paramagnetic? What is the net number of σ and π bonds? What is the carbon-oxygen bond order?

9. Nitrogen, N_2, can ionize to form N_2^+ or absorb an electron to give N_2^-. Compare these species with regard to (a) their magnetic character, (b) net number of π bonds, (c) bond order, (d) bond length, and (e) bond strength.

10. Which of the following molecules or molecule-ions should be paramagnetic? Assume the molecular orbital diagram for O_2 (Screen 10.11) applies to all of them.

(a) NO (d) Ne_2^+

(b) OF- (e) CN

(c) O_2^{2-}

CHAPTER 11
Bonding and Molecular Structure: Organic Chemistry

Screen 11.2 Carbon-Carbon Bonds

1. a) Benzene, C_6H_6, is a planar molecule (Screen 10.7SB). Here we see that another 6-carbon cyclic molecule, cyclohexane (C_6H_{12}), is not planar. Contrast the carbon atom hybridization in these two molecules.

   ___

   b) Why is π electron delocalization possible in benzene?

   ___

   c) Why is cyclohexane not planar with π electrons delocalized over the ring? Use valence bond theory (with hybrid atomic orbitals) to explain this difference.

   ___

   ___

   ___

2. What is the main difference between carbon and another Group 4A element, silicon, that explains why carbon can form long chains of atoms, whereas silicon forms only very short chains?

   ___

   ___

3. What is the hybridization of carbon in diamond? ________ In graphite? ________

Screen 11.3 Hydrocarbons

1. What is the primary structural difference between alkanes, alkenes, and alkynes?

   ___

   ___

2. What is the primary structural feature of aromatic compounds? (See the *Models* folder for more aromatic compounds such as benzene, naphthalene, phenanthrene, and anthracene.)

   ___

   ___

3. On Screen 10.7 we described the bonding of ethene, commonly called ethylene. To which class of hydrocarbons does this compound belong? ________

4. Consider the molecule styrene. To which class or classes of hydrocarbons does the compound belong? ____________________

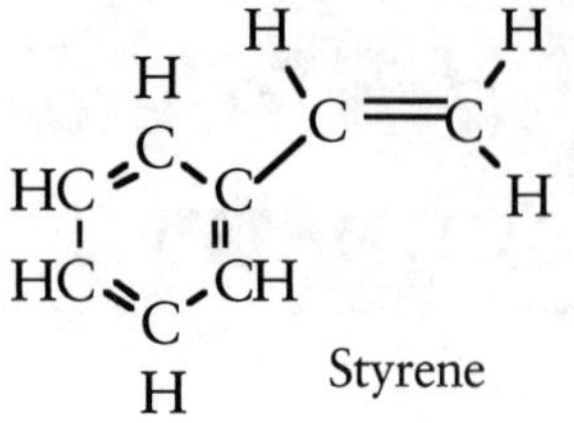

Styrene

Screen 11.4 Hydrocarbons and Addition Reactions

1. Why are addition reactions termed "addition?"

____

2. Many molecules other than those shown on this screen can be involved in addition reactions. Draw expanded structural formulas for the products of each of the following reactions:

   ethylene + HCl → ?

   ethylene + H_2O → ?

3. Alkenes can be hydrogenated to form alkanes. Alkynes can be hydrogenated as well. What class or classes of hydrocarbons can be formed from the hydrogenation of an alkyne?

   ____

4. Alkanes do not undergo addition reactions. Why not?

   ____

Screen 11.5 Functional Groups

1. Describe the hybrid orbitals used by the indicated atoms.

Atom	Hybridization
The O atom in an alcohol	
The C=O carbon in an aldehyde	
The C=O oxygen atom in a ketone	
The C=O carbon in a carboxylic acid	
The C—O—C oxygen atom in an ester	
The N atom in an amine	
The N and C=O atoms of an amide	

(Note that you may need to fill in the appropriate number of lone electron pairs on the atoms; these are not depicted in the diagrams on the screen.)

Screen 11.6 Reactions of Alcohols

1. What is the difference between a substitution reaction and an elimination reaction?

2. Compare the elimination reaction shown on this screen with the hydrogenation reaction shown on Screen 11.4.

Screen 11.7 Fats and Oils

1. What type of reaction is used to make a fat or oil from glycerol and a fatty acid: addition, substitution, or elimination?

2. a) What is the primary structural difference between fats and oils?

   b) What types of functional groups do each contain?

   c) What class of hydrocarbon fragments does each contain?

3. What structural feature of oil molecules prevents them from coiling up upon themselves as fat molecules do?

Screen 11.8 Amino Acids and Proteins

1. What two functional groups do amino acids contain?

2. The functional group in a peptide, —NH—CO—, is planar. Use hybrid orbitals and the concept of resonance to explain why this can be the case.

3. What type of reaction is the condensation reaction shown in the animation on this screen: addition, substitution, or elimination? ____________________

4. The peptide shown in the animation is composed of three amino acids. Could this "tripeptide" further react with one or more additional amino acid molecules?

5. Draw the tripeptide that can be formed by linking together three amino acids called alanine.

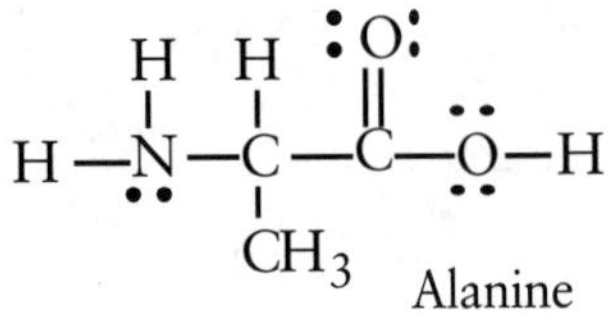

H H :O:
H—N—C—C—O—H
CH$_3$

Alanine

Screen 11.9 Addition Polymerization

1. What is the primary structural feature of the molecules used to form addition polymers?

2. Consider the animation of a polymerization reaction shown on this screen. The polymer made here has a chain of 14 carbon atoms. Could the chain that was formed have been shorter than this or longer than this?

3. What controls the length of the polymer chains formed?

4. Can the polymerization reaction be classified as one of the reaction types we studied earlier: addition, substitution, or elimination?

Screen 11.10 Condensation Polymerization

1. What is the primary structural feature necessary for a molecule to be useful in a condensation polymerization reaction?

2. The polyester on this screen is formed from ethylene glycol and terephthalic acid.

   a) Draw the structure of one unit of the condensation polymer.

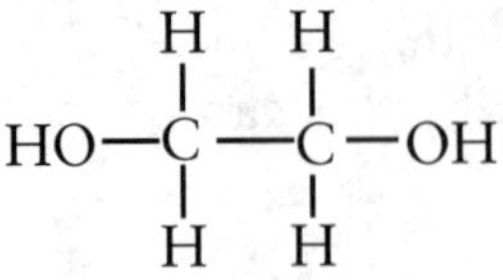

HO—CH$_2$—CH$_2$—OH

Ethylene glycol

HO—C(=O)—C$_6$H$_4$—C(=O)—OH

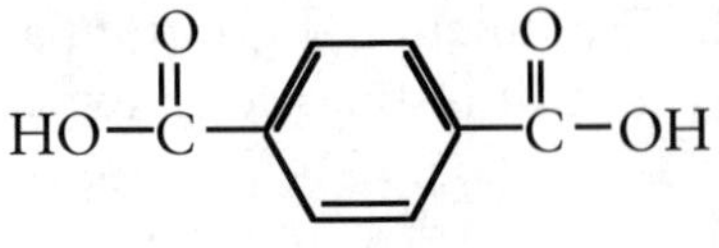

Terephthalic acid

   b) What molecule is eliminated during the condensation process? ______

c) Why is the reaction called a condensation reaction?

3. In what way are the reactions shown on this screen dissimilar to the addition polymerization reactions shown on the previous screen?

4. Shown below are the structures of the two reactants used to make nylon in the video on this screen. In this reaction, HCl is eliminated instead of H_2O. Draw a segment of the structure of nylon.

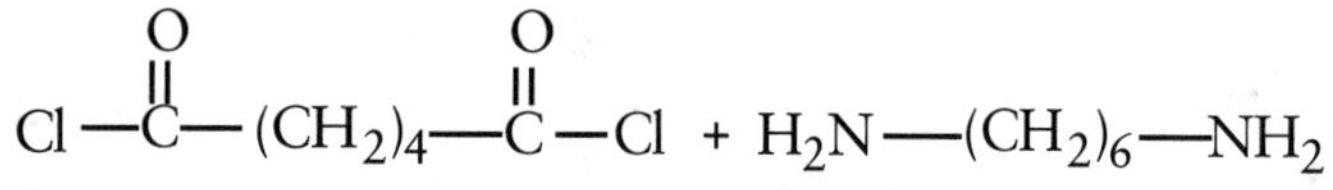

$$Cl-\overset{\overset{\displaystyle O}{\|}}{C}-(CH_2)_4-\overset{\overset{\displaystyle O}{\|}}{C}-Cl + H_2N-(CH_2)_6-NH_2$$

Screen 11.11 Return to Puzzler

1. The structures of butadiene and of styrene are shown here. They undergo an addition polymerization reaction similar to that shown on Screen 11.9. Draw a segment of the structure of the SBR rubber formed.

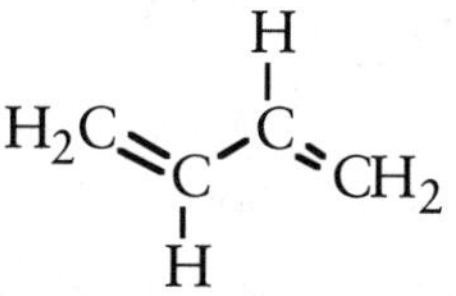

$H_2C{=}CH{-}CH{=}CH_2$

1,3-Butadiene

$H_2C{=}CH{-}C_6H_5$

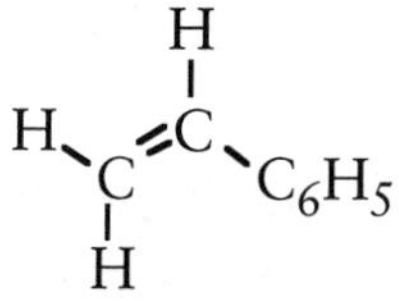

Styrene

Thinking Beyond

1. Many people speculate that a version of life could exist based on the chemistry of silicon, which is in the same periodic group as carbon. On Screen 11.2, we noted that silicon does not form long chains of silicon atoms. In addition, silicon very rarely forms double bonds, which are an important feature of the chemistry of carbon. Speculate as to the reason silicon is much less able to form π bonds than than carbon.
2. When ethylene reacts with itself, it forms long chains of singly bonded carbon atoms, the polymer polyethylene. Explain why the following reaction does not occur instead:

   $2\ H_2C{=}CH_2 \rightarrow$ cyclobutane

3. Consider the addition reaction of propene, $H_2C{=}CHCH_3$ and HCl. Draw structures for the two isomers that are possible from this reaction.

STUDY QUESTIONS

In doing these questions, it may be helpful to review Screens 3.4 (Isomers) and 3.6 (Alkanes)

1. Which of the following is an alkane? Which could be a cycloalkane?

   (a) C_2H_4, (b) C_5H_{10}, (c) $C_{14}H_{30}$, and (d) C_7H_8?

2. Isomers are molecules that have the same formula but different atom-to-atom connections. The two molecules below are so-called structural isomers of hexane, and there are others having this same formula. Draw at least two more isomers of hexane, C_6H_{14}.

   $CH_3—CH_2—CH_2—CH_2—CH_2—CH_3$
   Hexane

   $CH_3—CH(CH_3)—CH(CH_3)—CH_3$
   2,3-Dimethylbutane

3. In addition to the structural isomerism of alkanes, other types of isomers are found in organic chemistry. For example, dimethyl ether (CH_3OCH_3) and ethanol (CH_3CH_2OH) are isomers; they have the same molecular formula but their chemical functionality is quite different.

   (a) Draw all of the isomers possible for C_3H_8O. Tell into what class of compounds each fits.

   (b) Draw the structural formula for an aldehyde and a ketone with the molecular formula C_4H_8O.

4. Complete the following reactions.

   (a) $CH_3CH_2CH{=}CH_2 + Br_2 \rightarrow$

   (b) $HC{\equiv}CCH_3 + H_2$ (in the presence of a catalyst) $\rightarrow$

   (c) $HC{\equiv}CCH_3 + 2\ H_2$ (in the presence of a catalyst) $\rightarrow$

   (d) $CH_3CH{=}CHCH_3 + H_2O \rightarrow$

5. Describe a simple chemical test to tell the difference between $CH_3CH_2CH_2CH{=}CH_2$ and its isomer cyclopentane.

CHAPTER 12
Gases

Screen 12.2 Properties of Gases

1. List the four principal measurements we use to describe a gas along with the units with which we describe them.

   a) ____________________ c) ____________________

   b) ____________________ d) ____________________

2. Explain how each of the parameters listed above changes as air is pumped into the tire during the video on this screen.

   a) ____

   b) ____

   c) ____

   d) ____

3. Does the nature of the particular compounds making up a gas seem to matter?

   ____

4. The two most common units of gas pressure are atmospheres and millimeters of mercury.

   a) What pressure, in atmospheres, is represented by a pressure of 722 mm Hg? __________

   b) What pressure, in mm Hg, is represented by a pressure of 1.25 atm? __________

EXERCISE 12.1 *Units of Pressure*

Rank the following pressures in decreasing order of magnitude (largest first, smallest last): 75 kPa, 250 mm Hg, 0.83 bar, and 0.63 atm.

Screen 12.3 Gas Laws

1. For each of the following questions, assume that properties of the gas not mentioned remain constant:

   a) As the volume of a gas increases, its pressure ________________ .

   b) As the temperature of a gas increases, its volume ________________ .

   c) As the number of moles of a gas decreases, the volume of the gas ________________ .

   d) As the volume of a gas decreases, its pressure ________________ .

   e) As the temperature of a gas decreases, its pressure ________________ .

   f) As the number of moles of gas increases, the pressure ________________ .

2. In the animation of Charles's Law, it is stated that if we extrapolate to a temperature of absolute zero a gas has, "in principle," no volume. Why can this experiment not be performed?

3. Consider the animation of Avogadro's Law. If we instead combined gas samples containing 12 H_2 molecules and 4 N_2 molecules, what would be the ratio of volumes before and after the following reaction took place?

$$3\ H_2(g) + N_2(g) \rightarrow 2\ NH_3(g)$$

EXERCISE 12.2 *The Gas Laws*

1. *Boyle's Law:* A sample of CO_2 has a pressure of 55 mm Hg in a volume of 125 mL. The sample is moved to a new flask in which the pressure of the gas is now 78 mm Hg. What is the volume of the new flask? (The temperature was constant throughout the experiment.)
2. *Charles's Law:* A balloon is inflated with helium to a volume of 45 L at room temperature (25 °C). If the balloon is inflated with the same quantity of helium on a very cold day (-10 °C), what is the new volume of the balloon? (Assume the pressure is the same in both cases.)
3. *Avogadro's Law:* Methane burns in oxygen to give the usual products, CO_2 and H_2O.

$$CH_4(g) + 2\ O_2(g) \rightarrow 2\ H_2O(g) + CO_2(g)$$

If 22.4 L of gaseous CH_4 is burned, what volume of O_2 is required for complete combustion? What volumes of H_2O and CO_2 are produced? Assume all gases are measured at the same temperature and pressure.

Screen 12.4 The Ideal Gas Law

1. The ideal gas law is a combination of the three laws described on Screen 12.3. It can be used to describe other relationships as well. Make up your own law (calling it, for instance, Erin's Law or Bob's Law) describing the relationship between temperature and pressure. Is the pressure of a gas directly or inversely related to the gas temperature?

2. What would be the value of the gas constant, R, if it were expressed in units of $L \cdot mm\ Hg/K \cdot mol$?

EXERCISE 12.3 *The Ideal Gas Law*

The balloon used by Jacques Charles in his historic flight in 1783 was filled with about 1300 moles of H_2. If the temperature of the gas was 20.0 °C, and its pressure was 750 mm Hg, what was the volume of the balloon?

Screen 12.5 Gas Density

1. The equation for gas density is given on this screen as d = PM/RT. Use the ideal gas law and the units of molar mass to derive this expression.

2. The masses shown on the scale on this screen (Click on "More") show values of 1.619 g for CH_4 and 1.988 g for SF_6. Assuming equal conditions, remark on what is represented by the masses seen. Are the masses of the balloons and the displaced air involved?

   ____

   ____

   ____

3. Calculate the densities of He(g) and of H_2(g) at 25 °C and 1.0 atm.

   Density of He = __________ g/L

   Density of H_2 = __________ g/L

   Why is helium used in dirigibles (blimps), despite the fact that it is more dense and much less abundant on earth than hydrogen?

   ____

EXERCISE 12.4 *Gas Density Calculation*

Calculate the density of dry air at 15.0 °C and 1.00 atm if its molar mass (average) under those conditions is 28.96 g/mol.

Screen 12.6 Using Gas Laws: Determining Molar Mass

1. Examine how the molar mass of sulfur hexafluoride is determined on this screen. An alternate method of determining the molar mass is to rearrange the formula on the previous screen (d = PM/RT). Perform the calculation in that manner and see if it matches that found by the method presented here.

EXERCISE 12.5 *Molar Mass from P, V, and T Data*

A 0.105-g sample of a gaseous compound has a pressure of 560. mm Hg in a volume of 125 mL at 23.0 °C. What is its molar mass?

Screen 12.7 Gas Laws and Chemical Reactions: Stoichiometry

1. Examine the calculation on this screen. What is assumed about the temperature of the sample during the reaction?

   ____

2. If the temperature increases during the reaction, would more or less sodium azide be required to inflate the bag to the required pressure? Explain.

   ____

3. Another substance that decomposes to form a gas is hydrogen peroxide.

   $$2\ H_2O_2(aq) \rightarrow 2\ H_2O(\ell) + O_2(g)$$

   What volume of an 8.8 M solution of hydrogen peroxide would be needed to fill the bag with oxygen gas in the same manner as described on this screen? (Note that the reaction of hydrogen peroxide is in fact too slow at normal temperatures to be useful in an air bag.)

EXERCISE 12.6 *Gas Laws and Stoichiometry*

Gaseous oxygen reacts with aqueous hydrazine to produce water and gaseous nitrogen according to the balanced equation

$$N_2H_4(aq) + O_2(g) \rightarrow 2\ H_2O(\ell) + N_2(g)$$

If a solution contains 180 g of N_2H_4, what is the maximum volume of O_2 that will react with the hydrazine if the O_2 is measured at a barometric pressure of 750 mm Hg and room temperature (21 °C)?

Screen 12.8 Gas Mixtures and Partial Pressures

Dry air at sea level is composed of 78.08% N_2, 20.95% O_2, 0.93% Ar, and 0.033% CO_2. Calculate the partial pressure of each of these gases (in units of mm Hg) if the total atmospheric pressure is exactly 1 atm.

$P(N_2)$ = __________

$P(O_2)$ = __________

$P(Ar)$ = __________

$P(CO_2)$ = __________

EXERCISE 12.7 *Partial Pressures*

Halothane has the formula $C_2HBrClF_3$. It is a nonflammable, nonexplosive, and nonirritating gas that is a commonly used inhalation anesthetic. Suppose you mix 15.0 g of halothane vapor with 23.5 g of oxygen gas. If the total pressure of the mixture is 855 mm Hg, what is the partial pressure of each gas?

Screen 12.9 The Kinetic Molecular Theory of Gases

1. What are all the pictures on this screen(of fish, coffee, garlic, blue cheese, and an orange) intended to make you think about? What does this have to do with the kinetic theory of gases?

2. If the absolute temperature of a gas doubles, by how much, on average, does the speed at which its molecules travel increase?

3. Do the tenets of the kinetic molecular theory include the shape, size, or chemical properties of gas molecules?

Screen 12.10 Gas Laws and the Kinetic Molecular Theory

1. What is the nature of pressure on the molecular level assumed on this screen?

2. This screen shows animations describing the following relationships on the molecular scale: P vs. n, P vs. T, and P vs. V. Sketch out a molecular scale animation for the relationship between n and V.

Screen 12.11 Distribution of Molecular Speeds

1. The video and plot of automobile speeds on this screen is an analogy to the distribution of gas speeds on the molecular scale. The effect of increasing temperature on the molecular scale is the same as that of moving cars from a slow city street to a fast interstate highway. Draw curves describing the distribution of automobile speeds in these two situations.

2. Examine the *Boltzman Distribution* tool on this screen.

   a) What trends do you observe about gas speeds and the molar mass of the gas?

   b) What trends are observed for changes in temperature?

   c) As temperature is increased, the curves shift to the right but also decrease in height. What does the decrease in height signify? Why does it happen?

Screen 12.12 Application of the Kinetic Molecular Theory

1. The presentation on this screen shows that NH_3 molecules diffuse through air more quickly than do HCl molecules. Why does this happen?

   Draw Boltzman distribution curves for each of these compounds to support your point.

2. What compound is formed when the two gases, NH_3 and HCl, meet? Why does it form a "smoke?"

3. On the sidebar to this screen, the separation of uranium isotopes is described. The process of diffusion of $UF_6(g)$ through a membrane needs to be repeated about a thousand times to enrich the ^{235}U from 0.7% to 4%. Why must the process be repeated so many times?

____

____

Thinking Beyond

1. The *Boltzman Distribution* tool on Screen 12.11 indicates that molecules in air travel hundreds of meters per second. When a bottle of a noxious chemical or a person wearing perfume or cologne enters a room, however, far more time is needed for the smell to permeate through the space. Why is this the case?
2. Consider the tenets of the kinetic molecular theory presented on screen 12.9. Which of these tenets is most likely *not* true?

STUDY QUESTIONS

1. A sample of CO_2 gas is placed in a 125-mL flask where it exerts a pressure of 67.5 mm Hg. What is the pressure of this gas sample when it is transferred to a 500.-mL flask at the same temperature?
2. You have 3.5 L of helium at a temperature of 22.0 °C. What volume would the helium occupy at 37 °C? (The pressure of the helium sample is constant.)
3. Water can be made by combining gaseous O_2 and H_2. If you begin with 1.0 L of H_2 gas at 380 mm Hg and 25 °C, how many liters of O_2 gas would you need for complete reaction if the O_2 gas is also measured at 380 mm Hg and 25 °C?
4. A 1.25-g sample of CO_2 is contained in a 850.-mL flask at 22.5 °C. What is the pressure of the gas?
5. To find the volume of a flask, it was first evacuated so that it contained no gas at all. Next, 4.4 g of CO_2 is introduced into the flask. On warming to 22 °C, the gas exerted a pressure of 635 mm Hg. What is the volume of the flask?
6. If you have a 150.-L tank of gaseous nitrogen, and the gas exerts a pressure of 41.8 mm Hg at 25 °C, how many moles of helium are there in the tank?
7. A 0.982-g sample of an unknown gas exerts a pressure of 700. mm Hg in a 450.-mL container at 23 °C. What is the molar mass of the gas?
8. A hydrocarbon, C_xH_y, is 82.66% carbon. Experiment shows that 0.218 g of the hydrocarbon has a pressure of 374 mm Hg in a 185-mL flask at 23 °C. What are the empirical and molecular formulas of the compound?
9. Forty miles above the earth's surface the temperature is 250 K and the pressure is only 0.20 mm Hg. What is the density of air (in g/L) at this altitude? (Assume the molar mass of air is 29 g/mol.)
10. If 12.0 g of O_2 is required to inflate a balloon to a certain size at 27 °C, what mass of O_2 is required to inflate it to the same size (and pressure) at 81 °C?

11. Iron reacts with acid to produce iron(II) chloride and hydrogen gas.

$Fe(s) + 2\ HCl(aq) \rightarrow FeCl_2(aq) + H_2(g)$

The H_2 gas from the reaction of 1.0 g of iron with excess hydrochloric acid is collected in a 15.0-L flask at 25 °C. What is the pressure of the H_2 gas in this flask?

12. Sodium azide, the explosive compound in automobile air bags, decomposes according to the equation

$2\ NaN_3(s) \rightarrow 2\ Na(s) + 3\ N_2(g)$

What mass of sodium azide is required to provide the nitrogen to inflate a 25.0-L bag to a pressure of 1.3 atm at 25 °C?

13. Hydrazine reacts with O_2 according to the equation

$N_2H_4(g) + O_2(g) \rightarrow N_2(g) + 2\ H_2O(\ell)$

Assume the O_2 needed for the reaction is in a 450-L tank at 23 °C. What must the oxygen pressure be in the tank to have enough oxygen to consume 1.00 kg of hydrazine completely?

14. Iron forms a series of compounds of the type $Fe_x(CO)_y$. If you heat the compounds in air, they decompose to Fe_2O_3 and CO_2 gas. After heating a 0.142-g sample of $Fe_x(CO)_y$, you isolate the CO_2 in a 1.50-L flask at 25 °C. The pressure of the gas is 44.9 mm Hg. What is the formula of $Fe_x(CO)_y$?

15. Helium (0.56 g) and hydrogen gas are mixed in a flask at room temperature. The partial pressure of He is 150 mm Hg and that of H_2 is 25 mm Hg. How many grams of H_2 are present?

16. A collapsed balloon is filled with He to a volume of 12 L at a pressure of 1.0 atm. Oxygen (O_2) is then added so that the final volume of the balloon is 26 L with a total pressure of 1.0 atm. The temperature, constant throughout, is equal to 20 °C.

(a) How many grams of He does the balloon contain?

(b) What is the final partial pressure of He in the balloon?

(c) What is the partial pressure of O_2 in the balloon?

17. Dichlorine oxide is a powerful oxidizing agent that is used to bleach wood pulp and to treat municipal water supplies. It is made by the reaction

$SO_2(g) + 2\ Cl_2(g) \rightarrow OSCl_2(g) + Cl_2O(g)$

If you put SO_2 in a flask so that its pressure is 125 mm Hg at 22 °C, and if you add Cl_2 gas to this same flask, what should the Cl_2 partial pressure be in order to have the correct stoichiometric ratio of SO_2 to Cl_2?

18. You are given two flasks of equal volume. Flask A contains H_2 at 0 °C and 1 atm pressure. Flask B contains CO_2 gas at 0 °C and 2 atm pressure. Compare these two gases with respect to each of the following:

(a) Average kinetic energy per molecule

(b) Average molecular velocity

(c) Number of molecules

(d) Mass of gas

19. Place the following gases in order of increasing average molecular speed at 25 °C: (a) Ar, (b) CH_4, (c) N_2, and (d) CH_2F_2.

CHAPTER 13
Bonding and Molecular Structure: Intermolecular Forces, Liquids, and Solids

Screen 13.2 Phases of Matter

1. List the similarities and differences between the three states of matter.

2. What holds molecules to one another in the liquid or solid phase?

Screen 13.3 Intermolecular Forces (1)

1. There are four recognized forces of nature: gravity, the coulombic force, the strong nuclear force, and the weak nuclear force. Which of these four is responsible for all intermolecular forces?

2. Watch the video on the problem screen associated with this screen. The highly exothermic reaction involves the decomposition of sucrose and the formation of water and carbon. What is primarily responsible for the release of energy during the reaction?

3. The table of Intermolecular Force Strengths (IMFs) on this screen indicates that substances with strong intermolecular forces are solids, those with moderate IMFs are liquids, and those with weak IMFs are gases. Explain this in terms of the molecular scale view of how molecules act in each of these phases.

Screen 13.4 Intermolecular Forces (2)

1. a) Watch the animation that plays when you study *Ion-Dipole Forces* on this screen. Why do you see an interaction between the Na^+ ions and the O atom of water molecules and not between Na^+ ions and the H atoms of water?

   b) Considering Coulomb's Law, explain the trend in enthalpy of hydration values for Group 1A cations.

   c) Which ion would you expect to have a greater enthalpy of hydration, Na^+ or Mg^{2+}?

2. a) Click on *Dipole-Dipole Forces* and then on the table of *Molar Masses and Boiling Points*. What is the connection between the strength of a compound's intermolecular forces and its boiling point?

   b) What two factors seem to control the strength of intermolecular forces, as reflected in the boiling point table?

   c) Which compound would you expect to have a higher boiling point, HF or Ne?

Screen 13.5 Intermolecular Forces (3)

1. What does the term "polarizable" mean?

2. Why are O_2 molecules attracted to molecules of water, despite the fact that O_2 is nonpolar and, as such, might be expected to be unaffected by coulombic forces?

3. What is the trend between the size of a molecule and the ease with which its electrons can be polarized? Why?

4. Can only nonpolar molecules be polarized and experience induced dipole forces, or can ions and polar molecules experience this effect as well?

5. Rank the halogens F_2, Cl_2, and Br_2 in terms of their expected boiling points. Explain briefly.

Screen 13.6 Hydrogen Bonding

1. What are the two reasons compounds containing F-H, O-H, and N-H bonds experience especially strong dipole-dipole forces (called hydrogen bonds)?

2. Which of the following compounds will undergo hydrogen bonding with another molecule of the same kind? You may need to examine the structures in the *Models* folder.
   a) urea, $(NH_2)_2CO$ (in the MISC_ORG folder within the ORGANIC folder)
   b) acetone, $(CH_3)_2CO$ (in the ALDE_KET folder within the ORGANIC folder)
   c) methanol, CH_3OH (in the ALCOHOLS folder within the ORGANIC folder)
   d) benzene, C_6H_6 (in the AROMATIC folder within the ORGANIC folder)

3. How might the structure of DNA be different if hydrogen bonding was not possible?

Screen 13.7 The Weird Properties of Water

This screen describes a number of unique properties of water and some reasoning behind those properties.

1. What is the primary reason water is a special substance?

2. Give a reason why CH_4 does not display "water-like" properties.

3. Give a reason why HF does not display "water-like" properties.

4. Which is usually more dense, the liquid or solid phase of a substance? What experiment could you do to determine which is more dense?

EXERCISE 13.1 *Intermolecular Forces*

Decide what type of intermolecular force is involved with (1) liquid O_2; (2) hydrated magnesium sulfate, $MgSO_4 \cdot 7\ H_2O$; and (3) O_2 in water.

Screen 13.8 Properties of Liquids (Enthalpy of Vaporization)

1. Describe the process of vaporization. What IMFs are broken during evaporation of a liquid such as water? Are any formed?

2. Consider the table of enthalpy of vaporization values. What two aspects of molecules seem to control enthalpy of vaporization values?

3. Can enthalpy of vaporization values ever be negative, that is, can evaporation ever be an exothermic process? Explain briefly.

Screen 13.9 Properties of Liquids (Vapor Pressure)

1. What relationship do you expect to exist between vapor pressure and enthalpy of vaporization for a series of compounds?

2. Look up the structures of propane and butane in the *Models* folder (located in the ALKANE folder within the ORGANIC folder or directory). Explain why propane tanks are made of steel whereas butane can be stored in weaker plastic containers. (It may be helpful to check the properties of these compounds in the Table on Screen 13.8.)

EXERCISE 13.2 *Enthalpy of Vaporization*

The molar enthalpy of vaporization of methanol, CH_3OH, is 35.21 kJ/mol at 64.6 °C. How much energy is required to evaporate 1.00 kg of this alcohol?

Screen 13.10 Properties of Liquids (Boiling Point)

1. What is the difference between evaporation and boiling?

2. Watch the series of photos on this screen and examine the attached problem screen. At what temperature would you expect water to boil when the external pressure is about 600 mm Hg? How will this affect the temperature at which pasta cooks in a city at high elevation, such as Lake Tahoe, California?

3. Ethanol, C_2H_5OH, with a normal boiling point of 78.5 °C, has weaker intermolecular forces than does water. Sketch plots of vapor pressure vs. temperature for water and ethanol. How do they differ?

4. Based on their boiling points, which liquid, water or ethanol, do you expect to have the greater enthalpy of vaporization?

Screen 13.11 Properties of Liquids (Surface Tension, Capillary Action, and Viscosity)

1. What controls a liquid's surface tension?

2. Think about toweling off after your daily shower. Explain this process in terms of adhesive and cohesive forces. Indicate which forces are stronger, the adhesive forces between the towel and the water, or between your skin and the water. Do you think your skin is more or less polar than the molecules in the towel?

3. Motor oil consists of long chain hydrocarbon molecules. (See models of molecules such as eicosane in the ALKANES folder in the ORGANIC folder of the *Models* folder.) Why is motor oil viscous?

____

____

4. Gasoline is composed of molecules that are very similar to those of motor oil, but they have much shorter carbon chains. Why is gasoline less viscous than motor oil?

____

____

EXERCISE 13.3 *Viscosity*

Glycerol, $CH_2(OH)CH(OH)CH_2(OH)$, is used in cosmetics. Is its viscosity greater or less than that of ethanol, CH_3CH_2OH? Why or why not? (A model of glycerol is included in the ALCOHOLS folder in the *Models* folder.)

Screen 13.12 Solid Structures (Crystalline and Amorphous Solids)

1. What is the difference between crystalline and amorphous solids?

____

____

2. What is a unit cell and why do we make use of them?

____

____

3. Watch the simple cubic unit cell animation on this screen. How many unit cells make up the rotating structure on the screen?

____

____

4. Consider the shapes of the following solids and decide if you think they are crystalline solids or amorphous solids:
   a) table salt _______________________________
   b) ice (think of snow flakes) _______________________________
   c) glass (think of how it breaks) _______________________________
   d) wood _______________________________
5. Examine the unit cells for gold and tungsten. They are found in the *Models* folder (and in the folder labeled SOLIDS). (Instructions for the use of these models, and for measuring distances, are given in Appendix A. Note that dimensions are given in Ångstrom units where 1 Å = 10^{-10} m = 100 pm.)

Name	Type of Unit Cell	Description	Edge Distance (in Å)
Gold	Face-centered cubic		
Tungsten	Body-centered cubic		

What is the difference between simple cubic, face-centered cubic, and body-centered cubic unit cells?

Screen 13.13 Solid Structures (Ionic Solids)

1. Watch the animation of a face-centered cubic (FCC) unit cell on this screen. Why is the sodium ion said to occupy "octahedral" holes?

2. What are the two attributes of the make up of an ionic compound that control its solid state structure?

3. The NaCl unit cell on this screen is "assembled" by placing Na^{+} ions in a face-centered cubic lattice of Cl^{-} ions. Examine the structure of NaCl using the model in the *Models* folder (in the SOLIDS folder). This was "assembled" from a face-centered cubic lattice of Na^{+} ions with Cl^{-} ions in the holes. Are there any differences between this model and the one on Screen 13.13? Do both lead to the same formula for salt?

4. Examine the structure of CaO using the model in the *Models* folder (in the SOLIDS folder).

a) Describe the structure. What type of lattice is described by the Ca^{2+} ions? What type of hole does the O^{2-} occupy in the lattice of Ca^{2+} ions? (If there are "lines" connecting the ions, these are not bonds; they can be removed by going to "Bond Shape" in the "View" menu and changing "cylinder radius" to 0.0 Å.)

b) How is the formula of CaO related to its unit cell structure? (How many Ca^{2+} ions and how many O^{2-} ions are "inside" the unit cell?)

c) How is the CaO structure related to the NaCl structure?

5. Examine the structure of CsCl using the model in the *Models* folder (in the SOLIDS folder).
   a) Describe the structure. What type of lattice is described by the Cs^+ ions (silver colored)? Where are the Cl^- ions (green) located?

   b) How is the formula of CsCl related to its unit cell structure? (How many Cs^+ ions and how many Cl^- ions are "inside" the unit cell?)

6. Examine the structure of ZnS using the model in the *Models* folder (in the SOLIDS folder or directory).
   a) Describe the structure. What type of lattice is described by the Zn^{2+} ions (silver color)? What type of holes do the S^{2-} ions (yellow color) occupy in the lattice of Zn^{2+} ions? How is the formula of ZnS related to its unit cell structure?

   b) Lead sulfide, PbS (commonly called galena) has the same type of formula as ZnS. Does it have the same solid structure? (See the model in the *Models* folder.) If different, how it is different? How is its unit cell related to its formula?

7. Examine the structure of CaF_2 using the model in the *Models* folder (in the SOLIDS folder or directory).

   a) Describe the structure. What type of lattice is described by the Ca^{2+} ions? What type of holes do the F^- occupy in the lattice of Ca^{2+} ions?

   b) How is the formula of CaF_2 related to its unit cell structure? (How many Ca^{2+} ions and how many F^- ions are "inside" the unit cell?)

   c) How is the CaF_2 structure related to the ZnS structure?

8. Examine the structure of TiO_2 using the model in the *Models* folder (in the SOLIDS folder or directory).

   a) Describe the structure. What type of lattice is described by the Ti^{4+} ions?

   b) How is the formula of TiO_2 related to its unit cell structure? (How many Ti^{4+} ions and how many O^{2-} ions are "inside" the unit cell?)

9. Examine the structure of NaCl using the structure in the *Models* folder labeled NACL/L.CSF (where L stands for "large") in the SOLIDS folder. How are the Na^+ and Cl^- ions arranged? What does this tell you about solid state structures in general? (Also examine the structure labeled RUTILE.CSF, which shows the extended lattice of TiO_2.) (Be sure to rotate these models to get a clear picture of the three-dimensional structure.)

Screen 13.14 Solid Structures (Molecular Solids)

1. Solid iodine, I_2, is held as a solid due to induced dipole-induced dipole forces between adjacent I_2 molecules. Draw a reasonable structure for the solid.

2. Sketch out what happens on the molecular scale for each of the following reactions involving I_2:

a) sublimation, $I_2(s) \rightarrow I_2(g)$

b) decomposition, $I_2(s) \rightarrow 2\ I(g)$

3. Draw a reasonable structure for solid HCl (which exists as a solid at low temperatures).

Screen 13.15 Solid Structures (Network Solids)

1. Both molecular solids (described on Screen 13.14) and network solids are composed of molecules held together by covalent bonds. What, then, is the difference between molecular solids and network solids?

____

____

2. In what way does the solid structure of diamond explain its great hardness?

____

____

3. Examine the model of the structure of quartz in the *Models* folder or directory (under SiO2.CSF in the SOLIDS folder). Which type of solid is quartz?

____

____

Screen 13.16 Silicate Minerals

1. Using the structure of amphibole, explain why silicate minerals are frequently both network and ionic solids.

____

____

2. Much of the study of the chemistry of the earth involves examining how atoms of one element can replace those of another element in solid structures. Which atom or ion do you think it would be easier to replace in the amphibole structure, O or Mg? Explain your reasoning.

____

Screen 13.17 Phase Diagrams

1. Examine the phase diagram for water shown on this screen. Notice that the line between liquid and solid slants up to the left as pressure is increased. What does this imply about the relative densities of liquid and solid water?

2. Consider moving from left to right across the phase diagram beginning at a very low temperature and at a pressure of about 0.5 atm. Explain each step in terms of what occurs on the molecular scale and whether each step involves an endothermic or an exothermic process?

Thinking Beyond

1. Water has many unique properties based on its strong intermolecular forces. One way in which water acts rather normally is its viscosity—it flows quite freely. Give an explanation for the fact that water, despite its strong IMFs, does not have a high viscosity.
2. You can calculate the density of a solid if you know its structure and the mass and volume of the unit cell.
   a) Calculate the density of gold. Find the structure of gold in the SOLIDS folder of the *Models* folder. Measure the unit cell dimensions and then calculate the volume of the unit cell. (Read the instructions for measuring distances on molecular models in Appendix A.) (Note that the cell dimensions are measured in Ångstrom units where $1\ \text{Å} = 10^{-10}\ \text{m} = 100\ \text{pm}$.)
   b) Calculate the density of ZnS. Find the structure of ZnS in the SOLIDS folder of the *Models* folder. Measure the unit cell dimensions and then calculate the volume of the unit cell.

STUDY QUESTIONS

1. When KCl dissolves in water, what type of attractive forces must be overcome in the liquid water? What type of forces must be overcome in solid KCl? What type of attractive forces allow KCl to dissolve in liquid water?
2. What type of intermolecular force must be overcome in converting each of the following from a liquid to a gas?
   (a) liquid O_2
   (b) mercury
   (c) CH_3I (methyl iodide)
   (d) CH_3CH_2OH (ethanol)
3. Explain why the boiling point of H_2S is lower than that of water.

4. Which of the following compounds would be expected to form intermolecular hydrogen bonds in the liquid state?

(a) CH_3OCH_3 (dimethyl ether)
(b) CH_4
(c) HF
(d) CH_3CO_2H (acetic acid)
(e) Br_2
(f) CH_3OH (methanol)

5. The simple hydrocarbons methane (CH_4) and ethane (C_2H_6) cannot be liquefied at room temperature, no matter how high the pressure. Propane (C_3H_8), the next compound in the series of simple alkanes, has a critical pressure of 42 atm and a critical temperature of 96.7 °C. Can this compound be liquefied at room temperature? Can you think of a place where you would find liquefied propane?

6. Outline a two-dimensional unit cell for the pattern shown here. If the black squares are labeled A and the white squares are B, what is the simplest formula for a "compound" based on this pattern?

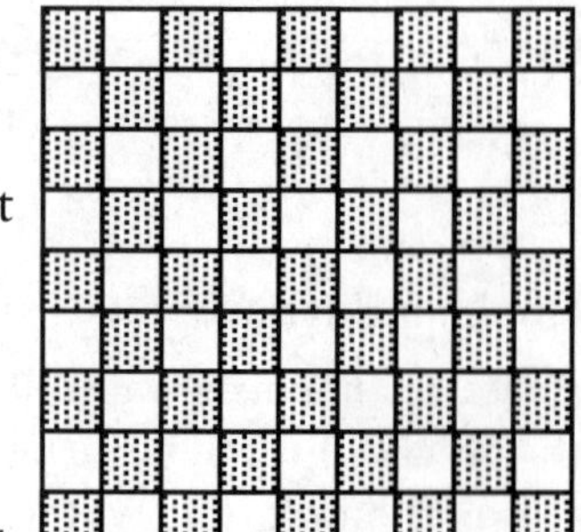

7. Acetone, $(CH_3)_2CO$, is a common laboratory solvent. However, it is usually contaminated with water. Why does acetone absorb water so readily? Draw molecular structures showing how water and acetone can interact. What intermolecular force(s) is (are) involved in the interaction?

8. Liquid ethylene glycol, $HOCH_2CH_2OH$, is one of the main ingredients in commercial antifreeze. Would you predict its viscosity to be greater or less than that of ethanol, CH_3CH_2OH?

9. Account for these facts:

(a) Although ethanol (C_2H_5OH) (bp, 80 °C) has a higher molar mass than water (bp, 100 °C), the alcohol has a lower boiling point.

(b) Salts of the ion HF_2^- are known.

(c) Mixing 50 mL of ethanol with 50 mL of water produces a solution with a volume less than 100 mL.

10. Can $CaCl_2$ have a unit cell like that of sodium chloride? Explain.

11. A unit cell of sodium oxide is shown here. How many sodium ions are there in the unit cell? How many oxide ions? What is the formula for the compound?

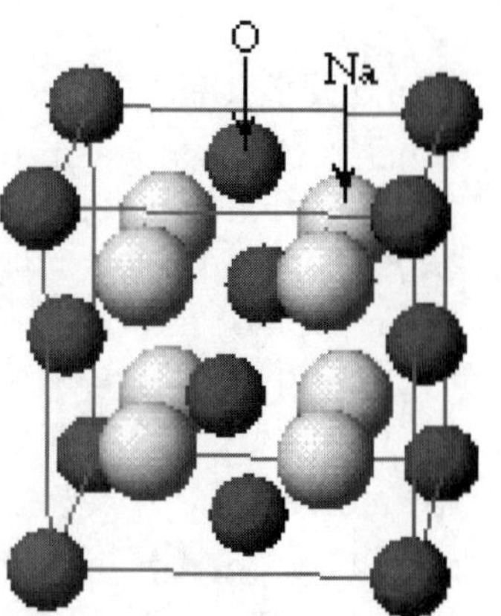

CHAPTER 14
Solutions and Their Behavior

Screen 14.2 Solubility

1. a) Examine the video that plays when the *Saturated* button is selected. Solutions of lead nitrate and potassium iodide are mixed together, resulting in a solid precipitate. Give the name and formula for the solid precipitate.

   ____

   b) Before the KI solution is added to the lead nitrate solution, is the KI solution saturated or unsaturated?

   ____

2. Examine the video that plays when the *Unsaturated* button is clicked. What is the evidence that the final solution of nickel chloride is unsaturated?

   ____

3. How can you test a solution to see if it is supersaturated?

   ____

4. Three alternate means of expressing concentration are introduced on this screen. Consider a solution that is formed by the dissolution of 4.88 g of $CaCl_2$ into 68.2 g of H_2O. Calculate the molality, mole fraction, and weight percent of $CaCl_2$ in this solution.

   a) Molality = ______________ mol/kg

   $$\text{Concentration(molality)} = \frac{\text{moles of solute}}{\text{kilograms of solvent}}$$

   b) Mole fraction = ______________

   $$X_{solute} = \frac{\text{moles of solute}}{\text{total moles of solute and solvent}}$$

   c) Weight percent = ______________

   $$\text{Weight \%} = \frac{\text{mass of solute}}{\text{total mass of solution}} \times 100\%$$

5. Examine the table of solubilities. Why do you think NH_3 is so much more soluble in water than either O_2 or N_2?

EXERCISE 14.1 *Mole Fraction, Molality, and Weight Percent*

If you dissolve 10.0 g or about 1 heaping teaspoonful of sugar, $C_{12}H_{22}O_{11}$, in a cup of water (250. g), what are the mole fraction, molality, and weight percentage of sugar?

Screen 14.3 The Solution Process

Examine the animation on this screen.

1. The process of converting the red liquid to a gas is depicted as endothermic. Do you think this is true of all liquids, or only some?

2. The condensation of the two types of molecules from the gaseous to the liquid state is depicted as exothermic. Do you think that all condensation processes are exothermic, or only some?

3. What, on the molecular scale, controls whether a dissolution reaction is endothermic or exothermic?

4. Examine the problem screen associated with this screen. What do you think would be found if the solubility of hexane (C_6H_{14}) were examined in these same two solvents, water and carbon tetrachloride? In which would hexane be more soluble?

Screen 14.4 Energetics of Solution Formation

1. What controls the solubility of an ionic compound: the strength of the ionic bonds holding the ions together, the forces of attraction between the ions and water molecules, or the relationship between these two?

2. Examine the animation dealing with the dissolution of KF on this screen. Why is the separation of the K^+ and F^- ions thought to be endothermic?

3. Examine the *Oil and Water* sidebar on this screen. What prevents water and oil from mixing intimately: strong attractions between oil molecules, strong attractions between water molecules, or weak attractions between oil and water molecules?

EXERCISE 14.2 *Enthalpy of Solution*

Use the data in the table on this screen to compare the enthalpies of solution for AgCl and RbF. Comment on the relation between the enthalpy of solution and solubility of these two salts.

Screen 14.5 Factors Affecting Solubility (1)

1. Henry's Law states that the solubility of a gas is proportional to the partial pressure of that gas. Calculate the solubility of O_2 at 25 °C if the partial pressure of O_2 is 0.21 atm. Calculate the solubility if the partial pressure is only 0.15 atm, as it would be if you were hiking high up in the mountains.

2. Why do different gases have different Henry's law constants?

Screen 14.6 Factors Affecting Solubility (2)

1. Sodium acetate becomes more soluble as temperature increases, whereas lithium sulfate becomes less soluble with increases in temperature. What is the difference in these two systems that leads to their differing behavior?

2. Many substances that are less soluble in warm water than cold water "appear" to dissolve more readily when added to warm water. How do our perceptions deceive us?

Screen 14.7 Colligative Properties (1)

1. What are colligative properties? Are they properties of the solvent in a solution or of the solute?

2. Why is the vapor pressure of a liquid lowered upon dissolution of a solute?

3. Why, on the molecular scale, is the vapor pressure of a solution proportional to the mole fraction of solvent?

4. What substance would have the greatest influence on the vapor pressure of water when added to 1000 g of the solvent: 10.0 g of sucrose ($C_{12}H_{22}O_{11}$), 10.0 g of ethylene glycol [$C_2H_4(OH)_2$], or 10.0 g of AgCl?

5. Eugenol, with an empirical formula of C_5H_6O, is the active component of oil of cloves. Suppose you dissolve 0.144 g of the compound in 10.00 g of benzene, C_6H_6. The vapor pressure of the benzene solution is 94.35 mm Hg at 25 °C, down from 95.00 mm Hg for pure benzene at this temperature.

   a) What is the mole fraction of eugenol in the solution?

   b) What quantity of eugenol (in moles) is dissolved in the solution?

   c) What is the molar mass of eugenol, and its molecular formula?

EXERCISE 14.3 *Using Raoult's Law*

Assume you dissolve 10.0 g of sugar ($C_{12}H_{22}O_{11}$) in 225 mL (225 g) of water and warm the water to 60 °C. What is the vapor pressure of the water over this solution? (The vapor pressure of pure water at 60 °C is 149.4 mm Hg.)

Screen 14.8 Colligative Properties (2)

1. In general, what does the addition of a solute do to the boiling point of a liquid? What does it do to the freezing point?

2. What does the addition of a solute do to the magnitude of the temperature range over which a compound exists as a liquid?

3. What is the boiling point of a solution in which 0.144 g of eugenol ($C_{10}H_{12}O_2$), which is found in oil of cloves, is dissolved in 10.0 g of benzene? (Answer: 80.32 °C)

4. What mass of ethylene glycol [$C_2H_4(OH)_2$] must be added to 5.50 kg of water to lower the freezing point of the water from 0.0 °C to -10.0 °C? (This is approximately the situation in your car.) (Answer: 1840 g)

EXERCISE 14.4 *Boiling Point Elevation*

What quantity of ethylene glycol, $C_2H_4(OH)_2$, must be added to 100. g of water to raise the boiling point by 1.0 °C? Express the answer in grams.

EXERCISE 14.5 *Determining Molar Mass by Boiling Point Elevation*

Crystals of the beautiful blue hydrocarbon, azulene (0.640 g), which has an empirical formula of C_5H_4, are dissolved in 100. g of benzene. The boiling point of the solution is 80.23 °C. What is the molecular formula of azulene?

EXERCISE 14.6 *Freezing Point Depression*

Is adding 500. g of ethylene glycol, $C_2H_4(OH)_2$, to 3.00 kg of water sufficient to keep the water from freezing at -25 °C?

Screen 14.9 Colligative Properties (3)

1. This screen explains the observations first seen on the Chemical Puzzler screen. Osmotic pressure is found to be responsible for the changes in the egg's size. What part of the egg acts as a semipermeable membrane?

2. If the egg were put in concentrated salt water, what would happen to its size?

3. Suppose you have two solutions separated by a semipermeable membrane. One contains 5.85 g of NaCl dissolved in 100. mL of solution and the other 8.88 g of KNO_3 dissolved in 100. mL of solution. If which direction will solvent flow, from the NaCl solution to the KNO_3 solution, or from KNO_3 to NaCl? Explain briefly.

4. Beta-carotene is the most important of the A vitamins. Its molar mass can be determined by measuring the osmotic pressure generated by a given mass of carotene dissolved in the solvent chloroform. A solution of 7.68 mg of carotene, dissolved in 10.0 mL of chloroform, has an osmotic pressure of 26.57 mm Hg at 25.0 °C.

   a) What is the concentration of the carotene?

   b) What is the molar mass of the carotene? (Answer = 537 g/mol)

EXERCISE 14.7 *Osmotic Pressure and Molar Mass*

Suppose 144 mg of aspartame, the artificial sweetener, is dissolved in exactly 25 mL of water. The osmotic pressure observed at 25 °C is 364 mm Hg. What is the molar mass of aspartame?

Screen 14.10 Colloids

Name three colloids not shown on this screen and identify the dispersing medium and the dispersed phase for each.

Screen 14.11 Surfactants

Explain how surfactants act to help oil and water form a solution.

____

____

____

Thinking Beyond

1. Silver halide salts are insoluble. The size and charge of a silver ion is approximately the same as that of a sodium ion (129 pm for Ag^+ and 119 pm for Na^+). Given this similarity, we expect the intermolecular forces experienced by AgCl and by NaCl to be similar. Why, then, is NaCl highly soluble whereas AgCl is very insoluble?
2. Ammonium nitrate dissolves in an endothermic process.

$$NH_4NO_3(s) \rightarrow NH_4NO_3(aq) \qquad \Delta H^\circ_{solution} = +25.7 \text{ kJ/mol}$$

The positive enthalpy change indicates that the products are less stable than the reactants, and it is necessary to add energy to allow the reaction to occur. If the products are less stable than the reactants, then why does the reaction occur in the first place?

STUDY QUESTIONS

1. Assume you dissolve 2.56 g of malic acid, $C_4H_6O_5$, in half a liter of water (500.0 g). Calculate the molarity, molality, mole fraction, and weight percent of acid in the solution.
2. You want to prepare a solution that is 0.200 m in $NaNO_3$; how many grams of the salt must you add to 250. g of water? What is the mole fraction of $NaNO_3$ in the resulting solution?
3. Hydrochloric acid is sold as a concentrated aqueous solution. If the molarity of commercial HCl is 12.0 mol/L and its density is 1.18g/cm^3, calculate (a) the molality of the solution and (b) the weight percent of HCl in the solution.
4. You make a saturated solution of NaCl at 25 °C. No solid is present in the beaker holding the solution. What can be done to increase the amount of dissolved NaCl in this solution? You know that NaCl has an endothermic enthalpy of solution. That is,

$$NaCl(s) + \text{heat} \rightarrow NaCl(aq) \quad \Delta H_{solution}(NaCl) = +3.9 \text{ kJ/mol}$$

   (a) Add more solid NaCl.

   (b) Raise the temperature of the solution.

   (c) Raise the temperature of the solution and add some NaCl.

   (d) Lower the temperature of the solution and add some NaCl.
5. A soda can has an aqueous CO_2 concentration of 0.0506 M at 25 °C. What is the pressure of CO_2 gas in the can? (Henry's law constant for CO_2 in water is 4.48×10^{-5} M/mm Hg.)
6. Some ethylene glycol ($C_2H_4(OH)_2$, 35.0 g) is dissolved in half a liter of water (500.0 g). The vapor pressure of water at 32 °C is 35.7 mm Hg. What is the vapor pressure of the water/glycol solution at 32 °C? (The glycol is assumed to be nonvolatile.)

7. Pure ethylene glycol [$C_2H_4(OH)_2$] is added to 2.00 kg of water in the cooling system of a car. The vapor pressure of the water in the system when the temperature is 90 °C is 457 mm Hg. How many grams of glycol are added? (Assume that ethylene glycol is not volatile at this temperature. The vapor pressure of pure water at 90 °C is 525.8 mm Hg.)
8. A quantity (10.0 g) of a nonvolatile solute is dissolved in 100. g of benzene (C_6H_6). The vapor pressure of pure benzene at 30 °C is 121.8 mm Hg, and that of the solution is 113.0 mm Hg at the same temperature. What is the molar mass of the solute?
9. What is the boiling point of a solution composed of 15.0 g of urea, $OC(NH_2)_2$, in 0.500 kg of water?
10. What is the boiling point of a solution composed of 15.0 g of $CHCl_3$ and 0.515 g of the nonvolatile solute acenaphthalene, $C_{12}H_{10}$, a component of coal tar?
11. BHA is used as an antioxidant in margarine and other fats and oils; it prevents oxidation and improves the shelf life of the food. What is the molar mass of BHA if 0.640 g of the compound, dissolved in 25.0 g of chloroform, produces a solution whose boiling point is 62.22 °C?
12. The empirical formula of anthracene, a hydrocarbon obtained from coal, is C_7H_5. To find its molecular formula you dissolve 0.500 g in 30.0 g of benzene. The boiling point of the pure benzene is 80.10 °C, whereas the solution has a boiling point of 80.34 °C. What is the molecular formula of anthracene?
13. A mixture of ethanol (C_2H_5OH) and water has a freezing point of -16.0 °C.
    (a) What is the molality of the alcohol?
    (b) What is the weight percent of alcohol in the solution?
14. An aqueous solution contains 0.180 g of an unknown, nonionic solute and 50.0 g of water. The solution freezes at -0.040 °C. What is the molar mass of the solute?
15. A solution contains 3.00% phenylalanine ($C_9H_{11}NO_2$) and 97.00% water by mass. Assume the phenylalanine is nonionic and nonvolatile. Find (a) the freezing point of the solution, (b) the boiling point of the solution, and (c) the osmotic pressure of the solution at 25 °C. In your view, which of these is most easily measurable in the laboratory?
16. An aqueous solution containing 1.00 g of bovine insulin (a protein) per liter has an osmotic pressure of 3.1 mm Hg at 25 °C. Calculate the molar mass of bovine insulin.
17. When solutions of $BaCl_2$ and Na_2SO_4 are mixed, a cloudy liquid is produced. After a few days, a white solid is observed on the bottom of the beaker with a clear liquid above it. (a) Write a balanced equation for the reaction that occurs. (b) Why is the solution cloudy at first? (c) What happens during the few days of waiting?
18. Water at 25 °C has a density of 0.997 g/cm^3. Calculate the molality and molarity of pure water at this temperature.
19. Hexachlorophene is used in germicidal soap. What is its molar mass if 0.640 g of the compound, dissolved in 25.0 g of chloroform, produces a solution whose boiling point is 61.93 °C?
20. Account for the fact that alcohols such as methanol (CH_3OH) and ethanol (C_2H_5OH) are quite miscible with water, whereas an alcohol with a long carbon chain such as octanol ($C_8H_{17}OH$) is poorly soluble in water.

CHAPTER 15
Principles of Reactivity: Chemical Kinetics

Screen 15.2 Rates of Chemical Reactions

1. What is the difference between an instantaneous rate and an average rate?

____

____

2. Observe the graph of food dye vs. time on this screen. (Click on the "tool" icon on this screen.) The plot shows the concentration of dye as the reaction progresses. What does the steepness of the plot at any particular time tell you about the rate of the reaction at that time?

____

____

____

3. As the reaction progresses, the concentration of dye decreases as it is consumed. What happens to the reaction rate as this occurs? What is the relationship between reaction rate and dye concentration?

____

____

____

EXERCISE 15.1 *Rate of Reaction*

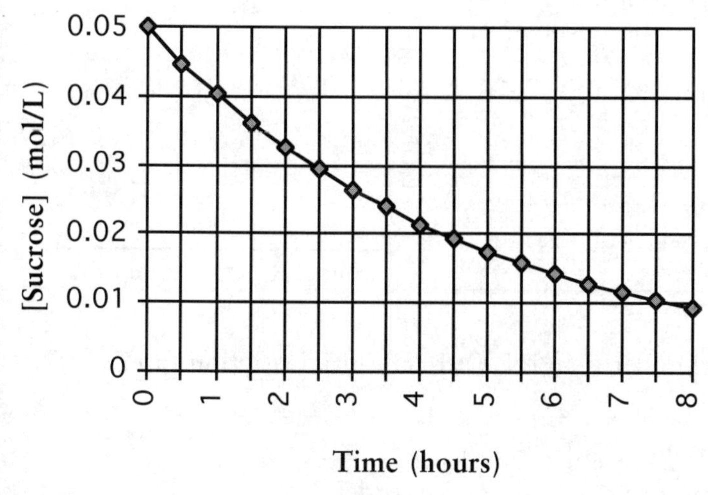

Sucrose decomposes to fructose and glucose in acid solution. A plot of the concentration of sucrose as a function of time is given here.

1. Is the resulting plot a straight line? ____________ (We shall see later why it is or is not.)
2. What is the rate of change of the sucrose concentration over the first two hours? ____________
3. What is the rate of change over the last two hours? ____________
4. Estimate the instantaneous rate at 4 h. ____________

Screen 15.3 Control of Reaction Rates (Surface Area)

1. What is the general relationship between the surface area of a solid reacting with a gas and the rate of the reaction?

2. The video on this screen is an illustration of what occurs in grain silo explosions. What should be avoided in grain silos to minimize the danger of an explosive reaction between grain dust and oxygen in the air?

Screen 15.4 Control of Reaction Rates (Concentration Dependence)

1. Watch the video on this screen. How does an increase in HCl concentration affect the rate of the reaction of the acid with magnesium metal?

2. On the second portion of this screen are data for rate measurements of the decomposition of N_2O_5 (click on "More"). The initial reaction rate is given for three separate experiments, each beginning with a different concentration of N_2O_5.

   a) Use the *Plotting Tool* to make a plot of initial reaction rate vs. initial $[N_2O_5]$. What do you observe? Is the plot linear or curved?

   Initial $[N_2O_5]$

   Initial Reaction Rate

   b) What is the mathematical relationship between $[N_2O_5]$ and initial reaction rate?

3. Explain the difference between a reaction rate and a reaction order.

Screen 15.5 Determination of Rate Equation (Method of Initial Rates)

This screen describes how to determine experimentally a rate law using the method of initial rates.

1. Why must the rate of reaction be measured at the very beginning of the process for this method to be valid?

2. The first experiment shows that the initial rate of NH_4NCO degradation is 2.2×10^{-4} mol/L · s when $[NH_4NCO] = 0.14$ M. Using the rate law determined on this screen, predict what the rate would be if $[NH_4NCO] = 0.18$ M.

EXERCISE 15.2 *Interpreting a Rate Law*

The rate equation for the reduction of NO to N_2 with hydrogen is

$$2\ NO(g) + 2\ H_2(g) \rightarrow N_2(g) + 2\ H_2O(g)$$

$$\text{Rate} = k\ [NO]^2[H_2]$$

1. What is the order of the reaction with respect to the NO? With respect to H_2?
2. If the concentration of NO is doubled, what happens to the reaction rate?
3. If the concentration of H_2 is halved, what happens to the reaction rate?

EXERCISE 15.3 *Using Rate Laws*

The rate constant k is 0.090/h for the reaction

$$Pt(NH_3)_2Cl_2(s) + H_2O(\ell) \rightarrow [Pt(NH_3)_2(H_2O)Cl]^+ + Cl^-$$

and the rate equation is "Rate = $k[Pt(NH_3)_2Cl_2]$." Calculate the rate of reaction when the concentration of $Pt(NH_3)_2Cl_2$ is 0.020 M. What is the rate of change in the concentration of Cl^- under these conditions?

EXERCISE 15.4 *Reaction Order*

In the following reaction, a Co—Cl bond is replaced by a Co—OH_2 bond.

$$[Co(NH_3)_5Cl]^{2+}(aq) + H_2O(\ell) \rightarrow [Co(NH_3)_5H_2O]^{3+}(aq) + Cl^-(aq)$$

$$\text{Rate} = k\ \{[Co(NH_3)_5Cl]^{2+}\}^m$$

Using the data in the table, find the value of *m* in the rate equation and calculate k.

Experiment	Initial Concentration of $[Co(NH_3)_5Cl]^{2+}$ (mol/L)	Initial Rate (mol/L · min)
1	1.0×10^{-3}	1.3×10^{-7}
2	2.0×10^{-3}	2.6×10^{-7}
3	3.0×10^{-3}	3.9×10^{-7}
4	1.0×10^{-2}	1.3×10^{-6}

Screen 15.6 Concentration-Time Relationships

1. In the problem on this screen, the equation $\ln([R]/[R]_o) = -kt$ is used to solve the problem. Why was that particular equation used instead of one of the others on this screen, such as $[R]_o - [R]_t = kt$?

2. If a 0.25 M sample of N_2O_5 is allowed to decompose in a first-order reaction, what will $[N_2O_5]$ be after 80. min?

3. Which of the three equations on this screen would be used to calculate $[NH_4NCO]$ at various times during its decomposition? Why did you choose this equation?

Screen 15.7 Determination of Rate Equation (Graphical Methods)

1. Construct a qualitative plot showing a linear mathematical relationship for the decomposition of sucrose. See the problem on Screen 15.6 for the additional information you need to solve this problem.

2. Phenyl acetate reacts with water according to the equation

$$CH_3CO_2C_6H_5 + H_2O \rightarrow CH_3CO_2H + C_6H_5OH$$

and the data in the following table have been obtained for $[CH_3CO_2C_6H_5]$ as the reaction progresses.

Time (min)	$[CH_3CO_2C_6H_5]$ (mol/L)
0	0.55
0.25	0.42
0.50	0.31
0.75	0.23
1.00	0.17
1.25	0.12
1.50	0.085

Construct appropriate plots using these data and the *Plotting Tool* and determine the rate equation for the reaction.

Screen 15.8 Half-Life: First-Order Reactions

1. What is meant by the term half-life?

____

____

2. Examine the graph and table of concentrations on the second portion of this screen.
   a) What will the concentration of H_2O_2 be after 3270 minutes have gone by? __________ After 3924 min? __________
   b) What fraction of the original concentration of H_2O_2 remains after each of these times? After 3270 minutes have gone by? __________ After 3924 min? __________
3. What is the half-life of the decomposition of sucrose in water at 25 °C? See the problem on Screen 15.6 for the additional information needed for this question.

EXERCISE 15.5 *Half-Life of a First-Order Process*

Cyclopropane, C_3H_6, has been used in a mixture with oxygen as an anesthetic. The compound is known to rearrange to propene, a different molecule of the same formula. What is the half-life of this reaction? What fraction of cyclopropane remains after 51.2 h? What fraction remains after 18.0 h? For this reaction the rate law is Rate = k [cyclopropane] where k = 5.4×10^{-2} h^{-1}.

H_2
C
H_2C—CH_2
Cyclopropane
→
$H_2C{-}C{=}CH_3$
H
Propene

Screen 15.9 Microscopic View of Reactions (1)

1. According to collision theory, what three conditions must be met for two molecules to react?

2. Examine the animations that play when numbers 1 and 2 are selected. One of these occurs at a higher temperature than the other. Which one? Explain briefly.

3. Examine the animations that play when numbers 2 and 3 are selected. Would you expect the reaction of O_3 with N_2

   $$O_3(g) + N_2(g) \rightarrow O_2(g) + ONN(g)$$

   to be more or less sensitive to requiring a proper orientation for reaction than the reaction displayed on this screen? Explain briefly.

Screen 15.10 Microscopic View of Reactions

1. Watch the animation on this screen. The transition state is the species with both F^- and Cl^- ions partially attached to the central C atom.

   a) How many valence electrons will the C be surrounded by at this stage of the reaction?

   b) Why is this transition state thought to be a high energy state? Explain briefly.

2. Examine the two animations on the *Reaction Coordinate Diagrams* sidebar to this screen.

   a) What is the difference between the way in which the two reactions occur?

   b) Draw the carbon-based intermediate that is formed during the two-step reaction.

Screen 15.11 Control of Reaction Rates (Temperature Dependence)

1. What is the general effect of an increase in temperature on reaction rates?

2. Describe the different portions of the Arrhenius equation and how they relate to the rate of a reaction.

3. The decomposition of N_2O_5 is described on Screen 15.4.

   $2\ N_2O_5(g) \rightarrow 4\ NO_2(g) + O_2(g)$

   The reaction has been studied at various temperatures and the following values of the rate constant, k, obtained.

T (K)	k (s^{-1})
338	4.87×10^{-3}
328	1.50×10^{-3}
318	4.98×10^{-4}
308	1.35×10^{-4}
298	3.46×10^{-5}
273	7.87×10^{-7}

   Use these data and the *Plotting Tool* to construct an Arrhenius plot like that shown on this screen to determine the activation energy for the decomposition of N_2O_5. Activation energy = __________ kJ/mol

Screen 15.12 Reaction Mechanisms

1. Examine the animation for the mechanism of the decomposition of ozone. Describe in words what occurs during each of the two steps of the reaction.

2. Do the same for the three steps involved in the formation of hydrazine.

3. Why is the second step in each mechanism said to be a bimolecular reaction.

Screen 15.13 Reaction Mechanisms and Rate Equations

1. What is the difference between an overall mechanism and an elementary step?

2. What is the relationship between the stoichiometric factors of the reactants in an elementary step and the rate law for that step?

3. What is the relationship between reactant concentrations and the rate law for that reaction?

4. What is the rate law for Step 2 of Mechanism 2?

5. The oxidation of iodide ion by hydrogen peroxide in acid solution occurs by the following mechanism.

   Step 1 (slow) $H_2O_2(aq) + I^-(aq) \rightarrow H_2O(\ell) + OI^-(aq)$

   Step 2 (fast) $H^+(aq) + OI^-(aq) \rightarrow HOI(aq)$

   Step 3 (fast) $HOI(aq) + H^+(aq) + I^-(aq) \rightarrow I_2(aq) + H_2O(\ell)$

   a) What is the rate-determining or rate-limiting step? ________

   b) What is the equation for the overall reaction? ________

   c) What is the rate law for the reaction? ________

6. Examine the *Isotopic Labeling* sidebar to this screen. If the reaction occurred in a single step transfer of an oxygen atom from NO_2 to CO, would any $N^{16}O^{18}O$ be found if the reaction is started using a mixture of $N^{16}O_2$ and $N^{18}O_2$? Why or why not?

EXERCISE 15.6 *Elementary Steps*

Nitrogen monoxide is reduced by hydrogen to give water and nitrogen,

$$2\ NO(g) + 2\ H_2(g) \rightarrow N_2(g) + 2\ H_2O(g)$$

and one possible mechanism to account for this reaction is

$$2\ NO(g) \rightarrow N_2O_2(g)$$
$$N_2O_2(g) + H_2(g) \rightarrow N_2O(g) + H_2O(g)$$
$$N_2O(g) + H_2(g) \rightarrow N_2(g) + H_2O(g)$$

What is the molecularity of each of the three steps? What is the rate equation for the third step? Show that the sum of these elementary steps is the net reaction.

Screen 15.14 Catalysis and Reaction Rate

1. What is the function of a catalyst?

2. Why does a catalyst not get consumed during a reaction?

3. Examine the mechanism for the iodide ion-catalyzed decomposition of H_2O_2. Explain how the mechanism shows that I^- is a catalyst.

4. How does the reaction coordinate diagram show that the catalyzed reaction is expected to be faster than the uncatalyzed reaction?

Screen 15.15 Return to the Puzzler

The animation on this screen shows that the reaction $2\ NO(g) \rightarrow N_2(g) + O_2(g)$ is accelerated by the presence of a metal surface. Do you think the adsorption of NO onto the metal surface strengthens or weakens the N-O bond? Explain briefly.

Thinking Beyond

1. All reactions go faster at higher temperatures (the value of the rate constant, k, increases with temperature) and yet many enzymatic reactions (enzymes are biochemical catalysts) show an increase in rate as temperature is increased and then upon further temperature increases show a decrease in rate. Propose how these two facts can be consistent.
2. Termolecular elementary steps, where three molecules meet simultaneously, have been observed but are extremely rare. Why are there so few examples of such reactions?

STUDY QUESTIONS

1. Experimental data are listed here for the hypothetical reaction $A \rightarrow 2\ B$.

Time (sec)	[A] (mol/L)
0.00	1.000
10.0	0.833
20.0	0.714
30.0	0.625
40.0	0.555

(a) Plot these data, connect the points with a smooth line, and calculate the rate of change of [A] for each 10-second interval from 0 to 40 seconds. Why does the rate of change decrease from one time interval to the next?

(b) How is the rate of change of [B] related to the rate of change of [A] in the same time interval? Calculate the rate of change of [B] for the time interval from 10 to 20 seconds.

2. The reaction between ozone and nitrogen dioxide has been studied at 231 K.

$$2\ NO_2(g) + O_3(g) \rightarrow N_2O_5(s) + O_2(g)$$

Experiment shows the reaction is first order in both NO_2 and O_3.

(a) Write the rate equation for the reaction.

(b) If the concentration of NO_2 is tripled, how does this affect the reaction rate?

(c) If the concentration of O_3 is halved, how does this affect the reaction rate?

3. Data are given at 660 K for the reaction $2\ NO(g) + O_2(g) \rightarrow 2\ NO_2(g)$.

Reactant Concentration (mol/L) [NO]	$[O_2]$	Rate of Disappearance of NO (mol/L·sec)
0.020	0.010	1.0×10^{-4}
0.040	0.010	4.0×10^{-4}
0.020	0.040	4.0×10^{-4}

(a) Write the rate equation for the reaction.

(b) Calculate the rate constant.

(c) Calculate the rate (in mol/L·sec) at the instant when [NO] = 0.045 M and $[O_2]$ = 0.025 M.

(d) At the instant when O_2 is reacting at the rate 5.0×10^{-4} mol/L·sec, what is the rate at which NO is reacting and NO_2 is forming?

4. The decomposition of N_2O_5 in nitric acid is a first-order reaction. It takes 4.26 minutes at 55 °C to decrease 2.56 mg of N_2O_5 to 2.50 mg. Find k in min^{-1} and sec^{-1}.

5. The rate constant for the decomposition of nitrogen dioxide

$$NO_2(g) \rightarrow NO(g) + 1/2\ O_2(g)$$

when irradiated with a laser beam is 3.40 L/mol·min. Find the time in seconds needed to decrease the concentration of NO_2 from 2.00 mol/L to 1.50 mol/L.

6. For a reaction with the rate equation "Rate = k[A]," and k = $3.33 \times 10^{-6}\ h^{-1}$, what is the half-life of the reaction? How long will it take for the concentration of A to drop from 1.0 M to 0.20 M?

7. The decomposition of phosphine, PH_3, proceeds according to the equation

$$4\ PH_3(g) \rightarrow P_4(g) + 6\ H_2(g)$$

It is found that the reaction has the rate equation "rate = k$[PH_3]$." If the half-life is 37.9 sec, how much time is required for three-fourths of the PH_3 to decompose?

8. Hypofluorous acid, HOF, is very unstable, decomposing to give HF and O_2 in a first-order reaction with a half-life of only 30. minutes at room temperature.

$$HOF(g) \rightarrow HF(g) + 1/2\ O_2(g)$$

If the partial pressure of HOF in a 1.00-L flask is initially 100. mm Hg at 25 °C, what is the total pressure in the flask and the partial pressure of HOF after 30. minutes? After 45 minutes?

9. The radioactive isotope copper-64 (^{64}Cu) is used in the form of copper(II) acetate to study Wilson's disease. The isotope decays in a first-order process with a half-life of 12.70 h. What quantity of copper(II) acetate remains after 2 days and 16 hours? After 5 days?

10. Common sugar, sucrose, reacts in dilute acid solution to give the simpler sugars glucose and fructose. Both of the simple sugars have the same formula, $C_6H_{12}O_6$.

$$C_{12}H_{22}O_{11}(aq) + H_2O(\ell) \rightarrow C_6H_{12}O_6(\text{glucose}) + C_6H_{12}O_6(\text{fructose})$$

The rate of this reaction has been studied in acid solution, and the following data were obtained.

Time (min)	$[C_{12}H_{22}O_{11}]$(mol/L)
0	0.316
39	0.274
80	0.238
140	0.190
210	0.146

(a) Plot the data above as ln[sucrose] versus time and l/[sucrose] versus time. What is the order of the reaction?

(b) Write the rate equation for the reaction and calculate the rate constant k.

(c) Estimate the concentration of sucrose after 175 minutes.

11. Data for the first-order decomposition of dinitrogen oxide, $2\ N_2O(g) \rightarrow 2\ N_2(g) + O_2(g)$, on a gold surface at 900 °C are given here. Find the rate constant by graphing these data in an appropriate manner. Write the rate equation and find the decomposition rate at 900 °C when $[N_2O] = 0.035$ mol/L.

Time (min)	$[N_2O]$ (mol/L)
15.0	0.0835
30.0	0.0680
80.0	0.0350
120.	0.0220

12. For the hypothetical reaction A + B → C + D, the activation energy is 32 kJ/mol. For the reverse reaction (C + D → A + B), the activation energy is 58 kJ/mol. Is the reaction A + B → C + D exothermic or endothermic?

13. The ozone, O_3, in the earth's upper atmosphere decomposes according to the equation $2\ O_3(g) \rightarrow 3\ O_2(g)$. The mechanism of the reaction is thought to proceed through an initial fast, reversible step and then a slow second step.

Step 1. Fast, reversible $O_3(g) \rightleftharpoons O_2(g) + O(g)$

Step 2. Slow $O_3(g) + O(g) \rightarrow 2\ O_2(g)$

(a) Which of the steps is the rate-determining or rate-limiting step?

(b) Write the rate equation for the rate-determining or rate-limiting step.

(c) What is the molecularity of each step?

14. Nitrogen oxides, NO_x (a mixture of NO and NO_2 collectively designated as NO_x), play an essential role in the production of pollutants found in photochemical smog. The average half-life for the removal of NO_x from smokestack emissions in a large city during daylight is 3.9 h.

(a) Starting with 1.50 mg in an experiment, what quantity of NO_x remains after 5.25 h? (Assume the reaction is first order.)

(b) How many hours of daylight must have elapsed to decrease 1.50 mg of NO_x to 2.50×10^{-6} mg?

15. Chlorine atoms are thought to lead to the destruction of the earth's ozone layer by the following sequence of reactions:

$$Cl(g) + O_3(g) \rightarrow ClO(g) + O_2(g)$$

$$ClO(g) + O(g) \rightarrow Cl(g) + O_2(g)$$

where the O atoms in the second step come from the decomposition of ozone by sunlight.

$$O_3(g) \rightleftharpoons O(g) + O_2(g)$$

What is the net equation obtained on adding the three equations? Why does this lead to ozone loss in the stratosphere? What is the role played by Cl in this sequence of reactions? What is the name given to species such as ClO?

16. The decomposition of dinitrogen pentaoxide

$$2\ N_2O_5(g) \rightarrow 4\ NO_2(g) + O_2(g)$$

has the rate equation "Rate = $k[N_2O_5]$." It has been found experimentally that the decomposition is 20% complete in 6.0 hours at 300. K. Calculate the rate constant and the half-life at 300. K.

CHAPTER 16
Principles of Reactivity: Chemical Equilibria

Screen 16.2 The Principle of Microscopic Reversibility

1. Describe the experiment shown on this screen. What is the objective of the experiment?

2. What is the principle of microscopic reversibility?

3. What does the double arrow in the equation $Pb^{2+}(aq) + 2\ Cl^{-}(aq) \rightleftharpoons PbCl_2(s)$ signify?

Screen 16.3 The Equilibrium State

1. Watch the reaction shown on this screen. The reaction does not go to completion; instead, only about 10% of the reactants go to form products. In what way would the solution's appearance differ if the reaction did go to completion?

2. What is meant by the term "dynamic equilibrium?"

3. Examine the *Isotopic Labeling* sidebar to this screen.
   a) What does this experiment illustrate?

   b) Could the experiment work using different isotopes of iron, instead of carbon? Explain briefly.

Screen 16.4 The Equilibrium Constant

1. The $Fe(SCN)^{2+}$ experiment seen on Screen 16.3 is used again on this screen. Examine the table of initial concentrations and the beakers of solutions formed by mixing these solutions. The quantity of iron used in all the experiments is the same, but additional SCN^- is used going from left to right. What effect does using more SCN^- have on the quantity of product formed?

____

____

2. Realizing that in each case the SCN^- is the limiting reactant, calculate the percent yield for each of the four reactions.

$[SCN^-]_{initial}(M)$	$[FeSCN^{2+}]_{final}(M)$	Percent Yield
2.00×10^{-4}	2.43×10^{-5}	________
4.00×10^{-4}	4.76×10^{-5}	________
6.00×10^{-4}	6.99×10^{-5}	________
1.00×10^{-3}	1.11×10^{-4}	________

What trend do you observe in the results? Does the reaction ever go nearly to completion?

____

____

3. Once equilibrium has been established, the rates of the forward and reverse reactions are equal,

$$\text{Rate } (Fe^{3+} + SCN^- \rightarrow FeSCN^{2+}) = \text{Rate } (FeSCN^{2+} \rightarrow Fe^{3+} + SCN^-)$$

and so $k_{forward}[Fe^{3+}][SCN^-] = k_{backward}[FeSCN^{2+}]$. Use these relationships to derive the form of the equilibrium constant expression. How do $k_{forward}$ and $k_{backward}$ relate to the equilibrium constant, K?

Screen 16.5 The Meaning of the Equilibrium Constant

1. In what way does the magnitude of the equilibrium constant relate to the tendency of reactants to form products?

____

2. Is $PbI_2(s)$ expected to dissolve to an appreciable extent? How is this reflected in the value of the reaction's equilibrium constant?

____

____

3. Would you expect the reaction $2\ NO_2(g) \rightleftharpoons 2\ NO(g) + O_2(g)$ to have a large or small equilibrium constant? Explain briefly.

4. Why can equilibrium constants not be negative?

Screen 16.6 Writing Equilibrium Expressions

Write equilibrium expressions for each of the following reactions:

1. $N_2(g) + 3\ H_2(g) \rightleftharpoons 2\ NH_3(g)$

2. $S_8(s) + 8\ O_2(g) \rightleftharpoons 8\ SO_2(g)$

3. $HCO_2H(aq) + H_2O(\ell) \rightleftharpoons HCO_2^-(aq) + H_3O^+(aq)$

EXERCISE 16.1 *Writing Equilibrium Constant Expressions*

Write equilibrium constant expressions for each of the following reactions:

1. $PCl_5(g) \rightleftharpoons PCl_3(g) + Cl_2(g)$
2. $Cu(OH)_2(s) \rightleftharpoons Cu^{2+}(aq) + 2\ OH^-(aq)$
3. $Cu(NH_3)_4{}^{2+}(aq) \rightleftharpoons Cu^{2+}(aq) + 4\ NH_3(aq)$
4. $CH_3CO_2H(aq) + H_2O(\ell) \rightleftharpoons CH_3CO_2^-(aq) + H_3O^+(aq)$

Screen 16.7 Manipulating Equilibrium Expressions

1. What is the general rule for changing an equilibrium constant when the stoichiometric factors of a reaction are multiplied by a constant?

2. What is the general rule for changing an equilibrium constant when the reaction is reversed, making the products the reactants, and making reactants the products?

3. What is the general rule for calculating an equilibrium constant when two reactions are added to give a third reaction?

EXERCISE 16.2 *Manipulating Equilibrium Constant Expressions*

1. The conversion of oxygen to ozone has a very small equilibrium constant.

   $3/2\ O_2(g) \rightleftharpoons O_3(g) \qquad K_c = 2.5 \times 10^{-29}$

   a) What is the value of K_c when the equation is written using whole-number coefficients?

   $3\ O_2(g) \rightleftharpoons 2\ O_3(g)$

   b) What is the value of K_c for the conversion of ozone to oxygen?

   $2\ O_3(g) \rightleftharpoons 3\ O_2(g)$

   (In each of these cases, K_c is based on the concentrations of reactants and products.)

2. The following equilibrium constants are given at 500 K:

   $H_2(g) + Br_2(g) \rightleftharpoons 2\ HBr(g) \qquad K_p = 7.9 \times 10^{11}$

   $H_2(g) \rightleftharpoons 2\ H(g) \qquad K_p = 4.8 \times 10^{-41}$

   $Br_2(g) \rightleftharpoons 2\ Br \qquad K_p = 2.2 \times 10^{-15}$

   Calculate K for the reaction of H and Br atoms to give HBr.

   $H(g) + Br(g) \rightleftharpoons HBr(g) \qquad K_p = ?$

   (In each of these cases, K_p is based on the partial pressures of reactants and products.)

Screen 16.8 Determining an Equilibrium Constant

1. Describe the process that must be followed to experimentally determine an equilibrium constant. What data must be obtained?

2. Often when making experimental measurements, the chemist realizes that there is the potential for experimental error to lead to inaccurate results. Considering the method used on this screen to determine the equilibrium constant of the reaction of formic acid with water, suggest some ways of minimizing the potential of obtaining an incorrect equilibrium constant.

Screen 16.9 Systems at Equilibrium

1. On the second part of this screen are shown three diagrammatic representations of butane/isobutane mixtures, one at equilibrium and two not at equilibrium. Sketch a diagram of another butane/isobutane mixture that would be at equilibrium.

2. Sketch a diagram of another mixture that would not be at equilibrium. For your mixture here, indicate which species, butane or isobutane, would react in order for the mixture to move towards equilibrium.

3. See the *Equilibrium and Reactions* sidebar to this screen. Is your body, at this very moment, at equilibrium? Explain briefly.

____

____

EXERCISE 16.3 *The Reaction Quotient*

At 2000 K the equilibrium constant, K_c, for the formation of NO(g)

$$N_2(g) + O_2(g) \rightleftharpoons 2\ NO(g)$$

is 4.0×10^{-4}. You have a vessel in which, at 2000 K, the concentration of N_2 is 0.50 mol/L, that of O_2 is 0.25 mol/L, and that of NO is 4.2×10^{-3} mol/L. Is the system at equilibrium? If not, predict which way the reaction will proceed to achieve equilibrium.

Screen 16.10 Estimating Equilibrium Concentrations

1. In a 1.00-L flask at 1000 K we place 1.00 mol of SO_2 and 1.00 mol of O_2. When equilibrium is achieved, 0.925 mol of SO_3 is formed.

   $$2\ SO_2(g) + O_2(g) \rightleftharpoons 2\ SO_3(g)$$

   a) What is the concentration of each reactant and product at equilibrium?

   $[SO_2]$ = ________ $[O_2]$ = ________ $[SO_3]$ = ________

b) What is K_c at this temperature? (Answer = 280)

2. A mixture of H_2 (9.838×10^{-4} mol) and I_2 (1.377×10^{-3} mol) is sealed in a quartz tube and kept at 350 °C for a week. During this time the reaction

$$H_2(g) + I_2(g) \rightleftharpoons 2\ HI(g)$$

comes to equilibrium. The tube is broken open, and 4.725×10^{-4} mol of I_2 is found.

a) Calculate the number of moles of H_2 and HI present at equilibrium.

b) Assume the volume of the tube is 10.0 mL. Calculate K_c for the reaction.

c) Is your value of K_c different if the volume of the tube is 20.0 mL?

EXERCISE 16.4 *Calculating a Concentration from an Equilibrium Constant*

A solution is prepared by dissolving 0.050 mol of diiodocyclohexane, $C_6H_{10}I_2$, in the solvent CCl_4. The total solution volume is 1.00 L. When the reaction

$$C_6H_{10}I_2 \rightleftharpoons C_6H_{10} + I_2$$

has come to equilibrium at 35 °C, the concentration of I_2 is 0.035 mol/L.

1. What are the concentrations of $C_6H_{10}I_2$ and C_6H_{10} at equilibrium?
2. Calculate K_c, the equilibrium constant.

Screen 16.11 Disturbing an Equilibrium (Le Chatelier's Principle)

1. State Le Chatelier's principle.

2. Examine the water tank animation on this screen. Describe how addition of water to the left-hand tank illustrates Le Chatelier's principle.

3. Of the three potential changes to an equilibrium system described on the screen, *Temperature Change, Addition or Removal of a Reactant or Product,* and *Volume Changes in Gas Phase Equilibria,* which does the tank demonstration illustrate?

4. a) What would you expect to occur in the tank equilibrium if the width of the left-hand tank were suddenly decreased to one half its present radius?

   b) Which of the three potential changes to an equilibrium system does this illustrate?

Screen 16.12 Disturbing an Equilibrium (Temperature Changes)

1. Watch the photographs shown on this screen. Describe the difference between the two states shown, both in terms of temperature and in terms of the concentrations of the species in the flask.

2. The dissolution of ammonium nitrate is an endothermic process.

   $NH_4NO_3(s) \rightleftharpoons NH_4^+(aq) + NO_3^-(aq)$ $\Delta H° = +25.7$ kJ/mol

   Do you expect the dissolution equilibrium to shift to the right or to the left if the temperature is increased? Explain briefly.

Screen 16.13 Disturbing an Equilibrium (Addition or Removal of a Reagent)

1. Watch the animated diagram on this screen. Sketch out what the graph would show if isobutane were added instead of butane.

2. On the previous screen, changes in temperature resulted in a change in the equilibrium constant, resulting in a shift in the equilibrium. Does the addition or removal of a reagent described on this screen result in a change in the equilibrium constant?

____

3. What is the difference between the reaction quotient, Q, and the equilibrium constant, K?

____

Screen 16.14 Disturbing an Equilibrium (Volume Changes)

1. For each of the following, decide if a decrease in volume will result in a shift left, shift right, or in no shift to the equilibrium:

   a) $2\ NO(g) + O_2(g) \rightleftharpoons 2\ NO_2(g)$ ______________________________

   b) $I_2(g) \rightleftharpoons 2\ I(g)$ ______________________________

   c) $NH_4I(s) \rightleftharpoons NH_3(g) + HI(g)$ ______________________________

EXERCISE 16.5 *Le Chatelier's Principle*

1. Does the concentration of SO_3 increase or decrease when the temperature increases?

   $2\ SO_2(g) + O_2(g) \rightleftharpoons 2\ SO_3(g)$ $\Delta H^\circ_{rxn} = -198$ kJ

2. Equilibrium exists between butane and isobutane when [butane] = 0.20 M and [isobutane] = 0.50 M. What are the equilibrium concentrations of butane and isobutane if 2.00 mol/L of isobutane are added to the original mixture?

3. The formation of ammonia from its elements is an important industrial process.

   $3\ H_2(g) + N_2(g) \rightleftharpoons 2\ NH_3(g)$

   a) Does the equilibrium shift to the left or the right when extra H_2 is added? When extra NH_3 is added?

   b) What is the effect on the position of the equilibrium when the volume of the system is increased? Does the equilibrium shift to the left or to the right, or is the system unchanged?

Screen 16.15 Return to the Puzzler

List the various ways that the equilibrium shown on this screen can be shifted toward the $Co(H_2O)_4Cl_2(aq)$ side.

____

____

Thinking Beyond

1. In the previous chapter we learned how to write rate expressions for mechanisms in which the first elementary step is the rate determining step. In many cases, however, the first step in a mechanism involves an equilibrium. This "preequilibrium" is then followed by a unidirectional, rate determining step. An example is the reaction of hydrogen and carbon monoxide to give formaldehyde, for which the proposed mechanism is as follows:

Step 1. Fast, equilibrium $H_2(g) \rightleftharpoons 2\ H(g)$

Step 2. Slow $H(g) + CO(g) \rightarrow HCO(g)$

Step 3. Fast $HCO(g) + H(g) \rightarrow HCOH(g)$

What is the rate expression expected for this mechanism?

2. Question 3 written for Screen 16.4 implies that the method for determining equilibrium constant expressions results directly from setting the rate equations for the forward and reverse reactions equal to each other. What does this assume about the mechanism of the reaction? Is this a valid assumption?

STUDY QUESTIONS

1. Write equilibrium constant expressions for the following reactions. For gases use either pressures or concentrations.
   (a) $2\ H_2O_2(g) \rightleftharpoons 2\ H_2O(g) + O_2(g)$
   (b) $PCl_3(g) + Cl_2(g) \rightleftharpoons PCl_5(g)$
   (c) $CO(g) + 1/2\ O_2(g) \rightleftharpoons CO_2(g)$
   (d) $C(s) + CO_2(g) \rightleftharpoons 2\ CO(g)$
   (e) $FeO(s) + CO(g) \rightleftharpoons Fe(s) + CO_2(g)$
2. Consider the following equilibria involving $SO_2(g)$ and their corresponding equilibrium constants. What is the relationship between their equilibrium constants, K_1 and K_2?
   $SO_2(g) + 1/2\ O_2(g) \rightleftharpoons SO_3(g)$ K_1
   $2\ SO_3(g) \rightleftharpoons 2\ SO_2(g) + O_2(g)$ K_2
3. Calculate K_c for the reaction
   $Fe(s) + H_2O(g) \rightleftharpoons FeO(s) + H_2(g)$
   given the following information:
   $H_2O(g) + CO(g) \rightleftharpoons H_2(g) + CO_2(g)$ $K_c = 1.6$
   $Fe(s) + CO_2(g) \rightleftharpoons FeO(s) + CO(g)$ $K_c = 1.5$
4. The equilibrium constant, K_c, for the reaction
   $2\ NOCl(g) \rightleftharpoons 2\ NO(g) + Cl_2(g)$
   is 3.9×10^{-3} at 300 °C. A mixture contains the gases at the following concentrations: $[NOCl] = 5.0 \times 10^{-3}$ mol/L, $[NO] = 2.5 \times 10^{-3}$ mol/L, and $[Cl_2] = 2.0 \times 10^{-3}$ mol/L. Is the reaction at equilibrium? If not, in which direction does the reaction move to come to equilibrium?
5. A mixture of SO_2, O_2, and SO_3 at 1000 K contains the gases at the following concentrations: $[SO_2] = 3.77 \times 10^{-3}$ mol/L, $[O_2] = 4.30 \times 10^{-3}$ mol/L, and $[SO_3] = 4.13 \times 10^{-3}$ mol/L. Calculate the equilibrium constant, K_c, for the reaction $2\ SO_2(g) + O_2(g) \rightleftharpoons 2\ SO_3(g)$
6. Hydrogen and carbon dioxide react at a high temperature to give water and carbon monoxide: $H_2(g) + CO_2(g) \rightleftharpoons H_2O(g) + CO(g)$.
   (a) At 986 °C there are 0.11 mol each of CO and water vapor and 0.087 mol each of H_2 and CO_2 at equilibrium in a 1.0-L container. Calculate the equilibrium constant for the reaction at 986 °C.

(a) At 986 °C there are 0.11 mol each of CO and water vapor and 0.087 mol each of H_2 and CO_2 at equilibrium in a 1.0-L container. Calculate the equilibrium constant for the reaction at 986 °C.

(b) If there were 0.050 mol of each of H_2 and CO_2 in a 2.0-L container at equilibrium at 986 °C, what amounts of CO(g) and H_2O(g), in moles, would be present?

7. At a very high temperature water vapor is 10.% dissociated into H_2(g) and O_2(g). (That is, 10.% of the original water has been transformed into products, and 90.% remains.)

   $H_2O(g) \rightleftharpoons H_2(g) + 1/2\ O_2(g)$

   Assuming a water concentration of 2.0 mol/L before dissociation, calculate the equilibrium constant, K_c.

8. The hydrocarbon C_4H_{10} can exist in two forms, butane and isobutane. The value of K_c for the interconversion of the two forms is 2.5 at 25 °C.

   Butane(g) $\rightleftharpoons$ Isobutane(g) $\quad K_c = 2.5$

   If you place 0.017 moles of butane in a 0.50-L flask at 25 °C and allow equilibrium to be established, what will be the equilibrium concentrations of the two forms of butane?

9. Carbonyl bromide, $COBr_2$, decomposes to CO and Br_2 with an equilibrium constant, K_c, of 0.190 at 73 °C.

   $COBr_2(g) \rightleftharpoons CO(g) + Br_2(g)$

   If there is 0.015 mol of $COBr_2$ in a 2.5-L flask at equilibrium, what are the concentrations of CO and Br_2 at equilibrium?

10. The equilibrium constant, K_c, for the reaction

    $N_2O_4(g) \rightleftharpoons 2\ NO_2(g)$

    at 25 °C is 5.88×10^{-3}. Suppose 20.0 g of N_2O_4 are placed in a 5.00-L flask at 25 °C. Calculate (a) the number of moles of NO_2 present at equilibrium and (b) the percentage of the original N_2O_4 that is dissociated.

11. K_c for the dissociation of iodine [$I_2(g) \rightleftharpoons 2\ I(g)$] is 3.76×10^{-3} at 1000 K. Suppose 1.00 mol of I_2 is placed in a 2.00-L flask at 1000 K. What are the concentrations of I_2 and I when the system comes to equilibrium?

12. K_p for the following reaction is 0.16 at 25 °C. The enthalpy change for the reaction at standard conditions is +16.1 kJ.

    $2\ NOBr(g) \rightleftharpoons 2NO(g) + Br_2(g)$

    Predict the effect of the following changes on the position of the equilibrium; that is, state which way the equilibrium will shift (left, right, or no change) when each of the following changes is made: (a) adding more Br_2(g); (b) removing some NOBr(g); (c) decreasing the temperature; and (d) increasing the container volume.

13. Calculate the equilibrium constant, K_c, at 25 °C for the reaction

    $2\ NOCl(g) \rightleftharpoons 2\ NO(g) + Cl_2(g)$

    using the following information. In one experiment 2.00 mol of NOCl were placed in a 1.00-L flask, and the concentration of NO after equilibrium was achieved was 0.66 mol/L.

CHAPTER 17
Principles of Reactivity: The Chemistry of Acids and Bases

Screen 17.2 Brønsted Acids and Bases

1. What is the difference between a Brønsted acid and a Brønsted base? How are the two related?

____

____

2. Give the formula of the conjugate base of each of the following Brønsted acids:
   a) HCN _______________
   b) HCO_2H _______________
   c) CH_3CO_2H _______________
3. Give the formula of the conjugate acid of each of the following Brønsted bases:
   a) CH_3NH_2 _______________
   b) HCO_3^- _______________
   c) NO_2^- _______________

EXERCISE 17.1 *Brønsted Acids and Bases*

1. Write a balanced equation for the reaction that occurs when H_3PO_4, phosphoric acid, donates a proton to water to form the dihydrogen phosphate ion.
2. Write a balanced equation for the reaction that occurs when the cyanide ion, CN^-, accepts a proton from water to form HCN. Is CN^- a Brønsted acid or base?
3. Write balanced equations for the two stepwise reactions that occur when $H_2C_2O_4$, oxalic acid, acts as a polyprotic acid (an acid that can donate more than one proton to the solution) in aqueous solution.
4. Identify the acid and base on the left side of the following reaction and their conjugate partners on the right side.

$$HSO_4^-(aq) + CO_3^{2-}(aq) \rightleftharpoons SO_4^{2-}(aq) + HCO_3^-(aq)$$

Screen 17.3 The Acid-Base Properties of Water

1. Explain how water acts as both an acid and a base in its autoionization reaction.

____

____

2. Why are the concentrations of H_3O^+ and OH^- both equal to 1.0×10^{-7} M in neutral water at 25 °C?

3. At temperatures above 25 °C, the value of K_w is greater than 1.0×10^{-14}. How will $[H_3O^+]$ in neutral water compare at, say, 50 °C with the value found at 25 °C?

Screen 17.4 The pH Scale

1. In other calculations involving species in solution, we use concentrations in units of moles per liter. Why do we bother with the seemingly more complicated concentration unit, pH?

2. In which solution is the concentration of H_3O^+ larger, one with a pH of 3 or one with a pH of 5?

3. If a solution has pH = 3.8, what are $[H_3O^+]$ and $[OH^-]$?

4. a) If a solution has $[H_3O^+] = 8.6 \times 10^{-4}$ M, what are the pH and pOH of the solution?

   pH = ________ pOH = ________

   b) Is this solution acidic or basic? ________

EXERCISE 17.2 *pH and Hydronium Ion Concentration*

The pH of a diet soda is 4.32 at 25 °C. What are the hydronium and hydroxide ion concentrations in the soda?

Screen 17.5 Strong Acids and Bases

1. What is the definition of a strong acid?

2. Consider a 0.026 M solution of HNO_3 at 25 °C. What are $[H_3O^+]$, $[OH^-]$, and the pH of this solution?

3. Consider a 4.8×10^{-3} M solution of KOH. What are $[H_3O^+]$, $[OH^-]$, and the pH of this solution?

4. How does the experiment shown on this screen prove that the hydride ion, H^-, is a strong base?

____

____

Screen 17.6 Weak Acids and Bases

1. What is the difference between a strong acid and a weak acid?

____

____

2. Examine the table of acids and bases on this screen.

   a) How does the acid strength change on moving from the top of the table to the bottom?

   ____

   b) How does the strength of bases change on moving from the top to the bottom of the table? ______________

3. The acids and their conjugate bases are listed in the same row in the table on this screen. How do the values of K_b for the bases change as the values of K_a for the acids decrease going down the table?

____

____

4. K_w is the product of the ionization constant for a weak acid (K_a) and the constant for its conjugate base (K_b), that is, $K_a \cdot K_b = K_w$. Use the equilibrium expressions for K_a and K_b for HF and F^- to show that this is true.

5. Explain how acid-base chemistry influences intermolecular forces and leads to absorption of aspirin into the cells that line the stomach.

____

____

EXERCISE 17.3 *Weak Acids and Bases*

Lactic acid, $CH_3CHOHCO_2H$, has $K_a = 1.4 \times 10^{-4}$. Where does this fit in the table of acids and bases on Screen 17.6? Name an acid stronger than lactic acid. Name a weaker acid. What is K_b for the conjugate base of lactic acid, the lactate ion ($CH_3CHOHCO_2^-$)?

Screen 17.7 Determining K_a and K_b Values

1. The method shown on this screen for determining K_a values can be used for other acids. The pH of a 0.10 M lactic acid solution is 2.43. Calculate the value of K_a for lactic acid.

____

____

2. Examine the sidebar to this screen. Which acid would you expect to have a greater value of K_a, CH_3CO_2H or FH_2CCO_2H? Explain briefly.

____

____

3. Structures for the series of acids $CX_nH_{3-n}CO_2H$ (X = F, Cl) are included in the folder labeled ACIDS inside the ORGANIC folder in the *Models* (or CAChe) folder or directory on this disc. Examine these acids, looking in particular at the partial charges on the atoms.
   a) Which H atom is donated to solution when these function as acids, the H atom on the carbon or on the oxygen? Explain briefly.

   ____

   ____

   ____

   ____

   b) Can these structures assist us in deciding what happens to the acid strength as a function of structure?

   ____

   ____

   ____

EXERCISE 17.4 *Calculating a K_a Value From a Measured pH*

A solution prepared from 0.10 mole of propanoic acid dissolved in sufficient water to give 1.0 L of solution has a pH of 2.94. Determine K_a for propanoic acid. The acid ionizes according to the balanced equation

$$CH_3CH_2CO_2H(aq) + H_2O(\ell) \rightleftharpoons H_3O^+(aq) + CH_3CH_2CO_2^-(aq)$$

Screen 17.8 Estimating the pH of Weak Acid Solutions

Examine the pH measurement simulation on this screen.

1. What is the general effect of increasing acid concentration on solution pH?

____

____

2. Determine the approximate pH values for each of the following:
   a) 0.003 M HCl ________
   b) 0.003 M CH_3CO_2H ________
   c) 0.003 M C_6H_5OH ________
   d) Rank these three acids in terms of acid strength: lowest K_a on the left, highest on the right.

   ____________ ____________ ____________

   weakest acid strongest acid

3. Estimate the pH of a 0.24 M solution of HCN. You will need to look up the K_a value on Screen 17.6.

EXERCISE 17.5 *Calculating Equilibrium Concentrations and pH from K_a*

What are the equilibrium concentrations of acetic acid, the acetate ion, and H_3O^+ for a 0.10 M solution of acetic acid? What is the pH of the solution?

Screen 17.9 Estimating the pH of Weak Base Solutions

Examine the pH measurement simulation on this screen.

1. What is the general effect of increasing base concentration on solution pH?

____

____

2. Determine the approximate pH values for each of the following:
   a) 0.003 M NaOH ________
   b) 0.003 M NH_3 ________
   c) 0.003 M $C_6H_5NH_2$ ________
   d) Rank these three bases in terms of base strength: lowest K_b on the left, highest on the right.

   ____________ ____________ ____________

   weakest base strongest base

3. Estimate the pH of a 0.24 M solution of NaCN. You will need to look up the K_b for CN^- on Screen 17.6.

EXERCISE 17.6 *Calculating the pH of an Aqueous Solution of a Weak Base*

What is the pH of a 0.025 M solution of ammonia, NH_3?

Screen 17.10 Acid-Base Properties of Salts

Predict the acid-base properties of each salt. Describe it as an acid, a base, or a neutral salt.

Name	Ions Produced in Aqueous Solution	Behavior in Aqueous Solution
NH_4Cl		
NaCl		
NaCN		
NH_4CN		
$NaNO_2$		
$(NH_4)_2HPO_4$		
$AlCl_3$		

1. Explain why you decided that aqueous NH_4Cl is acidic, basic, or neutral.

2. Explain why you decided that aqueous NaCN is acidic, basic, or neutral.

3. Why is NH_4CN basic, even though it contains the weak acid, NH_4^+?

4. Why does $AlCl_3$ behave as you have predicted?

5. How did you decide that aqueous $(NH_4)_2HPO_4$ will be acidic, neutral, or basic?

EXERCISE 17.7 *Calculating the pH of an Aqueous Salt Solution*

The ammonium ion is the conjugate acid of the weak base ammonia.

$$NH_4^+(aq) + H_2O(\ell) \rightarrow H_3O^+(aq) + NH_3(aq)$$

What is the pH of a 0.50 M solution of ammonium chloride?

EXERCISE 17.8 *Predicting the pH of Salt Solutions*

For each of the following salts, predict whether the pH will be greater than, less than, or equal to 7: 1. LiBr; 2. $FeCl_3$; 3. NH_4NO_3; 4. Na_2HPO_4.

Screen 17.11 Lewis Acids and Bases

1. What is the difference between a Lewis acid and a Brønsted acid?

___

2. Can water act as (choose all that apply)

   a) a Brønsted acid?
   b) a Brønsted base?
   c) a Lewis acid?
   d) a Lewis base?

   Explain your choices briefly, and use Lewis electron structures in your discussion.

___

___

___

___

3. On Screen 17.5 you learned that the hydride ion, H^-, is a strong base. Is it also a Lewis base? Use the electron dot structure of H^- to prove your point.

___

___

___

Screen 17.12 Cationic Lewis Acids

1. Explain how Cu^{2+} ion acts as a Lewis acid on this screen.

___

___

2. Name another compound that could take the part of ammonia in this reaction.

___

3. Examine the *Blood and CO* sidebar to this screen. Why is carbon monoxide a poison?

___

4. Are ligands Lewis acids or Lewis bases? Explain briefly.

___

___

Screen 17.13 Neutral Lewis Acids

This screen describes how certain compounds—oxides of the nonmetals—can give acidic solutions, even though they contain no hydrogen atoms. Show how the compound SO_2 can result in acidic solutions when dissolved in water.

Screen 17.14 Return to the Puzzler

1. Give the primary reason why most foods are acidic.

____

2. Give a reason why most cleaners are basic.

____

Thinking Beyond

1. Both acid-base reactions and redox reactions involve electron donation and acceptance. What is the difference between these reactions?

STUDY QUESTIONS

1. Write the formula and give the name of the conjugate base of each of the following acids:
   (a) HCN (b) HSO_4^- (c) HF (d) HNO_2 (e) HCO_3^-
2. What are the products for each of the following acid-base reactions? Indicate the acid and its conjugate base and the base and its conjugate acid.
   (a) $HNO_3 + H_2O \rightarrow$
   (b) $HSO_4^- + H_2O \rightarrow$
   (c) $H_3O^+ + F^- \rightarrow$
3. Dissolving K_2CO_3 in water gives a basic solution. Write a balanced equation showing how the carbonate ion is responsible for this effect.
4. In each of the following acid-base reactions, identify the Brønsted acid and base on the left, and their conjugate partners on the right.
   (a) $HCO_2H(aq) + H_2O(\ell) \rightleftharpoons HCO_2^-(aq) + H_3O^+(aq)$
   (b) $H_2S(aq) + NH_3(aq) \rightleftharpoons HS^-(aq) + NH_4^+(aq)$
   (c) $HSO_4^-(aq) + OH^-(aq) \rightleftharpoons SO_4^{2-}(aq) + H_2O(\ell)$
5. Several acids are listed here with their respective equilibrium constants.
   $HF(aq) + H_2O(\ell) \rightleftharpoons H_3O^+(aq) + F^-(aq)$ $K_a = 7.2 \times 10^{-4}$
   $HS^-(aq) + H_2O(\ell) \rightleftharpoons H_3O^+(aq) + S^{2-}(aq)$ $K_a = 1.3 \times 10^{-13}$
   $CH_3CO_2H(aq) + H_2O(\ell) \rightleftharpoons H_3O^+(aq) + CH_3CO_2^-(aq)$ $K_a = 1.8 \times 10^{-5}$
   (a) Which is the strongest acid? Which is the weakest?

(b) What is the conjugate base of the acid HF?

(c) Which acid has the weakest conjugate base?

(d) Which acid has the strongest conjugate base?

6. State which of the following ions or compounds has the strongest conjugate base and briefly explain your choice: (a) HSO_4^-, (b) CH_3CO_2H, and (c) HClO.

7. A certain table wine has a pH of 3.40. What is the hydronium ion concentration of the wine? Is it acidic or basic?

8. What is the pH of a 0.0013 M solution of HNO_3? What is the hydroxide ion concentration of the solution?

9. What is the pH of a 0.0015 M solution of $Ca(OH)_2$?

10. A 2.5×10^{-3} M solution of an unknown acid has a pH of 3.80 at 25 °C.

    (a) What is the hydronium ion concentration of the solution?

    (b) Is the acid a strong acid, a moderately weak acid (K_a of about 10^{-5}), or a very weak acid (K_a of about 10^{-10})?

11. A 0.025 M solution of hydroxylamine has a pH of 9.11. What is the value of K_b for this weak base?

    $$H_2NOH(aq) + H_2O(\ell) \rightleftharpoons H_3NOH^+(aq) + OH^-(aq)$$

12. The ionization constant of a very weak acid HA is 4.0×10^{-9}. Calculate the equilibrium concentrations of H_3O^+, A^-, and HA in a 0.040 M solution of the acid.

13. If you have a 0.025 M solution of HCN, what are the equilibrium concentrations of H_3O^+, CN^-, and HCN? What is the pH of the solution?

14. Benzoic acid, $C_6H_5CO_2H$, has a K_a of 6.3×10^{-5}, while that of a derivative of this acid, 4-chlorobenzoic acid ($ClC_6H_4CO_2H$), is 1.0×10^{-4}.

    (a) Which is the stronger acid?

    (b) For a 0.010 M solution of each of these monoprotic acids, which will have the higher pH?

15. If each of the salts listed here were dissolved in water to give a 0.10 M solution, which solution would have the highest pH? Which would have the lowest pH?

    (a) Na_2S
    (b) Na_3PO_4
    (c) NaH_2PO_4
    (c) NaF
    (d) $NaCH_3CO_2$ (sodium acetate)
    (e) $AlCl_3$

16. Sodium cyanide is the salt of the weak acid HCN. Calculate the concentration of H_3O^+, OH^-, HCN, and Na^+ in a solution prepared by dissolving 10.8 g of NaCN in 500. mL of pure water at 25 °C.

17. Decide if each of the following substances should be classified as a Lewis acid or base.

    (a) Mn^{2+}
    (b) CH_3NH_2
    (c) H_2NOH in the reaction $H_2NOH(aq) + HCl(aq) \rightarrow [H_3NOH]Cl(aq)$
    (d) SO_2 in the reaction $SO_2(g) + BF_3(g) \rightleftharpoons O_2S\text{—}BF_3(s)$
    (e) $Zn(OH)_2$ in the reaction: $Zn(OH)_2(s) + 2\ OH^-(aq) \rightleftharpoons Zn(OH)_4^{2-}(aq)$

18. Given the following solutions:

0.1 M NH_3
0.1 M NH_4Cl
0.1 M Na_2CO_3
0.1 M $NaCH_3CO_2$ (sodium acetate)
0.1 M NaCl
0.1 M $NH_4CH_3CO_2$ (ammonium acetate)
0.1 M CH_3CO_2H

(a) Which of the solutions are acidic?
(b) Which of the solutions are basic?
(c) Which of the solutions is most acidic?

19. Arrange the following 1.0 M solutions in order of increasing pH.

(a) NaCl
(b) NH_3
(c) NaCN
(d) HCl
(e) NaOH
(f) CH_3CO_2H

CHAPTER 18
Principles of Reactivity: Reactions Between Acids and Bases

Screen 18.2 Acid-Base Reactions

1. What types of acid-base reactions are being described in this chapter, those between Brønsted acids and bases, or those between Lewis acids and bases?

2. Examine the table of acid-base reaction types on this screen. What seems to be more important in controlling the acidity of the solution after reaction, strong acids or weak acids?

Screen 18.3 Acid-Base Reactions (Strong Acids + Strong Bases)

1. Explain why a neutral solution results when equal molar amounts of a strong acid and a strong base are mixed.

2. Watch the video shown on this screen. Is the initial solution acidic or basic? Explain your choice.

3. When the reaction is complete, the solution pH is quite low. What does this imply about the amount of HCl added relative to the amount of NaOH originally in the beaker?

Screen 18.4 Acid-Base Reactions (Strong Acids + Weak Bases)

1. Explain why the solution resulting from reaction of a strong acid and a weak base is acidic, even though all the strong acid is consumed by the base.

2. What do the reaction equations shown at the bottom of the screen tell us about the extent of reaction between a strong acid and a weak base? Do they react fully or only to a limited extent?

3. Estimate the pH of a solution resulting from mixing 100. mL of a 0.15 M HCl solution with 100. mL of a 0.15 M NH_3 solution.

EXERCISE 18.1 *A Strong Acid/Weak Base Reaction*

Aniline, $C_6H_5NH_2$, is a weak organic base. If you mix exactly 50 mL of 0.20 M HCl with 0.93 g of aniline, are the acid and base completely consumed? What is the pH of the resulting solution? (The K_a for $C_6H_5NH_3^+$ is 2.4×10^{-5}).

$$HCl(aq) + C_6H_5NH_2(aq) \rightarrow C_6H_5NH_3^+(aq) + Cl^-(aq)$$

Screen 18.5 Acid-Base Reactions (Weak Acids + Strong Bases)

1. Explain why the solution resulting from reaction of a strong base and a weak acid is basic, even though all the strong base is consumed by the acid.

___

___

2. What do the reaction equations shown at the bottom of the screen tell us about the extent of reaction between a strong base and a weak acid? Do they react fully or only to a limited extent?

___

___

3. Estimate the pH of a solution resulting from mixing 100. mL of a 0.25 M NaOH solution with 100. mL of a 0.25 M HNO_2 solution. (Note: The K_a for HNO_2 is 2.22×10^{-11}.)

EXERCISE 18.2 *pH at the Equivalence Point of a Strong Base/Weak Acid Reaction*

What volume of 0.100 M NaOH, in milliliters, is required to react completely with 0.976 g of the weak, monoprotic acid benzoic acid ($C_6H_5CO_2H$)? What is the pH of the solution after reaction? (See Screen17.7 for the acid-base table giving the K_b for the benzoate ion, $C_6H_5CO_2^-$.)

Screen 18.6 Acid-Base Reactions (Weak Acids + Weak Bases)

1. What controls the acidity of a solution resulting from mixing solutions of a weak acid and a weak base?

2. Do all weak acid-weak base reactions go nearly to completion?

Screen 18.7 The Common Ion Effect

1. This screen shows how adding acetate ion to a solution containing acetic acid results in a shift towards a more basic solution. Using the reaction equation for the reaction of acetic acid and water to give acetate ion and hydronium ion, show how this illustrates Le Chatelier's principle.

2. Consider the plot of concentrations vs. time shown on the second part of this screen. What would the plot look like if we began with an acetate solution and then added acetic acid? Could this be considered a case of the common ion effect, even though no ionic compound would be added?

EXERCISE 18.3 *The Common Ion Effect*

Assume you have a 0.30 M solution of formic acid (HCO_2H) and add enough sodium formate ($NaHCO_2$) to make the solution 0.10 M in the salt. Calculate the pH of the formic acid solution before and after adding sodium formate.

Screen 18.8 Buffer Solutions

1. What are the main components of a buffer solution?

2. What are some practical uses of buffer solutions? How do the videos on this screen illustrate these uses?

3. Write chemical equations to show how a buffer composed of the weak acid sodium bicarbonate, $NaHCO_3$, and its conjugate base sodium carbonate, Na_2CO_3, can neutralize either added acid or added base.

Screen 18.9 pH of Buffer Solutions

1. Consider the Henderson-Hasselbalch equation shown when you press the *General Rule* button on this screen. Clearly the pH of a buffer solution depends on the pK_a of the weak acid and on the relative concentrations of the conjugate acid and base. Which of these two is more important in controlling buffer pH?

2. Examine the sidebar to this screen. Suggest a reason why the body does not use the acetic acid/acetate buffer system for holding blood pH near 7.4.

3. Estimate the pH of a buffer solution composed of 5.8 g CH_3CO_2H (acetic acid) and 12.8 g $NaCH_3CO_2$ (sodium acetate) dissolved in 100. mL of solution.

EXERCISE 18.4 *pH of a Buffer Solution*

You dissolve 15.0 g of $NaHCO_3$ and 18.0 g of Na_2CO_3 in enough water to make 1.00 L of solution. Calculate the pH of the solution. (You can consider this as a solution of the weak acid HCO_3^- with CO_3^{2-} as its conjugate base.)

Screen 18.10 Preparing Buffer Solutions

1. The first rule in preparing a buffer solution is to choose an acid-base conjugate pair with a pK_a near the desired pH. Suggest a reason for this rule based on the Henderson-Hasselbalch equation examined on the previous screen.

____

____

2. Describe how a buffer solution with pH = 3.9 would be prepared using nitrous acid, HNO_2, and sodium nitrite, $NaNO_2$.

EXERCISE 18.5 *Preparing a Buffer Solution*

Using an acetic acid/sodium acetate buffer solution, what ratio of acid to conjugate base will you need to maintain the solution's pH at 5.00? Explain how you would prepare such a solution.

Screen 18.11 Adding Reagents to Buffer Solutions

Consider 100. mL of a buffer solution composed of 0.10 M acetic acid and 0.10 M acetate ion.

1. What is the pH of the buffer?

2. What reaction will occur if 5.0 mL of 0.10 M HCl is added?

3. What will the solution pH be after the acid is added?

4. What reaction would have occurred if 5.0 mL of 0.10 M NaOH had been added to the original buffer solution?

5. What would the resulting pH be in this case?

6. Calculate the expected pH for each of the above examples if HCl or NaOH were added to pure water instead of the buffer solution.
   a) 5.0 mL of 0.10 M HCl added to 100.0 mL of pure water: pH = __________

   b) 5.0 mL of 0.10 M NaOH added to 100.0 mL of pure water: pH = __________

Screen 18.12 Titration Curves

The *Titration Tool* on this screen allows you to titrate two different weak acids, acetic acid and hypochlorous acid (HOCl), in two different concentrations with the strong base NaOH.

1. Titrate 50.0 mL of 0.1 M acetic acid ($K_a = 1.8 \times 10^{-5}$; $pK_a = 4.75$) with 0.2 M NaOH in 0.5 mL increments.
   a) What should the pH of the solution be before NaOH is added?

   ____

   b) What quantity of 0.2 M NaOH is required to reach the equivalence point?

   ____

   c) Proceed to the point in the titration where one-half of the acid has been neutralized. (This is called the *half-neutralization point* or *midpoint*) What is the pH at this point in the titration? What is its relation to the pK_a of the acid?

   ____

   ____

   d) Proceed to the equivalence point. What is the pH at the equivalence point? __________
   e) What are the species in solution at the equivalence point?

   ____

   f) Which of these determines the pH at the equivalence point? Explain briefly.

   ____

   ____

2. Titrate 50.0 mL of 0.1 M hypochlorous acid with 0.2 M NaOH in small increments.

a) How does this titration curve differ from the one for acetic acid?

b) Determine the pH at the half-neutralization point. What is the pK_a and K_a for HOCl?

c) What is the pH at the equivalence point? ________ Demonstrate, by a calculation, that this pH is the expected value.

3. Alpha plots are useful to determine the makeup of the solution of a weak acid (or base) at various pH values. Examine the alpha plot that is displayed following the acetic acid titration.

a) What is depicted by the line that begins at an alpha of 1.0 and slopes downward as the pH increases?

b) What is depicted by the line that begins at an alpha of 0.0 and slopes upward?

c) What is the composition of the solution when the two lines cross? How is this related to the titration with NaOH? How is this related to the pK_a of the acid?

d) How are alpha plots related to buffer solutions?

e) What is the pH of a solution that contains 0.90 M acetic acid and 0.10 M acetate ion?

f) Which species predominates in a solution that has a pH of 5, acetic acid or acetate ion? Explain briefly.

Screen 18.13 Return to the Puzzler

1. Explain why hyperventilating can cause the pH of blood to become dangerously high.

2. Why does the loss of $CO_2(g)$ during hyperventilation change blood pH even though CO_2 contains no H atoms?

Thinking Beyond

1. Why does solution pH change much more quickly near the end point of a titration than either before or after the end point?
2. Will a solution composed of a weak acid and the conjugate base of a *different* weak acid act as a buffer solution?

STUDY QUESTIONS

1. Calculate the hydronium ion concentration and pH of the solution that results when 22.0 mL of 0.10 M acetic acid, CH_3CO_2H, is mixed with 22.0 mL of 0.10 M NaOH.
2. For each of the following cases, decide whether the pH is less than 7, equal to 7, or greater than 7.
   (a) Equal volumes of 0.10 M acetic acid, CH_3CO_2H, and 0.10 M KOH are mixed
   (b) 25 mL of 0.015 M NH_3 is mixed with 25 mL of 0.015 M HCl
   (c) 100. mL of 0.0020 M HNO_3 is mixed with 50. mL of 0.0040 M NaOH
3. For each of the following cases, decide whether the pH is less than 7, equal to 7, or greater than 7.
   (a) 25 mL of 0.45 M H_2SO_4 is mixed with 25 mL of 0.90 M NaOH
   (b) 15 mL of 0.050 M formic acid, HCO_2H, is mixed with 15 mL of 0.050 M NaOH
   (c) 25 mL of 0.15 M $H_2C_2O_4$ (oxalic acid) is mixed with 25 mL of 0.30 M NaOH. (Both H^+ ions of oxalic acid are titrated.)
4. Does the pH of the solution increase, decrease, or stay the same when you
   (a) add solid ammonium chloride to a dilute aqueous solution of NH_3?
   (b) add solid sodium acetate to a dilute aqueous solution of acetic acid?
   (c) add solid KCl to a dilute aqueous solution of KOH?
5. Does the pH of the solution increase, decrease, or stay the same when you
   (a) add 10.0 mL of 0.10 M HCl to 25.0 ml of 0.10 M NH_3?
   (b) add 25.0 mL of 0.050 M NaOH to 50.0 mL of 0.050 M acetic acid?
6. What is the pH of a buffer solution that is 0.20 M with respect to ammonia, NH_3, and 0.20 M with respect to ammonium chloride, NH_4Cl?

7. What mass of ammonium chloride, NH_4Cl, would have to be added to exactly 500 mL of 0.10 M NH_3 solution to give a solution with a pH of 9.0?
8. If a buffer solution is made of 12.2 g of benzoic acid ($C_6H_5CO_2H$) and 7.20 g of sodium benzoate ($NaC_6H_5CO_2$) in exactly 250 mL of solution, what is the pH of the buffer? If the solution is diluted to exactly 500 mL with pure water, what is the new pH of the solution?
9. Arrange the following solutions in order of increasing pH (all reagents are 0.10 M).

(a) NaCl	(d) HCl
(b) NH_3	(e) NH_4Cl
(c) $CH_3CO_2H/NaCH_3CO_2$	(f) CH_3CO_2H

10. The pH of human blood is controlled by several buffer systems, among them the reaction

    $H_2PO_4^-(aq) + H_2O(\ell) \rightarrow H_3O^+(aq) + HPO_4^{2-}(aq)$

    Calculate the ratio $[H_2PO_4^-]/[HPO_4^{2-}]$ in normal blood having a pH of 7.40.
11. Two acids, each approximately 10^{-2} M in concentration, are titrated separately with a strong base. The acids show the following pH values at the equivalence point: HA, pH = 9.5 and HB, pH = 8.5.
    (a) Which is the stronger acid, HA or HB?
    (b) Which of the conjugate bases, A^- or B^-, is the stronger base?
12. During active exercise, lactic acid ($K_a = 1.4 \times 10^{-4}$) is produced in the muscle tissues. At the pH of the body (pH = 7.4), which form will be primarily present: unionized lactic acid $[CH_3CH(OH)CO_2H]$ or the lactate ion $[CH_3CH(OH)CO_2^-]$?
13. The pH/Volume curve for the titration of 0.10 M NH_3 with 0.10 M HCl is depicted here.
    (a) What is the pH of the solution before HCl is added?
    (b) What is the pK_b for ammonia? Is this value reflected in the titration?
    (c) What species are in solution at the equivalence point?
    (d) What is the pH at the equivalence point?
    (e) Compare the NH_3/HCl curve with that for the titration of acetic acid with NaOH. Why are these curves different?

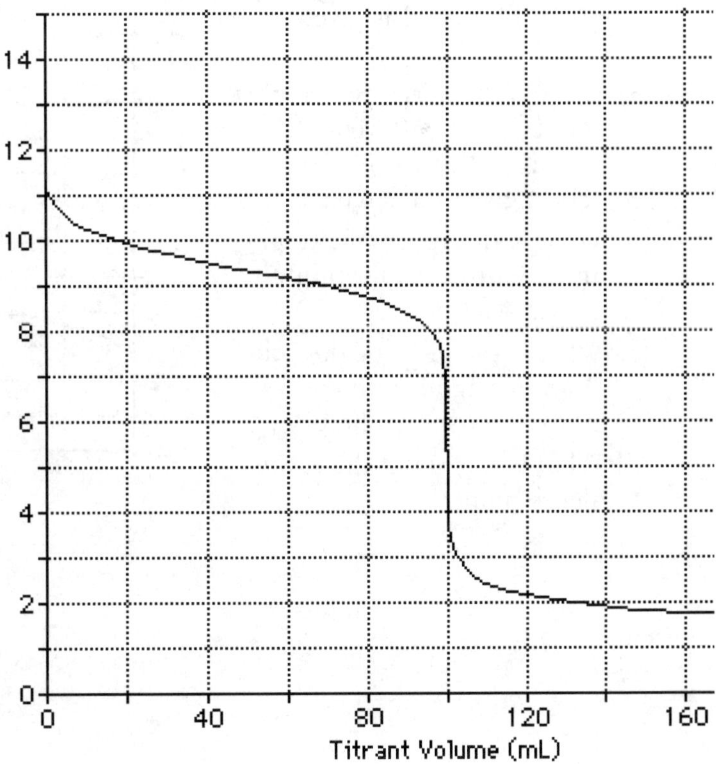

14. An alpha plot for formic acid, HCO_2H, is given here.

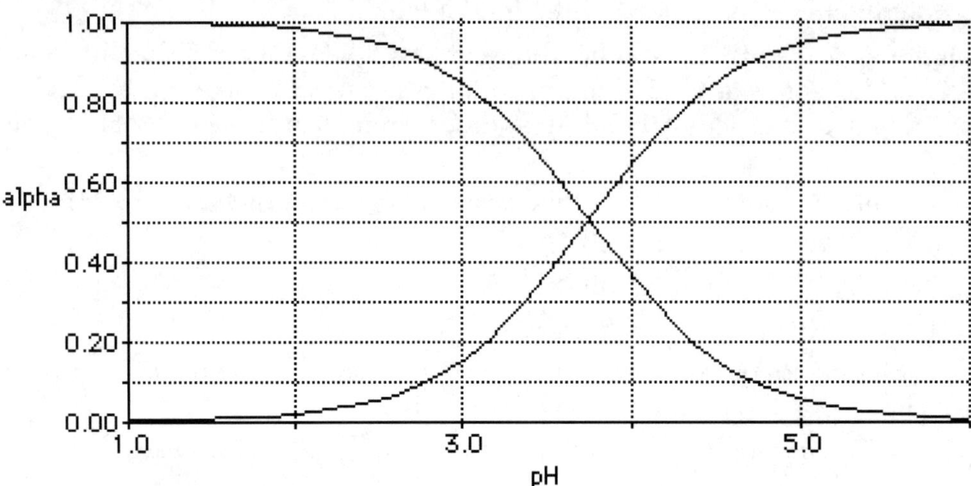

(a) Use the alpha plot to estimate the pK_a for the acid. Look up the acid on Screen 17.6 and determine if your estimate agrees with the known pK_a.

(b) Describe how you would make a buffer solution that has a pH of 4.00 if you have formic acid and sodium formate, $NaHCO_2$.

15. The pH/Volume curve for the titration of 100.0 mL of 0.050 M benzoic acid, $C_6H_5CO_2H$, with 0.10 M NaOH is given here.

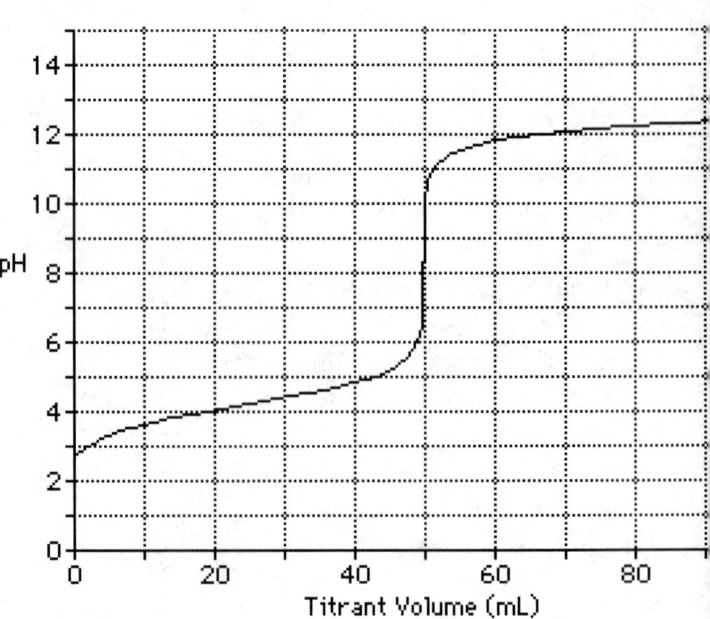

(a) What is the initial pH? Is this correct for 0.050 M acid? (The K_a value for benzoic acid is given on Screen 17.6.)

(b) What is the approximate pH at the half-neutralization point?

(c) What is the pK_a for the acid? Does this agree with the experiment shown here?

(d) What is the pH at the equivalence point?

CHAPTER 19
Principles of Reactivity: Precipitation Reactions

Screen 19.2 Precipitation Reactions

1. Draw a diagram to illustrate what occurs on the molecular scale when solid $CaCl_2$ dissolves in water.

2. This screen describes the dissolution of ionic compounds in water. How does this differ from the way a nonionic compound, such as ammonia, NH_3, dissolves in water?

Screen 19.3 Solubility

1. Explain how the series of photos on this screen illustrates the principle of solubility.

2. Sodium chloride is considered a soluble compound. What does this mean? If one keeps adding more and more NaCl to a sample of water, will the salt keep dissolving?

3. Name three common substances that are insoluble in water.

Screen 19.4 Solubility Product Constant

1. Why does the K_{sp} expression shown on the screen for the dissolution of $PbCl_2$ not have a denominator, such as those of other equilibrium reactions we have studied?

2. Explain the difference between the K_{sp} value of a salt and the salt's solubility.

3. Examine the table of K_{sp} values. Do you see any trends? For example, do salts that dissolve to give one cation and one anion (such as AgCl) tend to have larger or smaller K_{sp} values than those that break up into more ions (such as Ag_2CrO_4)?

EXERCISE 19.1 *Writing K_{sp} Expressions*

Write K_{sp} expressions for the following insoluble salts.

(a) $BaSO_4$ (b) BiI_3 (c) Ag_2CO_3

Screen 19.5 Determining K_{sp}

1. In the example on this screen, we determine the K_{sp} value for $PbCl_2$ by measuring the concentration of Pb^{2+} ions in solution. Why do we not need to measure the concentration of Cl^- ions in this case?

2. Watch the video on this screen describing the use of an atomic absorption spectrometer. Explain how the spectrometer works. It may be helpful to draw a diagram.

EXERCISE 19.2 *K_{sp} from Solubility Measurements*

The barium ion concentration, [Ba^{2+}], in a saturated solution of barium fluoride is 7.5×10^{-3} M. Calculate the K_{sp} for BaF_2.

$$BaF_2(s) \rightleftharpoons Ba^{2+}(aq) + 2\ F^-(aq)$$

Screen 19.6 Estimating Salt Solubility

1. This screen describes how the solubility of $BaSO_4$ is estimated using its K_{sp} value. The calculation assumes that the ions produced, Ba^{2+} and $SO_4{}^{2-}$, do not react further in solution. What would happen to the solubility of $BaSO_4$ if some of the sulfate ion reacted with acid in solution?

$$SO_4{}^{2-}(aq) + H_3O^+(aq) \rightarrow HSO_4{}^-(aq) + OH^-(aq)$$

____

____

____

2. We can compare the solubilities of different ionic compounds by directly comparing their K_{sp} values only if two salts break up into the same number of ions. For different types of salts (e.g., AgCl and $PbCl_2$) we cannot make a direct comparison. Explain briefly.

____

____

3. Calculate the solubility of AgI in water at 25 °C.

EXERCISE 19.3 *Calculating Solubility from K_{sp}*

Calculate the solubility in water of CuI ($K_{sp} = 5.1 \times 10^{-12}$) and of $Mg(OH)_2$ ($K_{sp} = 1.5 \times 10^{-11}$).

EXERCISE 19.4 *Comparing Salt Solubilities*

Using K_{sp} values (Screen 19.4), tell which salt in each pair is the more soluble.

1. AgCl or AgI
2. $CaCO_3$ or $SrCO_3$
3. CaF_2 or $CaCO_3$

Screen 19.7 Can a Precipitation Reaction Occur?

1. What is the difference between the reaction quotient, Q, and K_{sp}?

2. What is the connection between the relative values of Q and K_{sp} and the likelihood that a precipitation reaction will occur?

3. Three cases are given on this screen: $Q < K_{sp}$, $Q > K_{sp}$, and $Q = K_{sp}$. Which of these represents a system at equilibrium?

4. Can a precipitation system be at equilibrium if there is no solid present? Explain briefly.

EXERCISE 19.5 *Solubility and the Reaction Quotient*

Solid PbI_2 ($K_{sp} = 8.7 \times 10^{-9}$) is placed in a beaker of water. After a period of time, the lead(II) ion concentration is measured and found to be 1.1×10^{-3} M. Has the system yet reached equilibrium? That is, is the solution saturated? If not, will more PbI_2 dissolve?

EXERCISE 19.6 *Deciding Whether a Precipitate Will Form*

The concentration of strontium ion in a solution is 2.5×10^{-4} M. Will $SrSO_4$ precipitate if enough of the soluble salt Na_2SO_4 is added to make the solution 2.5×10^{-4} M in SO_4^{2-}? K_{sp} for $SrSO_4$ is 2.8×10^{-7}.

EXERCISE 19.7 *Ion Concentrations Required to Begin a Precipitation*

What is the minimum concentration of I^- that can cause precipitation of PbI_2 from a 0.050 M solution of $Pb(NO_3)_2$? K_{sp} for PbI_2 is 8.7×10^{-9}.

Screen 19.8 The Common Ion Effect

1. State the common ion effect in your own words.

2. The animation on this screen illustrates the common ion effect for the case of adding extra chloride ion to an equilibrium system containing $PbCl_2(s)$, $Pb^{2+}(aq)$, and $Cl^-(aq)$. Explain the changes you see in terms of the solubility product constant expression for this system.

3. Can a common "ion" effect occur for a solubility reaction involving a molecular solid, which does not break up into ions as it dissolves but simply leads to discrete molecules in solution?

4. Explain how the common ion effect is an example of Le Chatelier's principle. Use the dissolution of a salt such as $PbBr_2$ as an example.

EXERCISE 19.8 *The Common Ion Effect*

Calculate the solubility of $BaSO_4$ (i) in pure water and (ii) in the presence of 0.010 M $Ba(NO_3)_2$. K_{sp} for $BaSO_4$ is 1.1×10^{-10}.

Screen 19.9 Using Solubility

Examine the flowchart for the separation of the ions Ag^+, Pb^{2+}, and Cu^{2+} in aqueous solution.

1. What is the initial step in the separation and why does it work?

2. In what way does the solubility of $PbCl_2$ differ from that of AgCl?

3. How do we know that Pb^{2+} ions were in solution?

4. Give a possible reason for our interest in designing ways to separate different ions from one another in aqueous solution.

Screen 19.10 Simultaneous Equilibria

1. The video on this screen shows that the white solid $PbCl_2$ can be converted into the yellow solid $PbCrO_4$. It might appear that the reaction occurs in the solid state, even though it actually occurs in aqueous solution. Write chemical equations to show how it occurs in solution.

2. Explain why the experiment on this screen is good evidence for the fact that chemical equilibria are dynamic as opposed to static.

3. What effect do you think adding more chloride ion to the solution would have on the amount of product formed? Would the effect be significant?

EXERCISE 19.9 *Simultaneous Equilibria*

Silver forms many insoluble salts. Which is more soluble, AgCl or AgBr? (See Screen 19.4 for K_{sp} values.) If you add sufficient bromide ion to an aqueous suspension of AgCl(s), can AgCl be converted to AgBr? To answer this, derive the value of the equilibrium constant for

$$AgCl(s) + Br^-(aq) \rightarrow AgBr(s) + Cl^-(aq)$$

Screen 19.11 Solubility and pH

This screen describes the effect of pH on solubility equilibria.

1. Explain how the example on this screen, the solubility of $Co(OH)_2(s)$, is an example of Le Chatelier's principle.

2. Notice that the solubility of the compound increases as the pH decreases. What effect do you think a pH decrease would have on the solubility of $MgCO_3$, considering the fact that carbonate ion, CO_3^{2-}, is a weak base?

3. Examine the *pH and Solubility* table. Explain why the solubility of $Co(OH)_2$ increases by 100 for each 1.0 decrease in pH.

4. Do you expect the solubility of $Fe(OH)_3$ to be more or less sensitive to pH than that of $Co(OH)_2$?

Screen 19.12 Complex Ion Formation and Solubility

1. Watch the video of the effect ammonia has on silver chloride solubility. In what way is this an example of Le Chatelier's principle?

2. Examine the sidebar to this screen. How does the chemistry of floor wax support the idea that reactions can be reversible?

Thinking Beyond

Why do metal sulfides (except those of the alkali metals) generally not dissolve in water? As a hint, consider assumptions you make about the type of bonding in "ionic" solids.

STUDY QUESTIONS

1. When 1.55 g of solid thallium(I) bromide is added to 1.00 L of water, the salt dissolves to a small extent.

   $TlBr(s) \rightleftharpoons Tl^+(aq) + Br^-(aq)$

   The thallium(I) and bromide ions in equilibrium with TlBr(s) each have a concentration of 1.8×10^{-3} M. What is the value of K_{sp} for TlBr?
2. Calcium hydroxide, $Ca(OH)_2$, dissolves in water to the extent of 0.93 g per liter. What is the K_{sp} of $Ca(OH)_2$?

   $Ca(OH)_2(s) \rightleftharpoons Ca^{2+}(aq) + 2\ OH^-(aq)$
3. At 25 °C, 34.9 mg of Ag_2CO_3 will dissolve in 1.0 L of pure water.

   $Ag_2CO_3(s) \rightleftharpoons 2\ Ag^+(aq) + CO_3^{2-}(aq)$

   What is the solubility product constant for this salt?
4. When 1.234 g of solid $Ca(OH)_2$ is placed in 1.00 L of pure water at 25 °C, the pH of the solution is found to be 12.40. Estimate the K_{sp} for $Ca(OH)_2$.
5. Estimate the solubility of silver cyanide in (a) moles per liter and (b) grams per liter in pure water at 25 °C.

   $AgCN(s) \rightleftharpoons Ag^+(aq) + CN^-(aq)$ (K_{sp} for AgCN = 1.2×10^{-16})

6. Use K_{sp} values to decide which compound in each of the following pairs is the more soluble.
   (a) AgBr ($K_{sp} = 3.3 \times 10^{-13}$) or AgSCN ($K_{sp} = 1.0 \times 10^{-12}$)
   (b) $SrCO_3$ ($K_{sp} = 9.4 \times 10^{-10}$) or $SrSO_4$ ($K_{sp} = 2.8 \times 10^{-7}$)
   (c) MgF_2 ($K_{sp} = 6.4 \times 10^{-9}$) or CaF_2 ($K_{sp} = 3.9 \times 10^{-11}$)
   (d) AgI ($K_{sp} = 2.5 \times 10^{-16}$) or HgI_2 ($K_{sp} = 4.0 \times 10^{-29}$)
7. Rank the following compounds in order of increasing solubility in water: Na_2CO_3, BaF_2 ($K_{sp} = 1.7 \times 10^{-6}$), $BaCO_3$ ($K_{sp} = 8.1 \times 10^{-9}$), and Ag_2CO_3 ($K_{sp} = 8.1 \times 10^{-12}$).
8. Sodium carbonate is added to a solution in which the concentration of Ni^{2+} ion is 0.0024 M. Will precipitation of $NiCO_3$ occur when the concentration of the carbonate ion is (a) 1.0×10^{-6} M or (b) when it is 100 times greater (or 1.0×10^{-4} M)? The equilibrium process involved is $NiCO_3(s) \rightleftharpoons Ni^{2+}(aq) + CO_3^{2-}(aq)$ (K_{sp} for $NiCO_3 = 6.6 \times 10^{-9}$)
9. If the concentration of Mg^{2+} in seawater is 1350 mg per liter, what OH^- concentration is required to precipitate $Mg(OH)_2$? (K_{sp} for $Mg(OH)_2 = 1.5 \times 10^{-11}$)
10. If you mix 10. mL of 0.0010 M $Pb(NO_3)_2$ with 5.0 mL of 0.015 M HCl, will $PbCl_2$ precipitate? (K_{sp} for $PbCl_2 = 1.7 \times 10^{-5}$)
11. Calculate the molar solubility of silver thiocyanate, AgSCN, in pure water and in water containing 0.010 M NaSCN. (K_{sp} for AgSCN = 1.0×10^{-12})
12. A solution contains 0.10 M iodide ion, I^-, and 0.10 M carbonate ion, CO_3^{2-}. If solid $Pb(NO_3)_2$ is slowly added to the solution, which salt will precipitate first, PbI_2 or $PbCO_3$? (K_{sp} for $PbCO_3 = 1.5 \times 10^{-13}$ and K_{sp} for $PbI_2 = 8.7 \times 10^{-9}$)
13. What is the equilibrium constant for the following reaction? (See Screen 19.4 for K_{sp} values.)

    $AgCl(s) + I^-(aq) \rightleftharpoons AgI(s) + Cl^-(aq)$

    Does the equilibrium lie predominantly to the left or right? Will AgI form if iodide ion, I^-, is added to a saturated solution of AgCl?
14. Explain how changing pH would affect the solubility of $Ni(OH)_2$.
15. Barium carbonate dissolves to some extent in the presence of carbon dioxide.

    $BaCO_3(s) + CO_2(g) + H_2O(\ell) \rightleftharpoons Ba^{2+}(aq) + 2\ HCO_3^-(aq)$ $K = 4.5 \times 10^{-5}$

    (a) How will the solubility of barium carbonate be affected by an increase in the pressure of CO_2?
    (b) How will the solubility of barium carbonate be affected by a decrease in the pH?

CHAPTER 20
Principles of Reactivity: Entropy and Free Energy

Screen 20.1 The Chemical Puzzler

1. The question being asked on this screen involves reactions occurring that are "energetically uphill." List three reactions you have seen on this CD-ROM that are "energetically downhill."

Screen 20.2 Reaction Spontaneity (Thermodynamics and Kinetics)

This screen is an expansion of Screen 6.3. You might wish to review that screen.

1. What is the relationship between the terms "spontaneous" and "product-favored?"

2. What is the relationship between a reaction's spontaneity and its rate?

3. If gaseous H_2 and O_2 are carefully mixed and left alone, they can remain intact for millions of years. Is this "stability" a function of thermodynamics or of kinetics?

4. Can a reaction be "driven" only by thermodynamics or only by kinetics?

Screen 20.3 Directionality of Reactions

1. This screen describes reactions as being product-favored if either matter or energy are dispersed. Is the "favorability" being discussed thermodynamic or kinetic in nature?

2. Which of the dispersal mechanisms involves the concept of enthalpy change?

3. Is energy dispersed during the process described in the chemical puzzler.

$NH_4NO_3(s) \rightarrow NH_4NO_3(aq)$

Screen 20.4 Entropy

1. Describe your concept of entropy.

2. Which of the dispersal mechanisms described on Screen 20.3 involved changes in a system's entropy?

3. Why does the entropy of a substance increase with temperature?

4. Review Screen 13.17, *Phase Changes*. Describe, in qualitative terms, the changes in entropy when taking solid water at -10 °C and 1 atm and heating it to 110 °C.

5. Examine the sidebar to this screen. Calculate the entropy change for the vaporization of ethanol, C_2H_5OH, at its boiling point of 78.5 °C. Ethanol has an enthalpy of vaporization of 39.3 kJ/mol.

EXERCISE 20.1 *Calculating ΔS Values for Phase Changes*

The enthalpy of vaporization of benzene (C_6H_6) is 30.9 kJ/mol at the boiling point of 80.1 °C. Calculate the entropy change for benzene going from (a) liquid to vapor and (b) vapor to liquid at 80.1 °C.

EXERCISE 20.2 *Entropy*

For each of the following processes, predict whether you would expect entropy to be greater for the products than for the reactants. Explain how you arrived at your prediction.

1. $CO_2(g) \rightarrow CO_2(s, \text{dry ice})$

2. $NaCl(s) \rightarrow NaCl(aq)$

3. $MgCO_3(s) + heat \rightarrow MgO(s) + CO_2(g)$

Screen 20.5 Calculating ΔS for a Chemical Reaction

1. Will reactions with products that are much more disordered than the reactants have a positive or a negative value of ΔS?

2. Which do you expect to have a greater entropy, potassium chloride as a solid, or potassium chloride dissolved in water?

EXERCISE 20.3 *Calculating ΔS for a Chemical Reaction*

Nitrogen dioxide is formed from nitrogen monoxide and oxygen in a product-favored reaction at 25 °C (Screen 4.17). Determine the standard entropy change, ΔS°, for the reaction, ΔS°_{rxn}.

THERMODYNAMIC DATA

Thermodynamic values needed to solve numerical problems in this chapter are found in Appendix D in this *Workbook*.

Screen 20.6 The Second Law of Thermodynamics

1. What is the second law of thermodynamics?

2. Consider the reaction of iron and carbon.

   $Fe(s) + Cl_2(g) \rightarrow FeCl_2(s)$

   a) Calculate ΔS° and ΔH° for the reaction.

   b) If a reaction of one mole each of iron and chlorine occurs at a temperature of 25 °C, calculate the entropy change for the system and for the surroundings.

c) According to the second law of thermodynamics, is this reaction product- or reactant-favored? Explain briefly.

EXERCISE 20.4 *Is a Reaction Product- or Reactant-Favored?*

Is the direct reaction of hydrogen and chlorine to give hydrogen chloride gas predicted to be product-favored or reactant-favored?

$H_2(g) + Cl_2(g) \rightarrow 2\ HCl(g)$

Answer the question by calculating the values for ΔS°_{system} and $\Delta S^\circ_{surroundings}$ at 25 °C and then summing them to determine $\Delta S^\circ_{universe}$.

Screen 20.7 Gibbs Free Energy

1. What is the use of calculating Gibbs free energy changes for reactions?

2. Do we ever calculate actual Gibbs free energies, or just changes in Gibbs free energies that occur during processes?

3. a) What are the two methods for calculating ΔG for reactions?

   b) Explain under what circumstances each equation would be used.

EXERCISE 20.5 *Calculating ΔG°_{rxn} From ΔH°_{rxn} and ΔS°_{rxn}*

Using values of ΔH°_f and S° to find ΔH°_{rxn} and ΔS°_{rxn}, respectively, calculate the free energy change for the formation of 1 mole of $NH_3(g)$ from the elements at standard conditions (and 25 °C): $1/2\ N_2(g) + 3/2\ H_2(g) \rightarrow NH_3(g)$.

EXERCISE 20.6 *Calculating ΔG°_{rxn} from ΔG°_f*

Calculate the standard free energy change for the combustion of 1.00 mole of benzene, $C_6H_6(\ell)$, to give $CO_2(g)$ and $H_2O(g)$.

Screen 20.8 Free Energy and Temperature

1. Are reactions that occur only at high temperature but not at low temperature enthalpy-favored, entropy-favored or both? Or, can't you tell?

____

2. Are reactions that occur only at low temperature but not at high temperature enthalpy-favored, entropy-favored or both? Or, can't you tell?

____

EXERCISE 20.7 *Temperature and Free Energy Change*

Is the reduction of magnesia, MgO, with carbon a spontaneous process at 25 °C? If not, at what temperature does it become spontaneous?

$MgO(s) + C(graphite) \rightarrow Mg(s) + CO(g)$

Screen 20.9 Thermodynamics and the Equilibrium Constant

1. What is the relationship between $\Delta G°$ for a reaction and the reaction's equilibrium constant, K? Does K increase or decrease as $\Delta G°$ becomes a more negative number?

____

____

2. Consider the isomerization of butane, presented originally on Screen 16.9.

butane $\rightleftharpoons$ isobutane   K = 0.40 at 25 °C

a) Calculate the value of $\Delta G°$ for the reaction at this temperature.

b) Is the reaction product-favored or reactant-favored? Explain briefly.

____

____

____

c) If we begin with a system where [butane] = 1.0 M and [isobutane] = 0 M, what will occur?

____

d) Are your answers to parts b and c consistent? Explain.

____

____

4. Calculate ΔH°, ΔS°, ΔG° and the equilibrium constant for the reaction to form ammonia from the elements at 25 °C.

$$2\ N_2(g) + 3\ H_2(g) \rightleftharpoons 2\ NH_3(g)$$

a) ΔH° = __________ kJ/mol

ΔS° = __________ J/K · mol

ΔG° = __________ kJ/mol

K = __________

b) One of the main problems in making ammonia commercially is the slowness of the reaction above. There are two potential remedies to this problem: run the reaction at a higher temperature to increase the reaction rate, or use a catalyst to increase the reaction rate. Which of these do you recommend if millions of dollars depend on your decision?

____

____

EXERCISE 20.8 *Calculating K from the Free Energy Change*

Calculate K at 298 K from the value of ΔG°_{rxn} for the reaction $CaCO_3(s) \rightarrow CaO(s) + CO_2(g)$.

Screen 20.10 Return to the Puzzler

When the question to the puzzler was posed, we asked how energetically unfavorable reactions can occur, and we used the dissolution of ammonium nitrate as our example. Based on what you have learned in this chapter, what is wrong with the way the question is worded? In particular, which word in the question is really ambiguous?

____

____

____

Thinking Beyond

1. The value of ΔS° for the dissolution of NaOH(s) in water is a negative value.

   $NaOH(s) \rightarrow NaOH(aq) \quad \Delta S^\circ = -16.36 \text{ J/K} \cdot \text{mol}$

   Give an explanation of what might occur on the molecular scale that could lead to the dissolution of a crystalline solid resulting in a decrease in the system's entropy.
2. In Chapter 14 we gave a lengthy explanation of how intermolecular forces controlled solubility. These arguments were based mainly on the relative strengths of intermolecular forces in the different compounds. We now know that reaction favorability is based on both enthalpy changes (based here on IMF strengths) and on entropy changes. How would you modify the "like-dissolves-like" rule to account for entropy? Do you think it needs modifying?
3. Does the formation of snowflakes from gaseous water in the atmosphere violate the second law of thermodynamics?

STUDY QUESTIONS

1. Which substance has the higher entropy in each of the following pairs?
   (a) A sample of dry ice (solid CO_2) at -78 °C or CO_2 vapor at 0 °C.
   (b) Sugar, as a solid or dissolved in a cup of tea.
   (c) Two 100-mL beakers, one containing pure water and the other containing pure alcohol, or a beaker containing a mixture of the water and alcohol.
2. By comparing the formulas or states for each pair of compounds, decide which is expected to have the higher entropy at the same temperature.
   (a) $KCl(s)$ or $AlCl_3(s)$
   (b) $CH_3I(\ell)$ or $CH_3CH_2I(\ell)$
   (c) $NH_4Cl(s)$ or $NH_4Cl(aq)$
3. The enthalpy of vaporization of liquid diethyl ether, $(C_2H_5)_2O$, is 26.0 kJ/mol at the boiling point of 35.0 °C. Calculate ΔS° for (a) liquid to vapor and (b) vapor to liquid at 35.0 °C.
4. Calculate the standard molar entropy change for the formation of gaseous propane (C_3H_8) at 25 °C.

   $3\ C(\text{graphite}) + 4\ H_2(g) \rightarrow C_3H_8(g)$
5. Calculate the standard molar entropy change for each of the following reactions at 25 °C.
   (a) $2\ Al(s) + 3\ Cl_2(g) \rightarrow 2\ AlCl_3(s)$
   (b) $C_2H_5OH(\ell) + 3\ O_2(g) \rightarrow 2\ CO_2(g) + 3\ H_2O(g)$
6. What are the signs of the enthalpy and entropy changes for the splitting of water to give gaseous hydrogen and oxygen, a process that requires considerable energy? Is this reaction likely to be product-favored or not? Explain your answer briefly.
7. Is the combustion of ethane, C_2H_6, likely to be a product-favored reaction?

   $C_2H_6(g) + 7/2\ O_2(g) \rightarrow 2\ CO_2(g) + 3\ H_2O(g)$

   Answer the question by calculating the value of $\Delta S_{universe}$. Required values of ΔH°_f and S° are in Appendix D. Does your calculated answer agree with your preconceived idea of this reaction?

8. Using values of ΔH°_f and S°, calculate ΔG°_{rxn} for each of the following reactions:
   (a) $Sn(s) + 2\ Cl_2(g) \rightarrow SnCl_4(\ell)$
   (b) $NH_3(g) + HCl(g) \rightarrow NH_4Cl(s)$

   Which of the values of ΔG°_{rxn} that you have just calculated corresponds to a standard free energy of formation, ΔG°_f? In those cases, compare your calculated values with the values of ΔG°_f tabulated in Appendix D. Which of these reactions is (are) predicted to be product-favored? Are the reactions enthalpy- or entropy-driven?
9. Write a balanced equation that depicts the formation of 1 mole of $Fe_2O_3(s)$ from its elements. What is the standard free energy of formation of 1.00 mole of $Fe_2O_3(s)$? What is the value of ΔG°_{rxn} when 454 g (1 pound) of $Fe_2O_3(s)$ are formed from the elements?
10. Using values of ΔG°_f, calculate ΔG°_{rxn} for each of the following reactions. Which are predicted to be product-favored?
    (a) $Ca(s) + Cl_2(g) \rightarrow CaCl_2(s)$
    (b) $NH_3(g) + 2\ O_2(g) \rightarrow HNO_3(aq) + H_2O(\ell)$
11. The formation of NO(g) from its elements

    $1/2\ N_2(g) + 1/2\ O_2(g) \rightarrow NO(g)$

    has a standard free energy change, ΔG°_f, of +86.57 kJ/mol at 25 °C. Calculate K_p at this temperature. Comment on the connection between the sign of ΔG° and the magnitude of K_p.
12. Sodium reacts violently with water according to the equation

    $Na(s) + H_2O(\ell) \rightarrow NaOH(aq) + 1/2\ H_2(g)$

    Without doing calculations, predict the signs of ΔH° and ΔS° for the reaction. Next, verify your prediction with a calculation.
13. For each of the following processes, give the algebraic sign of ΔH°, ΔS°, and ΔG°. No calculations are necessary; use your common sense.
    (a) The splitting of liquid water to give gaseous oxygen and hydrogen, a process that requires a considerable amount of energy.
    (b) The explosion of dynamite, a mixture of nitroglycerin, $C_3H_5N_3O_9$, and diatomaceous earth. The explosive decomposition gives gaseous products such as water, CO_2, and others; much heat is evolved.
    (c) The combustion of gasoline in the engine of your car, as exemplified by the combustion of octane.

    $2\ C_8H_{18}(g) + 25\ O_2(g) \rightarrow 16\ CO_2(g) + 18\ H_2O(g)$
14. A crucial reaction for the production of synthetic fuels is the conversion of coal to H_2 with steam.

    $C(s) + H_2O(g) \rightarrow CO(g) + H_2(g)$

    (a) Calculate ΔG°_{rxn} and K_p for this reaction at 25 °C assuming C(s) is graphite.
    (b) Is the reaction predicted to be product-favored under standard conditions? If not, at what temperature will it become so?

CHAPTER 21
Principles of Reactivity: Electron Transfer Reactions

Screen 21.2 Redox Reactions

1. What is the difference between an oxidizing agent and a reducing agent?

____

____

2. What is the difference between a direct redox reaction and an indirect redox reaction?

____

____

3. In the reaction between copper and silver nitrate displayed under Direct Redox Reactions, what is the oxidizing agent (_______________) and what is the reducing agent (_______________)?

Screen 21.3 Balancing Equations for Redox Reactions

1. When first learning about balancing equations for reactions in chapter 4, we learned that the number of atoms of each element in the products and reactants must be equivalent. What are some additional factors that must be taken into account when balancing equations for redox reactions?

____

____

2. What are half-reactions? What two aspects of these equations must be balanced?

____

____

3. Balance the equation for the reaction of aluminum metal with copper(II) ion.

   $Al(s) + Cu^{2+}(aq) \rightarrow Al^{3+}(aq) + Cu(s)$

   Half-reaction for oxidation: _______________

   Half-reaction for reduction: _______________

   Net balanced equation: _______________

   How many electrons are transferred from the half-reaction for oxidation to that for reduction? _______________

   The oxidizing agent is __________ and the reducing agent is __________. The substance reduced is __________ and the substance oxidized is __________.

4. a) Balance the equation for the reaction of oxalic acid with permanganate ion in acid solution.

   $H_2C_2O_4(aq) + MnO_4^-(aq) \rightarrow Mn^{2+}(aq) + CO_2(g)$

Half-reaction for oxidation: ______________________________

Half-reaction for reduction: ______________________________

Net balanced equation: ______________________________

b) How many electrons are transferred from the half-reaction for oxidation to that for reduction? ______________

c) The oxidizing agent is __________ and the reducing agent is __________ . The substance reduced is __________ and the substance oxidized is __________ .

EXERCISE 21.1 *Balancing an Equation for an Oxidation-Reduction Reaction*

Balance the equation

$Cr^{2+}(aq) + I_2(aq) \rightarrow Cr^{3+}(aq) + I^-(aq)$

Write the balanced half-reactions and the balanced net ionic equation. Identify the oxidizing agent, the reducing agent, the substance oxidized, and the substance reduced.

EXERCISE 21.2 *Balancing Equations for Oxidation-Reduction Reactions in Acid Solution*

Cobalt metal reacts with nitric acid to give a cobalt(III) salt and NO_2 gas. The unbalanced net ionic equation is

$Co(s) + NO_3^-(aq) \rightarrow Co^{3+}(aq) + NO_2(g)$

1. Balance the equation for the reaction in acid solution.
2. Identify the oxidizing and reducing agents and the substance oxidized and the substance reduced.

Screen 21.4 Electrochemical Cells

1. What electron transfer reaction occurs at the anode of an electrochemical cell?

____

2. What electron transfer reaction occurs at the cathode of an electrochemical cell?

____

3. Why can an indirect redox reaction not occur if the wire connecting the two compartments is not attached to the electrodes?

____

4. Draw a sketch like that on this screen for an electrochemical cell that uses the following reaction:

   $Cu(s) + 2\ Ag^+(aq) \rightarrow Cu^{2+}(aq) + 2\ Ag(s)$

The reaction at the anode is: ________________________________

The half-reaction at the cathode is: ________________________________

Electrons move in the wire from the __________ electrode to the __________ electrode.

If the salt in the salt bridge is $NaNO_3$, the NO_3^- ions must move from the __________ compartment to the __________ .

EXERCISE 21.3 *Electrochemical Cells*

A voltaic cell has been assembled with the net reaction

$Ni(s) + 2\ Ag^+(aq) \rightarrow Ni^{2+}(aq) + 2\ Ag(s)$

Give the half-reactions for this electron transfer process, indicate whether each is an oxidation or reduction, and decide which happens at the anode and which at the cathode. What is the direction of electron flow in an external wire connecting the two electrodes? If a salt bridge connecting the cell compartments contains KNO_3, what is the direction of flow of the nitrate ions?

Screen 21.5 Electrochemical Cells and Potentials

1. When the contents of the half-cells in the photos are changed, the standard potential for the cells also change. What does this imply about the tendency of different metals to either hold onto or release electrons?

____

____

2. a) Rank the three metals used on this screen, Ag, Cu, and Zn, in order of their tendency to release electrons—most likely on the left, least likely on the right.

__________ __________ __________

most likely least likely

b) Rank the three metal ions used in the reactions on this screen, Ag^+, Cu^{2+}, and Zn^{2+}, in order of their tendency to accept electrons- most likely on the left, least likely on the right.

__________ __________ __________

most likely least likely

c) What is the connection between the two rankings you have just made?

____

____

____

3. Consider the following reactions, which occur spontaneously in the direction written:

$Zn(s) + Ni^{2+}(aq) \rightarrow Zn^{2+}(aq) + Ni(s)$

$Zn(s) + Cu^{2+}(aq) \rightarrow Zn^{2+}(aq) + Cu(s)$

$Ni(s) + Cu^{2+}(aq) \rightarrow Ni^{2+}(aq) + Cu(s)$

a) Which metal is the best reducing agent? __________

b) Which is the best oxidizing agent? __________

c) Which metal ion is the best oxidizing agent? ______________

d) Which is the poorest oxidizing agent? ______________

4. Chemical reactions are controlled by thermodynamic and kinetic factors. The sidebar for Screen 21.5 describes the relationship between the free energy change for a reaction and its value of E°. What is this relationship?

____

____

EXERCISE 21.4 *The Relation Between E° and ΔG°_{rxn}*

The following reaction has an E° value of -0.76 V. Calculate ΔG°_{rxn} and tell whether the reaction is product- or reactant-favored.

$H_2(g) + 2\ H_2O(\ell) + Zn^{2+}(aq) \rightarrow Zn(s) + 2\ H_3O^+(aq)$

Screen 21.6 Standard Potentials

1. What is the value of E° for each of the following half-reactions? (Values are given in the *Standard Reduction Potential Table.*)

a) $Cu^{2+}(aq) + 2\ e^- \rightarrow Cu(s)$ E° = ____________ V

b) $Ni(s) \rightarrow Ni^{2+}(aq) + 2\ e^-$ E° = ____________ V

c) $Al(s) \rightarrow Al^{3+}(aq) + 3\ e^-$ E° = ____________ V

d) $2\ H_3O^+(aq) + 2e^- \rightarrow H_2(g) + 2\ H_2O(\ell)$ E° = ____________ V

e) $Cl^-(aq) + OH^-(aq) \rightarrow OCl^-(aq) + H_2O(\ell) + 2\ e^-$ E° = ____________ V

2. What is the best reducing agent in the *Reduction Potential Table?* __________

What is the best oxidizing agent in the *Reduction Potential Table?* __________

What is the poorest reducing agent in the *Reduction Potential Table?* __________

What is the poorest oxidizing agent in the *Reduction Potential Table?* __________

3. Which is the better reducing agent, Al or Fe? __________

Which is the best oxidizing agent, Fe^{2+}, I^-, or Cl^-? __________

4. Decide if each of the following reactions is product-favored in the direction written and calculate E°_{net}.

a) $2\ Al(s) + 3\ Sn^{4+}(aq) \rightarrow 2\ Al^{3+}(aq) + 3\ Sn^{2+}(aq)$

b) $2\ Cl^-(aq) + I_2(s) \rightarrow 2\ I^-(aq) + Cl_2(aq)$

5. Consider the following half-reactions:

HALF-REACTION	E°(V)
$Ce^{4+}(aq) + e^- \rightarrow Ce^{3+}(aq)$	+1.61
$Ag^+(aq) + e^- \rightarrow Ag(s)$	+0.80
$Hg_2^{2+}(aq) + e^- \rightarrow 2\ Hg(\ell)$	+0.79
$Sn^{2+}(aq) + 2\ e^- \rightarrow Sn(s)$	-0.14
$Ni^{2+}(aq) + 2\ e^- \rightarrow Ni(s)$	-0.25
$Al^{3+}(aq) + 3\ e^- \rightarrow Al(s)$	-1.66

a) Which is the weakest oxidizing agent in the list? ______________

b) Which is the strongest oxidizing agent? ______________

c) Which is the strongest reducing agent? ______________

d) Which is the weakest reducing agent? ______________

e) Will Sn(s) reduce $Ag^+(aq)$ to Ag(s)? ______________

f) Will $Hg(\ell)$ reduce $Sn^{2+}(aq)$ to Sn(s)? ______________

g) Name the ions that can be reduced by Sn(s). ______________

h) What metals can be oxidized by $Ag^+(aq)$? ______________

EXERCISE 21.5 *Constructing an Electrochemical Cell*

Draw a diagram of an electrochemical cell using the half-cells $Zn(s)/Zn^{2+}(aq)$ and $Al(s)/Al^{3+}(aq)$. What is the net reaction and its E° value? Show the direction of electron flow in the external wire and the directions of ion flow in the salt bridge. Tell which compartment is the anode and which is the cathode.

Screen 21.7 Electrochemical Cells at Nonstandard Conditions

1. Show how the Nernst equation supports Le Chatelier's principle.

____

____

____

2. Two factors are important in determining the potential of a cell: the nature of the reactants and the concentration of those reactants. Which of these plays the more important role? Explain your answer in terms of the Nernst equation.

____

____

EXERCISE 21.6 *Using the Nernst Equation*

Calculate E_{net} for the following reaction:

$2\ Ag^+(aq, 0.80\ M) + Hg(\ell) \rightarrow 2\ Ag(s) + Hg^{2+}(0.0010\ M, aq)$

Is the reaction product-favored or reactant-favored under these conditions? How does this compare with the reaction under standard conditions?

Screen 21.8 Batteries

1. How does a battery relate to the electrochemical cells we have been describing in the first half of this chapter?

2. In what practical way do the batteries shown differ from the "two-beaker" cells shown thus far?

3. What is the difference between a primary battery and a secondary battery?

Screen 21.9 Corrosion

1. When iron rusts, what is the reducing agent? What is the oxidizing agent?

2. What is responsible for the fact that aluminum does not "rust": thermodynamics or kinetics?

Screen 21.10 Electrolysis

1. Is the electrolysis of water a product-favored or a reactant-favored system? Explain briefly.

2. What materials that we use in our homes and in automobiles are produced by electrolysis?

3. In the production of aluminum by electrolysis, is the aluminum produced at the anode or the cathode? ________________ What is produced at the other electrode? ________________

Screen 21.11 Coulometry

1. In the description of the electrolytic production of chlorine bleach, it is noted that although the value of E° for the reaction is only 2.19 V, commercial cells are normally run at much higher voltage. Is the reason for this based on thermodynamics or on kinetics?

2. A current of 1.50 amperes was passed through a solution containing silver ions for 15.0 minutes. The voltage was high enough that silver deposited at the cathode.

$Ag^+(aq) + e^- \rightarrow Ag(s)$

From the half-reaction we know that if 1 mol of electrons passed through the cell then 1 mol of silver was deposited. To find the number of moles of electrons, we need to know the total electric charge passed through the cell. The charge can be calculated from experimental measurements of the current (in amps) and the time the current flowed. Remember that current, charge, and time are related by the equation

$$\text{Charge, I (amps)} = \frac{\text{electric charge (coulombs, C)}}{\text{time (seconds, s)}}$$

a) Calculate the charge (number of coulombs) passed in 15.0 min.

b) Calculate the number of moles of electrons.

c) Calculate the number of moles of silver and the mass of silver deposited.

EXERCISE 21.7 *Coulometry*

One of the half-reactions occurring in the lead storage battery is

$Pb(s) + SO_4^{2-}(aq) \rightarrow PbSO_4(s) + 2\ e^-$

If a battery delivers 1.50 amperes, and if its lead electrode contains 500. g, how long can current flow before the lead in the electrode is consumed?

Thinking Beyond

1. Examine the sidebar to Screen 21.10. Notice that before electricity became plentiful, aluminum was a very rare, expensive metal. Why do you think this is? Couldn't aluminum be made in the same way that iron is made, by reduction with carbon in a furnace (see Screen 20.8)?

2. Using the half-cell value of E° found in the table on Screen 21.6 and for the half-reaction below, calculate the solubility of AgCl at 25 °C.

$AgCl(s) + e^- \rightarrow Ag(s) + Cl^-(aq) \qquad E^\circ = +0.222\ V$

3. Consider the reduction potentials for the alkali metals.

$Na^+(aq) + e^- \rightarrow Na(s)$ $E° = -2.714$ V

$K^+(aq) + e^- \rightarrow K(s)$ $E° = -2.925$ V

$Li^+(aq) + e^- \rightarrow Li(s)$ $E° = -3.045$ V

One might think that K would be the best reducing agent of the three because it has the lowest ionization energy, while Li would be the weakest reducing agent of the three. What other thermodynamic effect has influenced the ordering of the reduction potentials?

STUDY QUESTIONS

1. Balance each of these half-reactions in acid solution. Describe each as an oxidation or a reduction.
   (a) $Cr(s) \rightarrow Cr^{3+}(aq)$
   (b) $AsH_3(g) \rightarrow As(s)$
   (c) $VO_3^-(aq) \rightarrow V^{2+}(aq)$
2. Balance each of these half-reactions in acid solution. Describe each as an oxidation or a reduction.
   (a) $Cr_2O_7^{2-}(aq) \rightarrow Cr^{3+}(aq)$
   (b) $CH_3CHO(aq) \rightarrow CH_3CO_2H(aq)$
   (c) $Bi^{3+}(aq) \rightarrow HBiO_3(aq)$
3. Balance each of these equations. Reaction (a) is in neutral solution. Reactions (b) and (c) are in acid solution.
   (a) $Cl_2(aq) + Br^-(aq) \rightarrow Br_2(aq) + Cl^-(aq)$
   (b) $Sn(s) + H^+(aq) \rightarrow Sn^{2+}(aq) + H_2(g)$
   (c) $Zn(s) + VO^{2+}(aq) \rightarrow Zn^{2+}(aq) + V^{3+}(aq)$
4. The reactions here are in acid solution. Balance them.
   (a) $Ag^+(aq) + HCHO(aq) \rightarrow Ag(s) + HCO_2H(aq)$
   (c) $H_2S(aq) + Cr_2O_7^{2-}(aq) \rightarrow S(s) + Cr^{3+}(aq)$
   (d) $Zn(s) + VO_3^-(aq) \rightarrow V^{2+}(aq) + Zn^{2+}(aq)$
5. In principle, the reaction of chromium and iron(II) ion can be used to build an electrochemical cell.
   $2Cr(s) + 3\ Fe^{2+}(aq) \rightarrow 2\ Cr^{3+}(aq) + 3\ Fe(s)$
   (a) Write the half-reactions involved.
   (b) Which half-reaction is an oxidation and which is a reduction?
   (c) Which half-reaction occurs in the anode compartment and which in the cathode compartment?
6. The standard potential for the reaction of Mg(s) with $I_2(s)$ is +2.91 V. What is the standard free energy change, $\Delta G°$, for the reaction?
7. Calculate the value of E° for each of the following reactions? Decide if each is product-favored in the direction written.

(a) $2\ I^-(aq) + Zn^{2+}(aq) \rightarrow I_2(g) + Zn(s)$

(b) $Zn^{2+}(aq) + Ni(s) \rightarrow Zn(s) + Ni^{2+}(aq)$

(c) $2\ Cl^-(aq) + Cu^{2+}(aq) \rightarrow Cu(s) + Cl_2(g)$

8. Consider the following half-reactions:

HALF-REACTION	E°(V)
$Cl_2(g) + 2\ e^- \rightarrow 2\ Cl^-(aq)$	+1.36
$I_2(g) + 2\ e^- \rightarrow 2\ I^-(aq)$	+0.535
$Pb^{2+}(aq) + 2\ e^- \rightarrow Pb(s)$	-0.126
$V^{2+}(aq) + 2\ e^- \rightarrow V(s)$	-1.18

(a) Which is the weakest oxidizing agent in the list?

(b) Which is the strongest oxidizing agent?

(c) Which is the strongest reducing agent?

(d) Which is the weakest reducing agent?

(e) Will Pb(s) reduce $V^{2+}(aq)$ to V(s)?

(f) Will $I^-(aq)$ reduce $Cl^-(aq)$ to $Cl_2(g)$?

9. Assume that you assemble an electrochemical cell based on the half-reactions $Zn^{2+}(aq)/Zn(s)$ and $Ag^+(aq)/Ag(s)$.

(a) Write the equation for the product-favored reaction that occurs in the cell and calculate E°_{net}.

(b) Which electrode is the anode and which is the cathode?

(c) Diagram the components of the cell.

(d) If you use a silver wire as an electrode, is it the anode or cathode?

(e) Do electrons flow from the Zn electrode to the Ag electrode, or vice versa?

(f) If a salt bridge containing $NaNO_3$ connects the two half-cells, in which direction do the nitrate ions move, from the zinc to the silver compartment, or vice versa?

10. Calculate the voltage delivered by an electrochemical cell using the following reaction if all dissolved species are 0.10 M:

$2\ Fe^{3+}(aq) + 2\ I^-(aq) \rightarrow 2\ Fe^{2+}(aq) + I_2(s)$

Does the voltage increase or decrease relative to E°_{net}, the standard voltage for the reaction?

11. A current of 2.50 amps is passed through a solution of $Ni(NO_3)_2$ for 2.00 hours. What mass of nickel is deposited at the cathode?

12. The basic reaction occurring in the cell in which Al_2O_3 and aluminum salts are electrolyzed is Al^{3+} + 3 e- → Al(s). If the cell operates at 5.0 V and 1.0×10^5 amps, how many grams of aluminum metal can be produced in an 8.0-hour day?

13. The reactions occurring in a lead storage battery are given on Screen 21.8. A typical battery might be rated at "50. ampere-hours." This means it has the capacity to deliver 50. amps for 1.0 hour (or 1.0 amp for 50. hours). If it does deliver 1.0 amp for 50. hours, how many grams of lead would be consumed to accomplish this?

14. The half-cells $Ni(s)/Ni^{2+}(aq)$ and $Cd(s)/Cd^{2+}(aq)$ are assembled into a battery.
   (a) Write a balanced equation for the reaction occurring in the cell.
   (b) What is oxidized and what is reduced? What is the reducing agent and what is the oxidizing agent?
   (c) Which is the anode and which is the cathode? What is the polarity of the Cd electrode?
   (d) What is E°_{net} for the cell?
   (e) What is the direction of electron flow in the external wire?
   (f) If the salt bridge contains KNO_3, toward which compartment will the NO_3^- ions migrate?
   (g) Calculate the equilibrium constant for the net reaction.
   (h) If the concentration of Cd^{2+} is reduced to 0.010 M, and $[Ni^{2+}]$ = 1.0 M, what is the voltage produced by the cell? Is the net reaction still the reaction given in part (a)?
   (i) If 0.050 amps are drawn from the battery, how long can it last if you begin with 1.0 L of each of the solutions and each was initially 1.0 M in dissolved species? The electrodes each weigh 50.0 g in the beginning.
15. Four metals, A, B, C, and D, exhibit the following properties:
   (a) Only A and C react with 1.0 M hydrochloric acid to give $H_2(g)$.
   (b) When C is added to solutions of the ions of the other metals, metallic B, D, and A are formed.
   (c) Metal D reduces B^{n+} to give metallic B and D^{n+}.

   Based on the information above, arrange the four metals in order of increasing ability to act as reducing agents.

Saunders Interactive General Chemistry Workbook

ActivChemistry™ Lessons

---

by Justin T. Fermann, Ph.D.
University of Massachusetts, Amherst

CHAPTER 1, Lesson 1

Introduction to Density

Did the iron float or sink? ____________________

List the combinations of substances that you compared	Which substance was denser?
________________________________	____________________
________________________________	____________________
________________________________	____________________
________________________________	____________________
________________________________	____________________
________________________________	____________________

Densest substance: ____________________

Least dense substance: ____________________

Thinking Beyond

Is the density of an iron boat, which floats, less than or greater than the density of water? Why?

Could you use this type of experiment to compare densities in space, where there is no gravity, and objects have no weight?

CHAPTER 1, Lesson 2

Quantitative Density Measurement

Iron		Copper	
Mass (g)	Volume (L)	Mass (g)	Volume (L)

d = m/V = ______________ d = m/V = ______________

Thinking Beyond

Are metals or nonmetals generally denser? Metals are on the left and center of the periodic table, nonmetals are on the right.

How would you measure the volume of a penny? of a cork? of the aluminum pull tab from the top of a can of soda?

CHAPTER 2, Lesson 1

Introduction to Atoms

Sketch how ActivChemistry depicts these substances on the molecular scale.

lead **iron** **neon**

ammonia **water** **acetic acid**

Sketch the products of the reaction between an atom of sodium and a molecule of water.

Thinking Beyond

What substances around you can you identify as compounds (made of more than one type of atom) and as elements (that have only one type of atom in them)?

CHAPTER 2, Lesson 2
The Periodic Table

Chromium:

Atomic weight: __________

Boiling point: __________

Melting point: __________

Mercury:

Atomic weight: __________

Boiling point: __________

Melting point: __________

Element with the highest density: __________ Atomic number: __________

density: __________ g/cm^3

Alkali metal with largest thermal conductivity: __________

Atomic number: __________

Thermal conductivity: __________ W/cm K

Group with best electrical conductors: __________

Thinking Beyond

What properties change in a regular manner, either increasing or decreasing from left to right across or from top to bottom down the periodic table?

CHAPTER 2, Lesson 3

Chemical Periodicity

Number of Cl atoms that bond to one atom of:

Mg ________ Na ________ Al ________

Ca ________ K ________ Ga ________

Sr ________ Rb ________ In ________

Sketch the molecule made with B and F atoms

What is the formula of a compound made by bonding hydrogen atoms to an atom of each of the following elements?

C: ________________ N: ________________

O: ________________ F: ________________

Thinking Beyond

What happens if you bond as many oxygen atoms as possible to C and to Si?

CHAPTER 3, Lesson 1

Polyatomic Ions

Ions present in a solution of

Potassium nitrate ________________

Lithium hydroxide ________________

Calcium acetate ________________

Zinc sulfate ________________

Ammonia ________________

Thinking Beyond

Why are most polyatomic ions anions, and why is ammonium the only common polyatomic cation?

CHAPTER 3, Lesson 2

Conductivity of Ionic Solutions

Molecular weight of NaCl ________ g/mol

g NaCl in 0.001 moles ________ g

Current measurements for NaCl solutions

g Na^+	g Cl^-	current (amps)

Current measurements for $ZnSO_4$ solutions

g Zn^{+2}	g SO_4^{2-}	current (amps)

Current measurements for Al_2O_3 and $Ca(OH)_2$in water:

g Al_2O_3	current (amps)	g $Ca(OH)_2$	current (amps)

Thinking Beyond

Does pure water conduct electricity?

Explain why sea water is a better conductor than rain water.

CHAPTER 3, Lesson 3

Percent Composition

_______ g Li in 2.0 g LiOH % Li ___________

_______ g Na in 2.0 g NaOH % Na ___________

_______ g K in 2.0 g KOH % K ___________

Formula weight of $CuSO_4$ ___________ predicted % Cu ___________

Formula weight of Cu_2SO_4 ___________ predicted % Cu ___________

g unknown used ___________

g Cu in unknown ___________

% Cu in unknown ___________

Identity of unknown ___________

Thinking Beyond

Some salts exist in a hydrated state, with several water molecules for each formula unit of salt. What effect does this have on the percent of each ion in a sample?

CHAPTER 4, Lesson 1

Conservation of Matter

$Na + H_2O$

Balanced equation:

___

Sketch the reactants on the molecular scale:

Sketch the products on the molecular scale:

$CH_4 + O_2$

Sketch the reactants on the molecular scale:

Sketch the products on the molecular scale:

Balanced equation:

___

Thinking Beyond

When you burn a candle, it gets lighter (in mass). Why does this not violate the conservation of matter?

CHAPTER 4, Lesson 2

Classification of Reactions

	Reactants	Products	Reaction Type
Zn and $CuSO_4$	__________	__________	__________
$CaCl_2$ and NaOH	__________	__________	__________
HCl and KCH_3CO_2	__________	__________	__________
Sn and HCl	__________	__________	__________

Thinking Beyond

What type of reaction occurs between solid (insoluble) $Mg(OH)_2$ and aqueous HCl?

What kind of reaction occurs between *gas phase* ammonia and *gas phase* HCl, producing *solid* ammonium chloride?

CHAPTER 5, Lesson 1

Limiting Reagents

Sketch the graph of Cd remaining vs. volume of acid added once the reaction is completed.

Label:

a) the equivalence point
b) region where Cd is in excess
c) the region where HCl is in excess.

Thinking Beyond

What would this graph look like if you added the Cd to a solution of HCl in small amounts?

CHAPTER 5, Lesson 2

Making a Solution by Direct Addition

Moles $CaCl_2$ needed to make
0.5 L of 0.2 M solution ____________________

Grams $CaCl_2$ needed ____________________

Mass of 0.5 L of solution ____________________

Concentration of solution made by mixing
11.32 g $CaCl_2$ in a 0.500 L solution ____________________

Thinking Beyond

Why is it easier to use this method to make concentrated solutions of high molecular weight solutes accurately?

CHAPTER 5, Lesson 3

Making a Solution by Dilution

Moles NaOH needed to make 1.00 L of 0.150 M solution ________________

Volume (mL) of 6.6275 M NaOH needed ________________

Concentration of solution made accurately by diluting 20.0 mL of 6.6275 M NaOH to 1.00 L ________________

Thinking Beyond

What kinds of solutions can only be made by dilution?

CHAPTER 5, *Lesson 4*

Titration

Moles of $Ca(OH)_2$ added ____________________

Moles HCl consumed ____________________

Volume HCl solution added ____________________

Concentration of HCl solution ____________________

Thinking Beyond

If you used 78.0 mL of this standardized HCl solution to dissolve an unknown sample of $Ca(OH)_2$, how many moles of $Ca(OH)_2$ were in the unknown sample?

CHAPTER 6, Lesson 1

Heat Capacity

	Time Heated (s)	Initial Temperature (°C)	Final Temperature (°C)	Change in Temperature (°C)
Water				
Copper				
Carbon				

Heat added ________________ J

Heat capacity of water ____________ J/ °C

Heat capacity of copper ____________ J/ °C

Heat capacity of carbon ____________ J/ °C

	Heat Added (J)	Change in Temperature (°C)	Heat Capacity J/ °C
500 g Water			
1000 g Water			
2000 g Water			

Thinking Beyond

Will a lighter or a heavier skillet get hot faster? Which will cool down less when food is added to it?

A common bit of folklore says that hot water freezes faster than cold water in an ice cube tray placed in a freezer. Use your discoveries here to explain whether this is right or wrong.

CHAPTER 6, Lesson 2
Specific Heat Capacity

Heat capacity of bottle	____________	J/ °C
Temperature change of bottle and 500 grams of water	____________	°C
Heat added to bottle and 500 grams of water	____________	J
Heat added to bottle	____________	J
Heat added to water	____________	J
Specific heat capacity of water	____________	J/g °C
Temperature change of bottle and 4000 g of Ni	____________	°C
Heat added to bottle and 4000 grams of Ni	____________	J
Heat added to bottle	____________	J
Heat added to Ni	____________	J
Specific heat capacity of Ni	____________	J/g °C

Thinking Beyond

How does the specific heat capacity of water compare with that of other substances? Why is a water solution used as the coolant in automobiles?

CHAPTER 6, Lesson 3

Heat Transfer

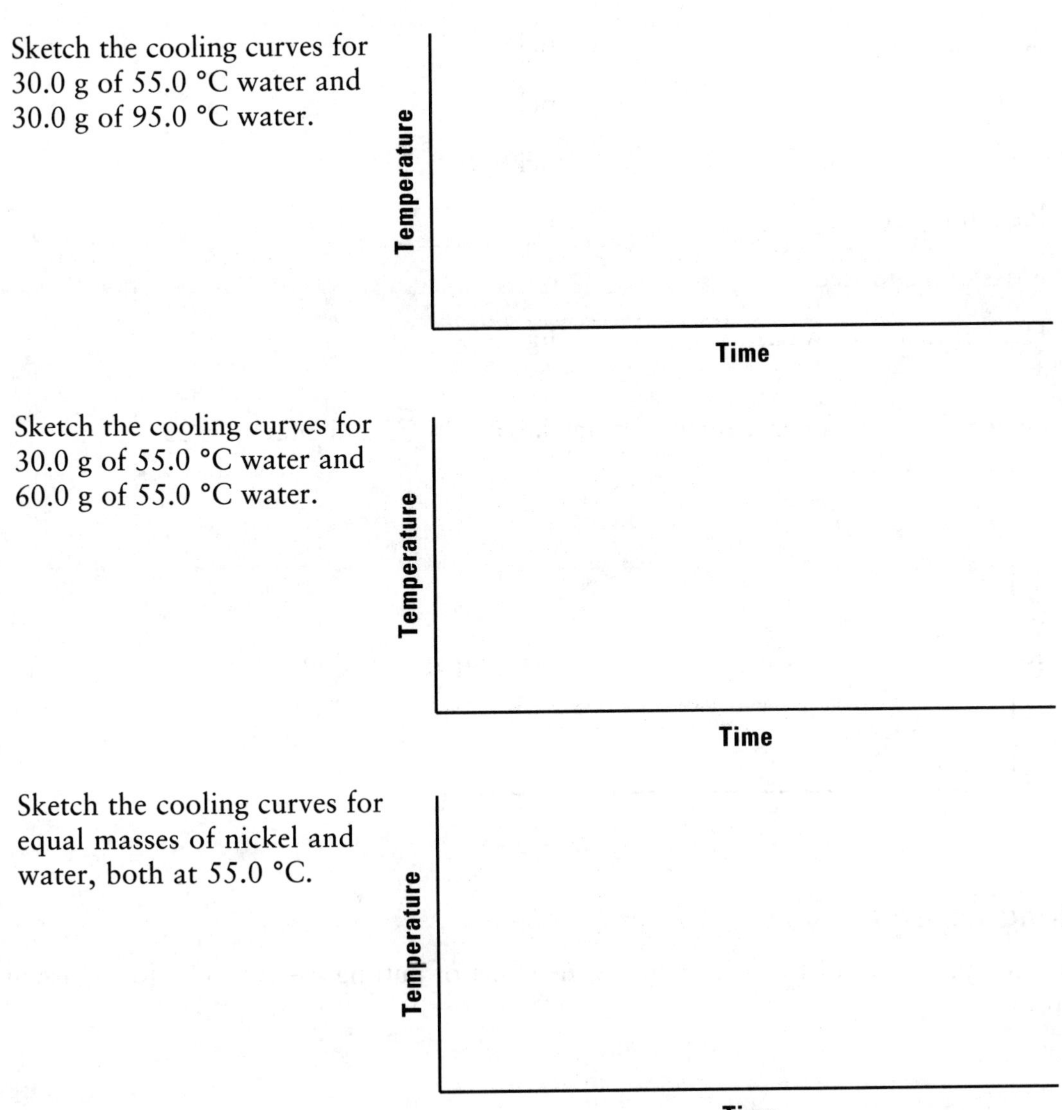
Sketch the cooling curves for
30.0 g of 55.0 °C water and
30.0 g of 95.0 °C water.
Temperature
Time
Sketch the cooling curves for
30.0 g of 55.0 °C water and
60.0 g of 55.0 °C water.
Temperature
Time
Sketch the cooling curves for
equal masses of nickel and
water, both at 55.0 °C.
Temperature
Time

Thinking Beyond

It takes the same amount of energy to melt ice whether or not it is chipped. Why does chipped ice melt faster?

CHAPTER 6, *Lesson 4*

Heat and Phase Changes

Starting time	____________	ticks
Ending time	____________	ticks
Total time to melt	____________	ticks
Heat transferred	____________	J
Mass of ice melted	____________	g
Heat of fusion of water	____________	J/g

Sketch the heating curve for water starting at -25 °C and ending at 25 °C.

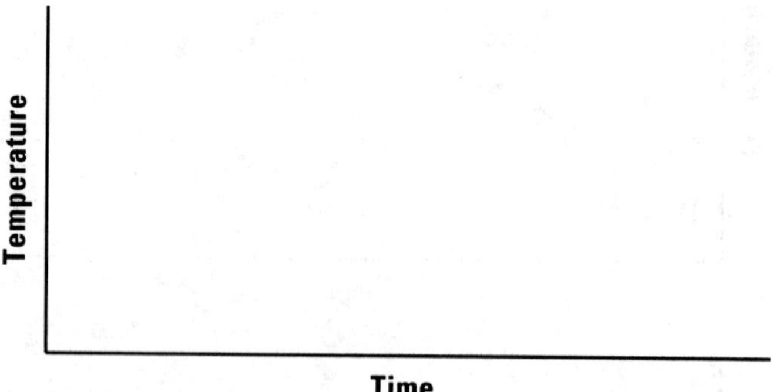

Thinking Beyond

Use the concept of heat of fusion to explain the effect of putting ice at 0 °C into a glass of warm water.

Why is it more dangerous to come in contact with steam at 100 °C than water at 100 °C?

CHAPTER 6, Lesson 5

Heat of Combustion

Heating water with methane

Trial	Change in Temperature (°C)

Absorbing all the heat

Change in temperature of water bath ______________ °C

Calculated heat capacity of water bath ______________ J/ °C

Heat absorbed by water ______________ J

Heat released by burning methane ______________ J

Enthalpy change of methane combustion

Trial	CH_4 consumed (g)	Enthalpy Change

Enthalpy change per gram of gas consumed

CH_4 ______________ C_2H_2 ______________ H_2 ______________

Thinking Beyond

Coal, composed mostly of large hydrocarbons, is much less expensive than gasoline, and provides similar quantities of energy. Why do automobiles burn gasoline?

CHAPTER 7, Lesson 1

Electromagnetic Radiation

λ that ejected an electron from a Na atom ____________ nm

Other λ used

____________ nm

____________ nm

____________ nm

____________ nm

____________ nm

Effect of λ on the speed of ejected electrons: ____________________________

Planck's Constant		
λ (nm)	E (J)	E · λ (J nm)
	Average:	

$h = E \cdot \lambda / C =$ ________________ J s

Thinking Beyond

Why do X-rays (light of very short wavelength) pass through soft body tissue, comprised mostly of water, and not through bone? Why does drinking a "barium shake" allow your digestive tract to show up in an X-ray scan?

CHAPTER 7, Lesson 2

Line Spectra

Wavelengths of photons emitted by an excited atom of hydrogen

__________ __________ __________ __________ __________

__________ __________ __________ __________ __________

__________ __________ __________ __________ __________

Thinking Beyond

Which of these wavelengths are in the visible range?

Sodium vapor lamps (used in some street lights) and fluorescent lights containing mercury vapor have both continuous and discrete spectra. Explain why this might be so.

CHAPTER 8, Lesson 1

Periodic Properties of Atoms

Sketch the depictions of Mg, Be, and Ca.

Sketch the depictions of C, Si, and Ge.

Which are the largest from each group? ______________ and ______________

What trend do atomic sizes exhibit across the periodic table? ______________

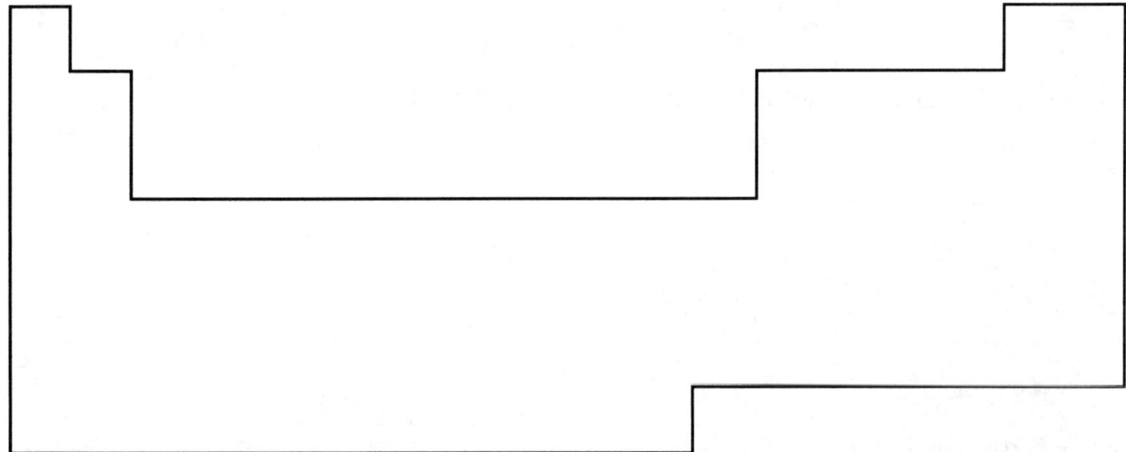

Label the region of this periodic table where the elements have the largest electronegativities, and the region where they have the smallest electronegativities.

Thinking Beyond

Why are the sizes of all the elements in one particular row of transition metals pretty much the same?

CHAPTER 8, Lesson 2

Ion Size

Depict the relative sizes of F, F^-, and Ne.

Depict the relative sizes of N^{+3} and Be.

Depict the relative sizes of N and Be^{-3}. (Note: Be^{-3} does not exist.)

Thinking Beyond

If you compare the sizes of O^{-2}, F^-, Ne, Na^+, and Mg^{+2} from a table of such values, Ne is anomalous. Why is this?

CHAPTER 8, Lesson 3

Ionization Energies

Calculated energy of a 50 nm photon _______________ J

Observed energy _____________ eV = _______________ J

Element	KE of Ejected Electron	Ionization Energy

Thinking Beyond

What atomic orbital is the electron removed from in all these examples?

Use an orbital-based argument to explain the exceptions in the trends you observed.

CHAPTER 9, Lesson 1

Bond Length

Sketch a plot of energy vs. the distance between two oxygen atoms as the atoms approach each other and form a chemical bond. Wait until the plot is complete in the simulation.

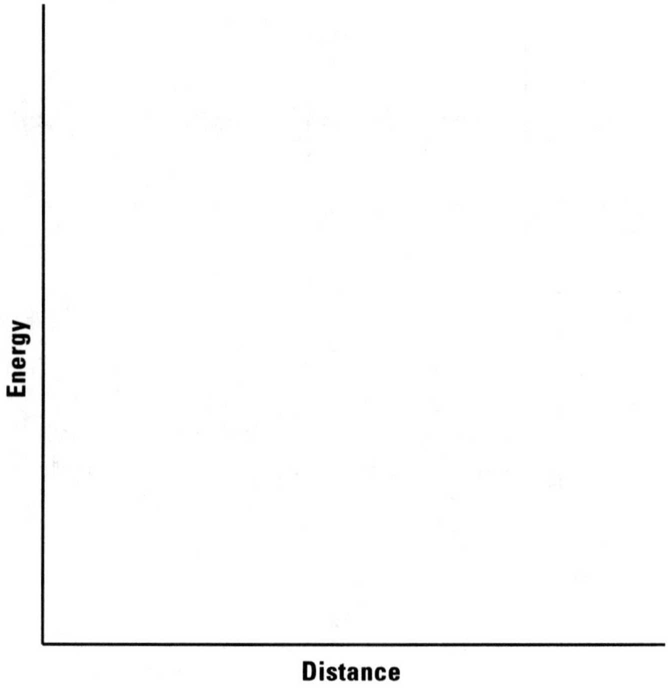

Energy

Distance

Label the bond length and the minimum energy.

Thinking Beyond

What chemical reactions do you know that release light when bonds are formed, as in this example?

What would cause the energy minimum, and hence the bond length, to move to a shorter distance between the nuclei? to a longer distance?

CHAPTER 9, Lesson 2

Bond Energies

Bond type	Bond Energy (eV)

Thinking Beyond

Add some other atoms from the small periodic table. Compare the strength of a NaCl (ionic) bond with the strength of a CCl (covalent) bond and the strength of a Cl_2 (pure covalent) bond.

What is the strongest bond you can find? the weakest?

CHAPTER 9, Lesson 3
Bond Energy and Reaction Enthalpy

Number of Br—Br bonds broken ____________________

Number of Al—Br bonds made ____________________

Energy of Br—Br bond ____________________ eV

Energy of Al—Br bond ____________________ eV

Total energy to break bonds ____________________ eV

Total energy released by bond formation ____________________ eV

Thinking Beyond

Use the "Open" command, under the "File" menu, to load the file named ch09s03b.sim and add atoms from the periodic table. Measure the energy of C—H, C_2=O, H—O, and O=O bonds. Use these measurements to explain why methane burns.

CHAPTER 12, Lesson 1

Properties of Gases

He

Trial	Mass (g)	Volume (L)	Temperature (°C)	Pressure (atm)

Thinking Beyond

f you have a slow chemical reaction that consumes the number of gas molecules, what effect vill this have on the properties you measure?

CHAPTER 12, Lesson 2

Gas Laws

0.100 mol He

Trial	Volume (L)	Pressure (atm)	Temper-ature (K)	Trial	Volume (L)	Pressure (atm)	Temper-ature (K)

Plot your data on good graph paper, and then sketch the result here.

Pressure

Temperature

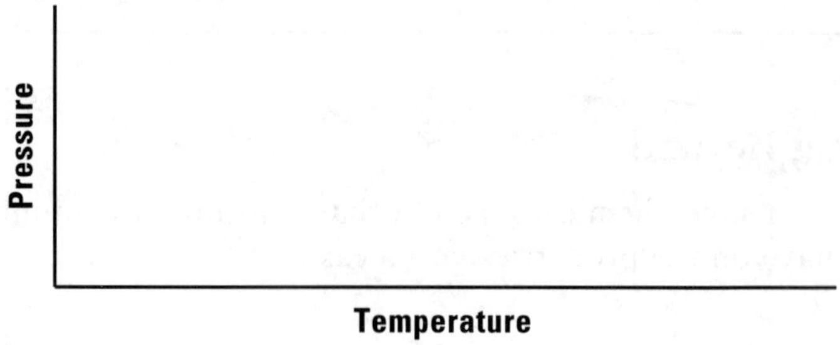

0.100 mol He

Trial	Pressure (atm)	Volume (L)	Temper-ature (K)	Trial	Pressure (atm)	Volume (L)	Temper-ature (K)

Plot your data on good graph paper, and then sketch the result here.

Volume

Temperature

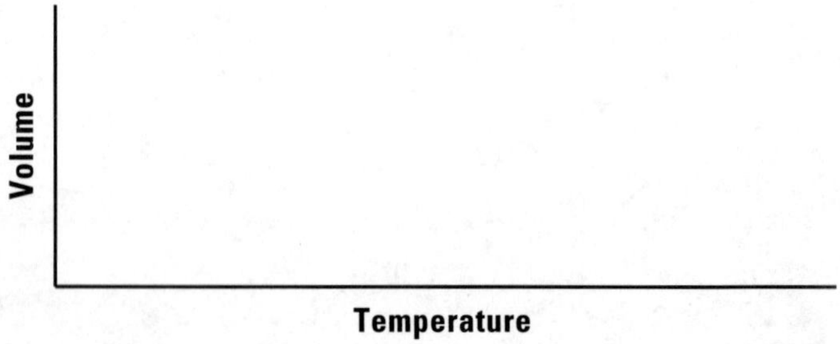

0.100 mol He

Trial	Pressure (atm)	Volume (L)	Temper-ature (K)	Trial	Pressure (atm)	Volume (L)	Temper-ature (K)

Plot your data on good graph paper, and then sketch the result here.

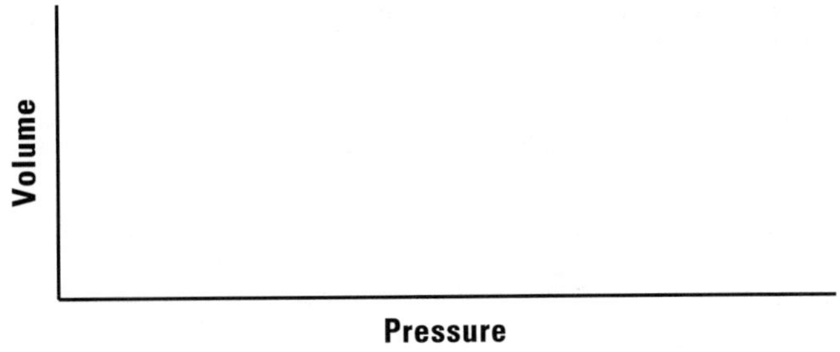

Volume

Pressure

Thinking Beyond

At what temperature would you extrapolate the volume of your sample to go to zero?

Give molecular scale interpretations of why and how the pressure of your sample depends on temperature and on volume (they are different).

CHAPTER 12, Lesson 3

Ideal Gas Law

0.100 mol He

P = ________ atm V = ________ L T = ________ K

R = PV/nT = ________ L atm / mol K

0.100 mol NH_3

Trial	Volume (L)	Temperature (°C)	Pressure (atm)	R (L atm/mol K)

0.100 mol H_2

P = ________ atm at 0.500 L and 300 K R = ________ L atm / mol K

0.100 mol H_2S

P = ________ atm at 0.500 L and 300 K R = ________ L atm / mol K

0.100 mol C_4H_{10}

P = ________ atm at 0.500 L and 300 K R = ________ L atm / mol K

Thinking Beyond

What assumption does the ideal gas model make about the molecules of a gas it is used to describe?

Would you expect the pressure of a gas with very large molecules to be larger or smaller than the pressure predicted by the ideal gas law? What about a gas with very polar molecules?

CHAPTER 12, Lesson 4

Gas Density

Xe

Trial	m (g)	V (L)	P (atm)	T (K)	Density (g/L)

H_2

Trial	m (g)	V (L)	P (atm)	T (K)	Density (g/L)

He

Trial	m (g)	V (L)	P (atm)	T (K)	Density (g/L)

H_2S

Trial	m (g)	V (L)	P (atm)	T (K)	Density (g/L)

NH_3 (ammonia)

Trial	m (g)	V (L)	P (atm)	T (K)	Density (g/L)

CHAPTER 12, Lesson 4

Gas Density *(continued)*

C_4H_{10} **(butane)**

Trial	m (g)	V (L)	P (atm)	T (K)	Density (g/L)

Thinking Beyond

Give a molecular level interpretation of how these gases differ in their behavior.

CHAPTER 13, Lesson 1

IMFs of Halogens and Noble Gases

Compound or Element	Melting Point (°C)	Boiling Point (°C)

Thinking Beyond

Air is a mixture of N_2 and O_2, with small amounts of Ar and other gases. How could you obtain pure samples of these three elements from liquefied air?

CHAPTER 13, *Lesson* 2

Vapor Pressure

H_2O		CCl_4	
Temperature (K)	Vapor Pressure (atm)	Temperature (K)	Vapor Pressure (atm)

Sketch the graph created by ActivChemistry™ for the vapor pressure of CCl_4 vs. temperature.

Vapor Pressure

Temperature

Sketch the graph created by ActivChemistry™ for the vapor pressure of H_2O vs. temperature.

Temperature

Thinking Beyond

How does vapor pressure play a role in the operation of a steam engine?

CHAPTER 13, Lesson 3

Boiling Point

Sealed Bottle		Open Bottle	
Temperature (K)	Pressure (atm)	Temperature (K)	Pressure (atm)

Thinking Beyond

Cooking instructions sometimes have "High Altitude" modifications and require longer boiling times. Why?

CHAPTER 14, Lesson 1
Solubility

Water as a solvent

Soluble Compounds	Insoluble Compounds

CCl_4 as a solvent

Soluble Compounds	Insoluble Compounds

Thinking Beyond

Why is water a good solvent for many ionic compounds, while ethanol, which also has a dipole, is not?

CHAPTER 14, Lesson 2

Colligative Properties

Pure Water

Freezing point

_______________ °C

Sketch the graph of temperature vs. time for freezing of water.

Temperature

Time

5 g NaCl in 200 g water

Freezing point

_______________ °C

Sketch the graph of temperature vs. time for freezing of this solution.

Temperature

Time

Temperature and contents of the bottle as the salt solution freezes:

	Temperature	Contents
At around 100 Ticks:		
At around 200 Ticks:		

	Pure Water	Water with 10 g NaCl	Water with 20 g NaCl
Freezing Point			

Thinking Beyond

Use what you have discovered to explain the use of salt to melt ice on roads and sidewalks in the winter.

Why is calcium chloride often more effective than sodium chloride at melting ice?

CHAPTER 16, Lesson 1

Determining an Equilibrium Constant

Equation for the dissociation of acetic acid

____

Quantities of solute

Solution	Volume of Solution (L)	CH_3CO_2H (g)	$CH_3CO_2^-$ (g)	H^+ (g)

Concentrations

Solution	$[CH_3CO_2H]$	H^+	$[CH_3CO_2^-]$

Equilibrium Constant ____________________

Thinking Beyond

This is a fairly weak acid. What is the dissociation constant for a very strong acid that dissociates nearly completely? For a very weak acid that hardly dissociates at all?

CHAPTER 16, Lesson 2

Estimating Equilibrium Concentrations

1.00 M acetic acid

Component:	Predicted concentration:	Actual concentration:

0.100 M acetic acid

Component:	Predicted concentration:	Actual concentration:

0.0100 M acetic acid

Component:	Predicted concentration:	Actual concentration:

Thinking Beyond

How would you make a solution that is 1.00 M in *undissociated* acetic acid? How would you prepare a solution that was exactly 0.010 M in H^+ ion?

CHAPTER 16, Lesson 3

Disturbing a Chemical Equilibrium

Solution Description	g CH_3CO_2H	g $CH_3CO_2^-$	g H^+	Ratio

Thinking Beyond

What would be the effect of adding equal numbers of moles of HCl and NaOH?

If you dilute vinegar, which is 5% acetic acid by weight, will the percentage of acetic acid that is dissociated increase or decrease? Why?

CHAPTER 17, Lesson 1

Conjugate Acid-Base Pairs

HCl

Reactants ______________________ Products ______________________

1st conjugate pair: Acid __________ Base __________

2nd conjugate pair: Base __________ Acid __________

CH_3CO_2H

Reactants ______________________ Products ______________________

1st conjugate pair: Acid __________ Base __________

2nd conjugate pair: Base __________ Acid __________

NH_3

Reactants ______________________ Products ______________________

1st conjugate pair: Acid __________ Base __________

2nd conjugate pair: Base __________ Acid __________

Thinking Beyond

Water has a conjugate base (OH^-) or a conjugate acid (H_3O^+) depending on what it reacts with. What other substances can act this way, either as an acid or as a base?

CHAPTER 17, Lesson 2

pH Scale

Quantity HCl Added (g)	H^+ in Solution (g)	Concentration of H^+	pH

Equation for determining pH: ______________________

Thinking Beyond

A solution with pH of 10 is basic. Are there any H^+ ions in that solution? If so, what is the concentration of H^+?

CHAPTER 17, Lesson 3
Strong Acids and Bases and pH

Equation for determining pH: ______________________________

Solution Description	pH	$[H^+]$

Equilibrium constant expression: ______________________________

Thinking Beyond

0.10 M HCl has a H^+ concentration of precisely 0.10 M, because the HCl molecules have dissociated completely. A 0.10 M solution of H_2SO_4 has a H^+ concentration greater than 0.10 M. Explain why.

CHAPTER 17, Lesson 4
pH of Weak Acid Solutions

0.10 M acetic acid

pH ________________

$[H^+]$ ________________

Percent dissociated ________________

Thinking Beyond

How does the percent of acetic acid that is dissociated change when you vary the concentration of the solution?

Workspace for Problems

CHAPTER 18, Lesson 1

Acid-Base Reactions

HCl + NaOH

Solution Description	pH	Contents of Bottle

CH_3CO_2H + NaOH pH = ________

NH_3 + HCl pH = ________

CH_3CO_2H + NH_3 pH = ________

Thinking Beyond

What is the result of equal numbers of moles of a polyprotic acid such as H_2SO_4 reacting with NaOH? What is the result of equal numbers of moles of H_2SO_4 reacting with $Ca(OH)_2$?

CHAPTER 18, Lesson 2

Buffer Solutions

pH of 0.10 M HCl ____________

pH after adding 15 mL of 0.10 M NaOH solution ____________

pH nearest to 5 you could make with HCl and NaOH ____________

Equation for the reaction between HCl and NaOH

____

Equation for the dissociation of acetic acid

____

125 mL 0.10 M acetic acid + 2.0 g potassium acetate

pH ____________

Adding HCl

	pH
1 mL	
2 mL	

Adding NaOH

	pH
2 mL	
3 mL	

Thinking Beyond

Most common stomach antacids contain the conjugate base of a weak acid. What do you think this does in the strongly acidic environment of a stomach?

CHAPTER 18, Lesson 3

pH Titrations

Sketch the graph of pH vs. volume of NaOH solution added

pH

Volume 0.10 M NaOH added

Unknown

Volume 0.10 M NaOH used ____________ L

Moles NaOH used ____________ mol

Moles HCl in unknown ____________ mol

Volume of unknown HCl solution used ____________ L

Concentration of unknown ____________ M

Thinking Beyond

KHP (potassium hydrogen phthalate, $KHC_8H_4O_4$) is a white acidic solid with a molecular weight of around 204 grams per mole. Why is KHP used to standardize NaOH solutions instead of HCl?

How could you use a solution with a known concentration of HCl to measure unknown quantities of some base?

CHAPTER 19, Lesson 1

Solubility

$CaCl_2$ + LiOH

Reactants

Products

$Ca(OH)_2$ **in water**

Final contents of bottle

$Mg(OH)_2$ **in water**

Final contents of bottle

Thinking Beyond

What would the pH of these solutions be?

How much water would it take to dissolve 2 g of $Ca(OH)_2$? 2 g of $Mg(OH)_2$?

CHAPTER 19, Lesson 2

Determining K_{sp}

Saturated solution of $Ca(OH)_2$

K_{sp} by measuring pH

pH ____________________

$[H^+]$ ____________________ M

$[OH^-]$ ____________________ M

$[Ca^{+2}]$ ____________________ M

K_{sp} ____________________

K_{sp} by direct measurement

bottle contents ____________________

volume ____________________

$[Ca^{+2}]$ ____________________ M

$[OH^-]$ ____________________ M

K_{sp} ____________________

Thinking Beyond

Is the K_{sp} of $Mg(OH)_2$ smaller or larger than that of $Ca(OH)_2$? Rationalize your answer based on previous solubility measurements.

CHAPTER 19, Lesson 3
Common Ion Effect

Contents of saturated solution of $Ca(OH)_2$:

Contents after adding more $Ca(OH)_2$:

Contents after adding $CaCl_2$:

Contents after adding LiOH:

Thinking Beyond

What would happen if you added an acid to this solution, *decreasing* the concentration of OH^-?

What would happen if you added $MgCl_2$?

CHAPTER 19, Lesson 4
pH and Solubility

Left bottle, $Ca(OH)_2$

pH	Volume (L)	Contents of Bottle

K_{sp} of $Ca(OH)_2$ ____________________

Right bottle, $Mg(OH)_2$

pH	Volume (L)	Contents of Bottle

K_{sp} of $Mg(OH)_2$ ____________________

After adding base:

Contents of $Ca(OH)_2$ solution   Concentration product ____________

Contents of $Mg(OH)_2$ solution   Concentration product ____________

After adding acid:

Contents of $Ca(OH)_2$ solution   Solubility ____________

Contents of $Mg(OH)_2$ solution   Solubility ____________

Thinking Beyond

Explain the effect of acid rain (dilute nitric and sulfuric acid) on limestone, $CaCO_3$

CHAPTER 21, Lesson 1
Electron Transfer Reactions

Cu + $ZnSO_4$

Products ____

Zn + $CuSO_4$

Products ____

Electrochemical Cell after 60 clock ticks

Zn Cell contents

Cu Cell contents

Thinking Beyond

You made, in essence, a battery. Explain now, based on this experiment, why batteries go dead and how some batteries might be recharged.

CHAPTER 21, Lesson 2

Electrochemical Cells and Potentials

1.0 M $ZnSO_4$ and 1.0 M $CuSO_4$

Initial cell voltage ____________________ V

1.0 M $ZnSO_4$ and 0.010 M $CuSO_4$

Initial cell voltage ____________________ V

0.010 M $ZnSO_4$ and 1.0 M $CuSO_4$

Initial cell voltage ____________________ V

Thinking Beyond

Describe the chemical reaction that would occur if you forced electricity to flow in the opposite direction through the wire.

APPENDIX A
CAChe Visualizer for Education

The *CAChe Visualizer for Education* is a tool for viewing and manipulating molecules. Over 300 molecular models have been created to accompany the application. The user can view these molecules and crystal structures and measure bond distances and bond angles. In addition, energy maps, molecular orbitals, and electron density surfaces can be visualized. In the Macintosh version it is also possible to view calculated infrared spectra. What follows are brief instructions for using the Macintosh and Windows version of this software. As you use this software, you may find additional features.

Macintosh Version

To open a molecule file, first locate the MODELS folder within the CAChe folder on the CD-ROM. Next, open the folder you are interested in, such as the ORGANIC folder. Find the molecule file you wish to open, such as the ETHANOL.CSF file. To open the file, choose one of two methods:

CAChe ball-and-stick model of ethanol, C_2H_5OH

Visualizer Icon

1. Open the Visualizer application (in the CAChe folder) by double-clicking its icon. Once that is open, use OPEN in the Visualizer FILE menu to open desired file.

or

2. Drag the file you wish to open on top of the Visualizer icon (holding the mouse button down). When the Visualizer icon turns black, release the mouse button, and the desired file will open.

Use the zoom box in the upper right corner of the window to enlarge the window so that it fills the screen.

Once you have a molecule file open, you have many options. Each of these options is described in detail below.

- Rotate, move, or scale the molecule.
- View a space-filling model of the molecule or view the partial charges on the atoms in the molecule.
- Determine bond lengths, bond angles, or dihedral angles in the molecule.
- View calculated data about the molecule, such as partial charge or bond order.
- View a two- or three-dimensional energy map of the molecule.
- View a calculated electron density surface or molecular orbital surface.

HOW TO ROTATE, MOVE, OR SCALE THE MOLECULE

Rotate (⌘ and shift)

With the molecule on the screen, hold down the command key (the key with the or ⌘ symbol) and the shift key at the same time. A "rotate circle" will appear on the screen. While still holding down the two keys, click and hold the mouse button, and drag the mouse inside the rotate circle to rotate the molecule about the x and y axes. Click and drag outside the rotate circle to rotate the molecule about the z axis. (The cursor will change from a straight arrow to a curved arrow.)

Scale (⌘ and control)

To scale a molecule (make it bigger or smaller), hold down the command (⌘) and control keys at the same time, click and hold the mouse button, and move the mouse. (The cursor will change from an arrow to a magnifying glass.)

Translate (⌘ and option)

To translate (move) a molecule, hold down the command and option keys at the same time, click and hold the mouse button, and move the mouse. (The cursor will change from an arrow to a hand.)

HOW TO CHANGE THE MOLECULE VIEW: SPACE-FILLING MODELS AND PARTIAL CHARGES

Space-filling models

The models on the disc were prepared as ball-and-stick models. Space-filling models may give a better idea of molecular shape, however. To view a space-filling model, you must first have a molecule file open. In the **View** menu, make sure **Atoms** and **Bonds** are checked, and then choose **Atom shape** from the **View** menu. The dialog box that appears allows you to decide what the atoms in the molecule will look like. Make sure the circles next to **Spheres** and **Shaded** are on and that the **Radius** pop-up box says "van der Waals." Change the value in the **Scale** box from "0.20" to a number closer to "1.00." Click the **OK** button. Now your molecule will look something like the picture shown here of a space-filling model of ethanol. To change back to a ball-and-stick model, open the **Atom shape** dialog box and change the value back to "0.20."

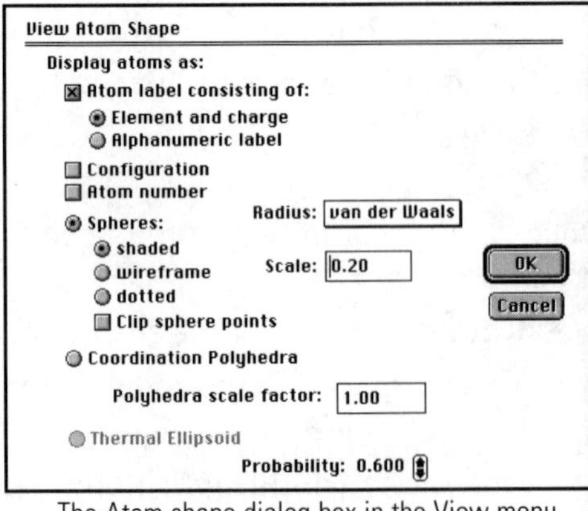

The Atom shape dialog box in the View menu.

A space-filling model of ethanol.

Note that the **Atom shape** dialog box allows you to turn on or off the atom labels and atom configuration or to display the atoms as shaded spheres, wireframe spheres, or dotted spheres, or as coordination polyhedra. You should experiment with the various options.

Partial Charge

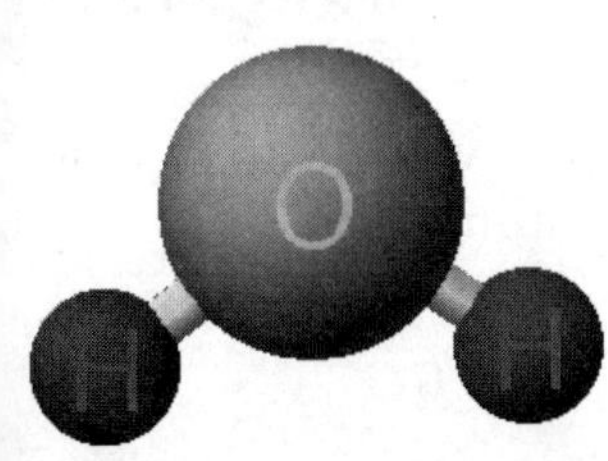

Model of H_2O showing relative partial charges on O and H atoms.

Partial charge values calculated by the CAChe software indicate how electron density is distributed within the molecule. Positive partial charge is shown as red spheres, and negative partial charge is shown as yellow spheres. The molecule files that have had partial charge calculations performed are in a PARTCHRG folder within the MODELS folder.

With a molecule file open, choose **Atom shape** from the **View** menu. Using the pop-up box, change **Van der Waal's radius** to **Partial charge.** You may want to increase the size from 0.20 to 0.6 or even larger. This can be changed again if the partial charge spheres are too large or too small. Click the **OK** button. Now your molecule will look something like the picture shown here of water. The sizes of the spheres surrounding each atom are proportional to the magnitude of the partial charge. Here the negative charge on the O atom is about twice as large as the positive charge on each of the H atoms. After viewing a partial charge model, you can return to the ball-and-stick model by changing the settings in the **Atom shape** dialog box to **Van der Waal's radius** and "0.20."

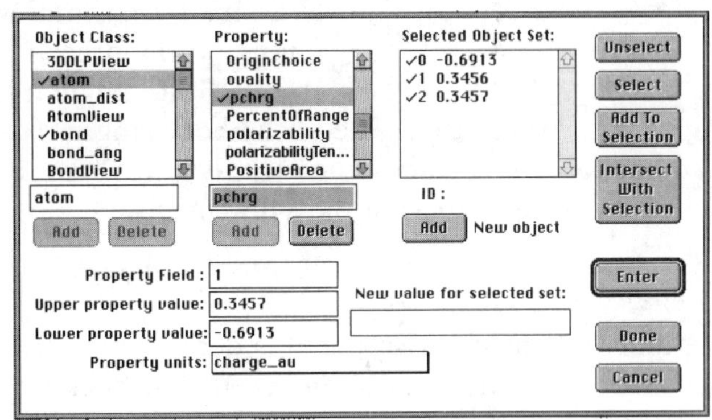

Internals dialog box (in Select menu) showing how to find the partial charges on atoms. The same box can be used to find calculated bond orders.

The calculated value of the partial charge can be found by going to the **Select** menu and choosing **Via Internals.** In this dialog box select **atom** (in Object Class) and then **pchrg** (in Property). In the window at the right you will see the values of the charges on the individual atoms. Here the O atom has a partial charge of about -0.7 whereas the H atoms are each about +0.35.

HOW TO DETERMINE BOND LENGTHS, BOND ANGLES, AND DIHEDRAL ANGLES

Bond length (atom distance)

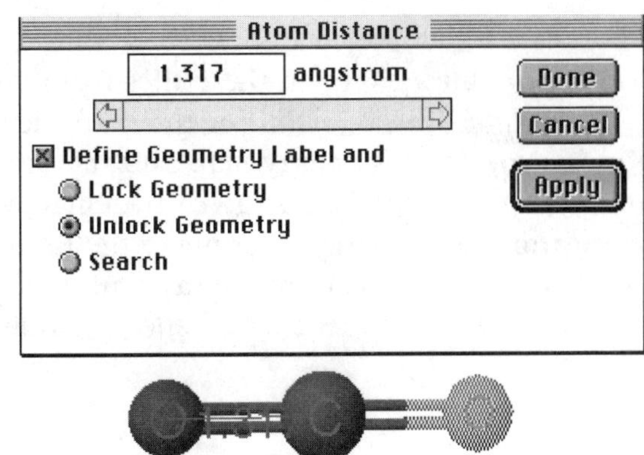

To determine the distance between two atoms, you must first select the two atoms. Select the first atom by placing the tip of the cursor on the center of the atom and clicking the mouse button. To select the second atom, hold down the shift key while you point and click on the second atom. Only the two atoms of interest should appear solid, and all of the other atoms and bonds should appear grayed out. If you make a mistake, simply click anywhere on the screen away from the molecule to deselect the entire molecule and begin again. With the two atoms selected, click on the **Geometry** menu and select **Atom distance.** A dialog box will appear, showing the distance between the two atoms. To place the measurement on the model, click **Define Geometry Label,** then **Apply** and **Done.** If you wish to remove the geometry labels later, click **Geometry labels** in the **View** menu.

Note that distances can be measured in solid state structures as well as in various inorganic and organic molecules. Also note that distances are measured in Ångstrom units, where 1 Å = 10^{-10} m or 100 pm.

Bond angle

To determine a bond angle, you must select three atoms. Select the first atom by clicking on it, then select the other two atoms by holding down the shift key and clicking on each one in succession. It is important to click on the atoms in the correct order. For example, to determine the H—O—H bond angle in water, first click on an H atom. Then, holding down the shift key, click the O atom, and finally on the other H atom. Once the three atoms are selected, choose **Bond angle** from the **Geometry** menu. A dialog box will appear, showing the value of the bond angle. To place the measurement on the model, click **Define Geometry Label,** then **Apply** and **Done.** If you wish to remove the geometry labels later, click **Geometry labels** in the **View** menu.

Dihedral angle

A dihedral angle (twist angle) is defined by four atoms. Select the four atoms of interest by clicking on the first atom and then shift-clicking on the remaining three. It is important to click on the atoms in the correct order. For example, you can determine a dihedral angle in ethanol by clicking on the following four atoms in order: H—O—C—C. When the four atoms are selected, choose **Dihedral angle** from the **Geometry** menu. A dialog box will appear with the value of the dihedral angle shown. Click the **Done** button to exit the dialog box.

HOW TO VIEW AN ENERGY MAP

There is a folder called ENER_MAP that contains two- and three-dimensional energy maps of some molecules. To open these files from within the CAChe application, choose **Open** from the **File** menu or drag the icon for the map file on top of the Visualizer icon. Two windows will appear, one for the energy map and the other for the molecule. The small gray ball on the energy map shows the energy of the current conformation of the molecule. To view a conformation of different energy, first make the map window active by clicking on it once, then click a point on the map. The small gray ball will move, and the molecule will change to show the conformation with that calculated energy value.

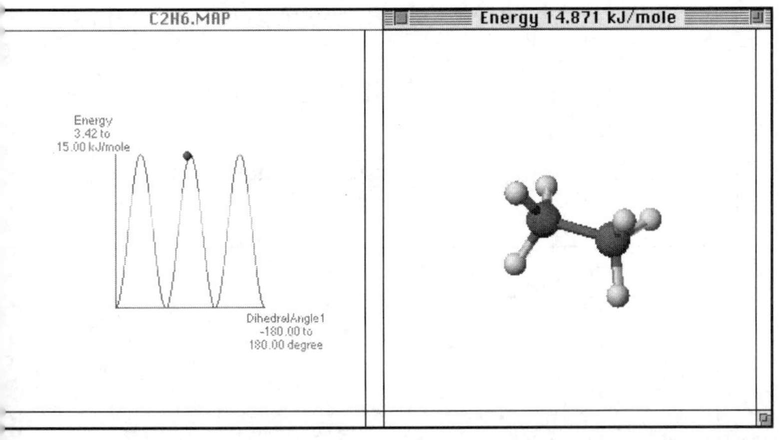

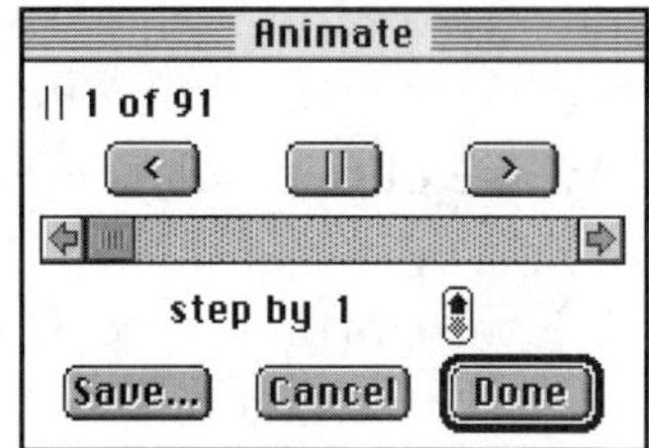

The energy maps can be animated to show how the energy changes as a function of the geometry label that was searched, such as a dihedral angle or a bond angle. With the map window active, choose **Animate** from the **Geometry** menu. A small dialog box will appear, which can be moved so that the map and molecule windows can be seen. Click on the button on the right side of the box to begin the animation. Click **Done** to exit the animation.

HOW TO VIEW ELECTRON DENSITY SURFACES AND MOLECULAR ORBITAL SURFACES

Electron density

Two types of surfaces have been calculated. One is called an "isovalue" urface. Every point on this surface has the same electron probability value. (The default calculation value was 0.01 $e^{-}/Å^{3}$.) The other type of urface is an "electron density gradient" surface. The electron density gradient surface shows the rate of change of electron probability. The greater he rate of change the more the color tends to red, green and light blue. **Surface Legend** in the **View** menu will show the numerical values. Click on the surface to make it active and then open **Surface Legend.**)

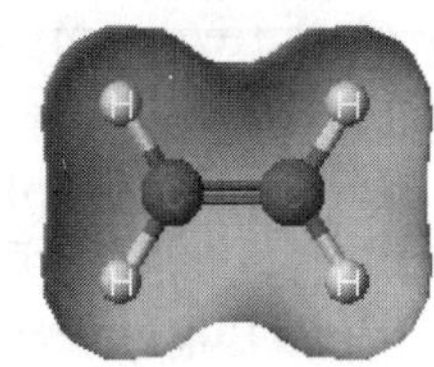

Isovalue surface for ethylene

To view these surfaces, go to the the SURFACES folder and double-click on a file named (molecule name).ISO or (molecule name).DEN. Open he file by dragging its icon onto the Visualizer icon. In both of the surfaces shown here we have removed the front of the surface to show the underlying molecule. This can be done by dechecking the item **Surface Front** in the **View** menu.

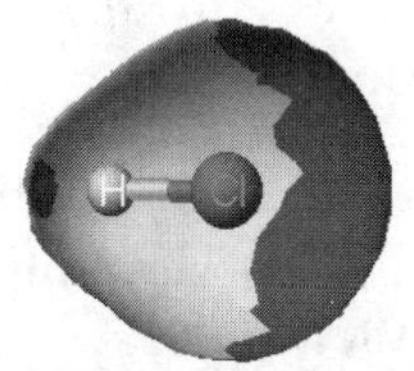

Density gradient surface for HCl

Molecular Orbitals

Molecular orbitals can be viewed by opening a file in the MOs folder within the ORBITALS folder. Go to the folder for a molecule and double-click on a file named (molecule name).MO# or (molecule name).M##. Alternatively, go the the **Open** menu, open a file named (molecule name).CSF within a folder in the ORBITALS folder. To view a molecular orbital surface, open the orbital surface using the **Open Surface** item in he **File** menu (or drag the orbital file icon on top of the Visualizer icon).

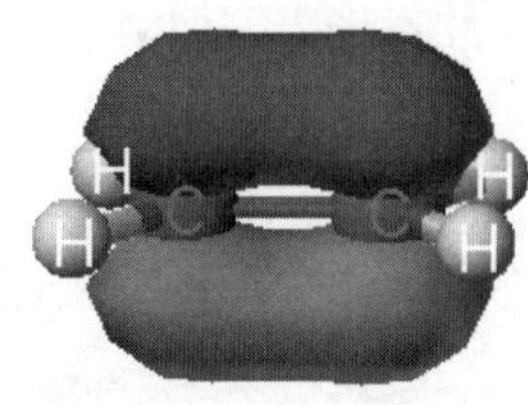

Ethylene π bonding molecular orbital.

- For each molecule the orbitals have a suffix of .MO# (or .M##); the molecular orbitals increase in energy as # increases.
- Any one of the molecular orbital surfaces may be opened, or all the orbitals for a molecule may be opened so that comparisons can be made.
- Molecular orbitals occupied by electrons are colored blue and green, whereas unoccupied orbitals are colored yellow and red.

HOW TO VIEW A CALCULATED INFRARED SPECTRUM

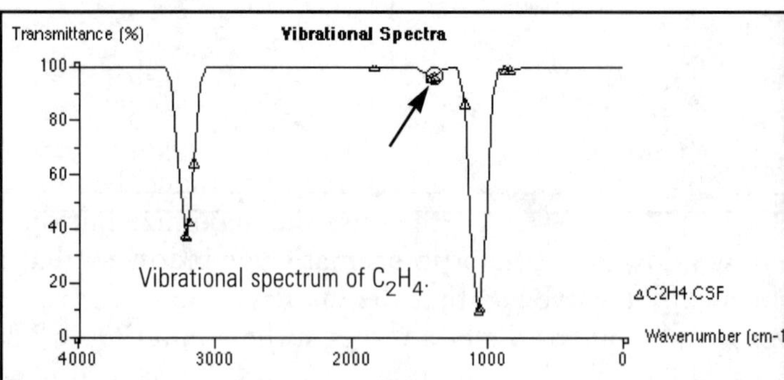

Vibrational spectrum of C_2H_4.

The full CAChe molecular modeling system allows one to compute the vibrational spectrum for a molecule, and a number of such spectra have been included in the folder marked INFRARED. To open one of these files, double click on the icon for one of the files in this folder or drag its icon on top of the Visualizer icon. A model of the molecule will appear on the screen. To view the infrared spectrum click on **Vibrational Spectra** in the **Analyze** menu, and the calculated spectrum will appear at the bottom of the screen.

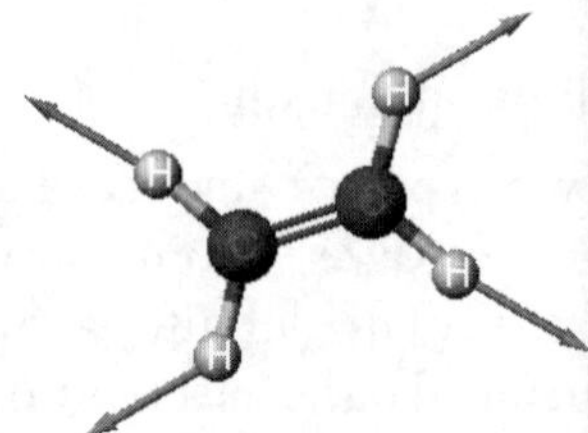
Vibration corresponding to band marked on spectrum.

- The energies of the calculated absorption bands observed may not agree completely with those observed in the laboratory. However, the order of bands and their relative intensities are often in good agreement.
- By clicking on an absorption band, vectors appear on the model of the molecule to show the vibration giving rise to the band.

OTHER ASPECTS OF THE VISUALIZER

1. If the model moves slowly on the screen, click **Faster Motion** in the **View** menu.
2. The background color for the models can be changed by clicking on **Color** in the **View** menu.
3. For molecules with chiral centers, clicking on **Chiral Centers** in the **View** menu will indicate whether the center is R or S.

CAChe Visualizer for Education

Windows Version

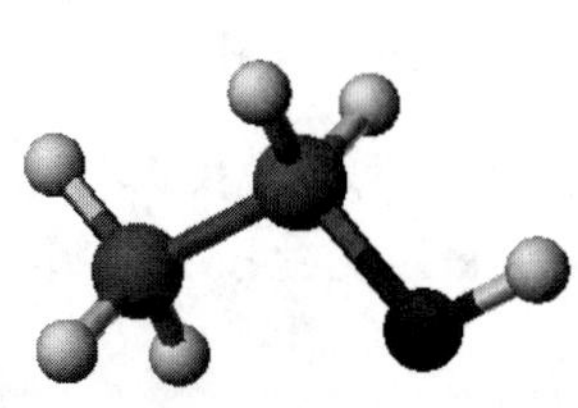

Ball-and-stick model of ethanol.

To open a molecule file, first locate the MODELS folder within the CAChe folder on the CD-ROM. Next, open the directory you are interested in, such as the ORGANIC directory. Find the molecule file you want to open, such as the ETHANOL.CSF file, within the ALCOHOLES folder. Double-click on the icon to open the file, and then use the maximize button in the upper right corner of the window to enlarge the window so that it fills the screen.

Once you have a molecule file open, you have many options. Each of these options is described in detail below.

1. Rotate, move, or scale the molecule.
2. View a space-filling model of the molecule or view the partial charges on the atoms in the molecule.
3. Determine bond lengths, bond angles, or dihedral angles in the molecule.
4. View a two- or three-dimensional energy map of the molecule.

How to Rotate, Move, or Scale the Molecule

Rotate

With the molecule on the screen, click on the rotate button in the toolbar on the left side of the screen. A "rotate circle" will appear on the screen. While holding down the mouse button, drag the mouse inside the rotate circle to rotate the molecule about the x and y axes. Drag outside the rotate circle to rotate the molecule about the z axis.

Scale

To scale a molecule (make it bigger or smaller), click on the scale button on the toolbar on the left side of the screen, click and hold the mouse button, and drag the mouse. (The cursor will change from an arrow to a magnifying glass.) Dragging to the left decreases the size, and dragging to the right increases the size.

Move

To move (translate) a molecule, click on the move button on the toolbar on the left side of the screen, click and hold the mouse button, and move the mouse. (The cursor will change from an arrow to a hand.)

SPACE-FILLING MODELS AND PARTIAL CHARGES

By default, molecules are viewed as ball-and-sticks when a molecule file is opened, but two other options are available.

 Space-filling models

Space-filling model of ethanol.

Space-filling models may give a better idea of molecular shape. To view a space-filling model, you must first have a molecule file open. Click on the space-filling model button on the toolbar at the top of the screen. Now your molecule will look something like the picture shown below of a space-filling model of ethanol. To change back to a ball-and-stick model, click on the ball-and-stick model button on the toolbar

Partial Charge

A model of H_2O showing relative partial charges on O and H atoms.

The partial charge values calculated by the CAChe software indicate how electron density is distributed within the molecule. Positive partial charge is shown as red spheres, and negative partial charge is shown as yellow spheres. The calculated partial charge values are shown on the atoms in the molecule. The molecule files that have had partial charge calculations performed are in a PARTCHRG directory. With a molecule file open, choose **Partial Chg & Calc. Bond Order** from the **View** menu. Your molecule will look something like the picture shown here of H_2O. You can return to the ball-and-stick model by clicking on the ball-and-stick model button in the top toolbar.

DETERMINING BOND LENGTHS, BOND ANGLES, AND DIHEDRAL ANGLES

Bond Length (atom distance)

To determine the distance between two atoms, you must first select the two atoms. Select the first atom by placing the tip of the cursor on the center of the atom and clicking the mouse button. To select the second atom, hold down the shift key while you point and click on the second atom. Only the two atoms of interest should appear solid, and all of the other atoms and bonds should appear grayed out. (If you make a mistake, simply click anywhere on the screen away from the molecule to deselect the entire molecule and begin again.) With the two atoms selected, click on the **Adjust** menu and select **Atom Distance**.

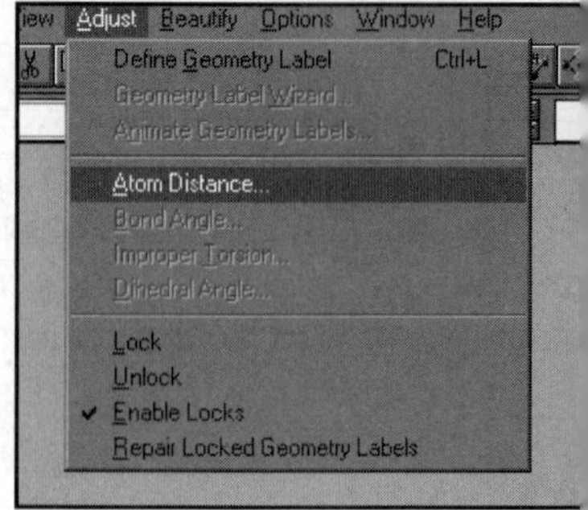

A dialog box will appear, showing the distance between the two atoms. Click the **OK** button to exit the dialog box.

Set Atom Distance
Distance: 1.541 angstrom
OK
Cancel
Apply
Define Geometry Label
Unlock Geometry
Lock Geometry
Search
between 1.386 and 1.695
using 6 steps of 0.052 angstrom
Help

Be sure to notice that bond lengths are given in Ångstrom units, where 1 Å = 10^{-10} m = 100 pm.

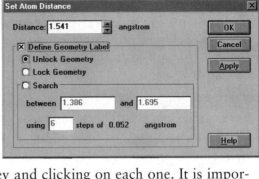

Bond Angle

To determine a bond angle, you must select three atoms. Select the first atom by clicking on it, then select the other two atoms by holding down the shift key and clicking on each one. It is important to click on the atoms in the correct order. For example, to determine the H—O—H bond angle in water, first click on an H atom, then shift-click on the O atom, and finally shift-click on the other H atom. Once the three atoms are selected, choose **Bond Angle** from the **Adjust** menu. A dialog box will appear, showing the value of the bond angle. Click the **OK** button to exit the dialog box.

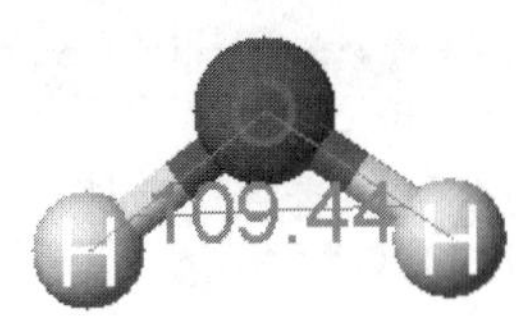

Dihedral Angle

A dihedral angle (twist angle) is defined by four atoms. Select the four atoms of interest by clicking on the first atom and then holding down the shift key and clicking on the remaining three. It is important to click on the atoms in the correct order. For example, you can determine a dihedral angle in ethanol by clicking on the following four atoms in order: H—O—C—C. When the four atoms are selected, choose **Dihedral Angle** from the **Adjust** menu. A dialog box will appear with the value of the dihedral angle shown. Click the **OK** button to exit the dialog box.

Energy Map

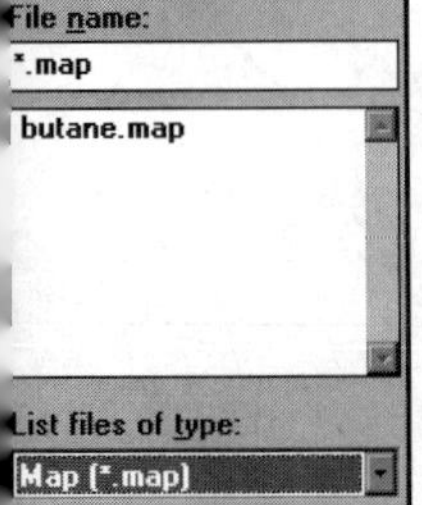

There is a directory called ENER_MAP (within the MODELS directory) that contains two- and three-dimensional energy maps of some molecules. To open these files from within the CAChe application, choose **Open** from the **File** menu, and change the **List files of type:** to ***.map**. Choose a map file, and click the **Open** button.

Two windows will appear, one for the energy map and the other for the molecule. The small gray ball on the energy map shows the energy of the current conformation of the molecule. To view a conformation of different energy, first make the map window active by clicking on it once, then click a point on the map. The small gray ball will move, and the molecule will change to show the conformation with that calculated energy value.

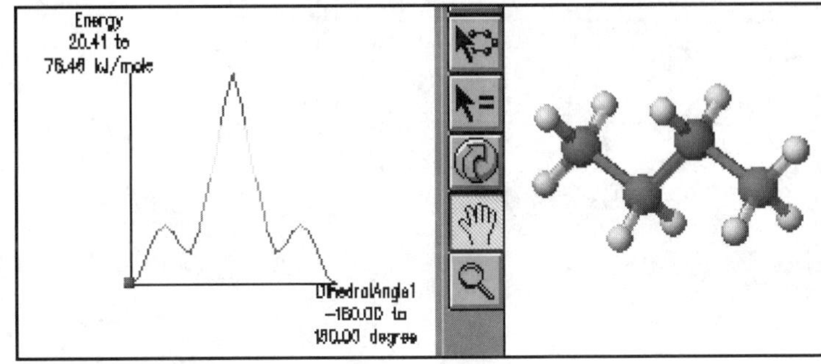

The energy maps can be animated to show how the energy changes as a function of the geometry label that was searched, such as a dihedral angle or a bond angle. With the map window active, choose **Animate Along Axis...** from the **Edit** menu. A small dialog box will appear, which can be moved so the map and molecule windows can be seen. Click on the ▶ button to start the animation.

Click the **Cancel** or **Done** button to exit the animation. Choose **Graph Attributes...** from the **View** menu to switch the axis that is animated. This is especially useful for three-dimensional energy maps.

HOW TO VIEW MOLECULAR ORBITALS

Molecular orbitals can be viewed by opening a file in the MOs directory within the ORBITALS directory. Go to the directory for a molecule and double-click on a file named (molecule name).CSF. When the molecule has appeared on the screen, go to the **View** menu and choose **Show Surface.**

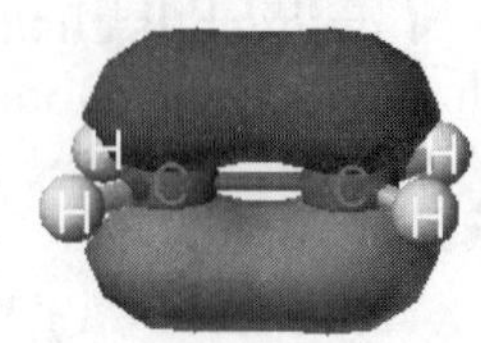

Ethylene π bonding molecular orbital.

- For each molecule the orbitals have a suffix of .MO# (or .M##); the molecular orbitals increase in energy as # increases.
- Any one of the molecular orbital surfaces may be opened, or all the orbitals for a molecule may be opened so that comparisons can be made..
- Molecular orbitals occupied by electrons are colored blue and green, whereas unoccupied orbitals are colored yellow and red.

OTHER ASPECTS OF THE VISUALIZER

For molecules with chiral centers, the R or S designation of the center can be found by going to the **View** menu. Then go to **Atom Attributes** and to **Label**. Choose **Options**, one of which is **Chirality**.

APPENDIX A (CONTINUED)
Using the Plotting Tool

Graphing is a basic tool for analyzing and presenting data. A plot shows the relationship between two variables and is a method of organizing data in order to find meaning. Once a plot has been generated, it can be used to find trends, compared with other graphs, or used to extrapolate and predict further data.

The Least Squares plotting tool, which is found in the TOOLS directory, is a general purpose x-y plotting program. It generates plots and performs least squares analyses as well as mathematical transformations. The steps to follow in making a plot are

1. Enter data.
2. Save data.
3. Manipulate data.
4. View plot on screen.
5. Print plot.

Instructions for each of these steps are given below. These instructions are for the Windows version of the program.

Input Data

1. To enter data, type values into the grid box.
   a) The column on the left is for x values (the horizontal axis), and the column on the right is for y values (the vertical axis).
   b) Corresponding data points (x,y) should be entered in the same row.
   c) Data points do not need to be in any order, they can be sorted by x or y values later.
   d) Use the up and down arrow key or the **Enter** key to move around the grid box (note that these keys move the active box across the rows, activating the x and y columns consecutively, instead of moving straight up or down one column).
   e) Holding the shift key down and pressing Enter changes the active column.
2. Data can be changed by typing over the previous values.
3. A data point (x,y) can be inserted or deleted by selecting the appropriate grid box (either the x or y) and then clicking on the **Edit** menu and selecting the desired operation.

Data Manipulation

1. To perform mathematical manipulations on the x or y values, click on the **X-transform** menu or **Y-transform** menu and select one of the nine operations.
   a) To subtract a constant, add a negative constant.
   b) To divide by a constant, multiply by its inverse.
   c) Raising to a power includes fractional powers and negative powers.
   d) Sorting by x or y values sorts the data from lowest to highest x or y value.

Plot Setup

1. To create labels for the x and y axes, type the **x-axis label** into the x-axis label box and type the y axis label into the **y-axis label** box.
2. To title the plot, type the title of the graph into the **Title** box.
3. The maximum and minimum values for the x and y axes and the tick spacing for the plot will be automatically set if the **Auto Axes** box is checked. If unselected, enter desired x-min, x-max, x-tick space, y-min, y-max, and y-tick space values in the appropriate boxes that appear at the right of the grid box.
4. To view your graph click on the **Plot** check box
   a) To add labels to the plot while viewing the graph, double click at the point you want the label and enter the desired text.
   b) To return to the main screen with grid box and menu options, simply click anywhere on the main screen. You may then also return to the plot by clicking anywhere on the plot screen.

Plot Options

The following plot options are available under the **Plot Options** menu.

1. Select **Grid Lines** to create a grid background for the plot.
2. Click on the **Least Squares Fit** for least squares analysis. Selecting the least squares fit option will cause a fit to be done on the points, and the slope, intercept and correlation coefficient will be displayed.
   a) The closer the correlation coefficient is to 1.0, the better the points fit a straight line.
3. Click on the **Connect Points** option to draw lines between the points on the graph.
   a) Note that the data points are connected in the order in which they appear in the data input grid. To have the connected line go from left to right, sort the data by x value using the **x-sort** option in the **X-transform** menu.
4. Selecting the **Set Label Precision** option allows you to change the number of digits in the numbers associated with each tick mark.
   a) The number you enter includes spaces for minus signs and decimal places.
5. Click on the **Plot Shape** option to select a shape for the data points.

File Management

1. To print your graph, click on the **Print Graph** option under the **File** menu. The plot printed is the one last viewed on the screen. If no plot has been viewed, or the last plot viewed was cleared, no printout can be obtained.
2. To save data, select **Save File** from the **File** menu.
   a) If you are using the Windows version, enter a valid DOS filename, preferably with a ".dat" extension.
   b) It is a good idea to save data immediately after entering it.

3. To retrieve data, select **Open File** from the **File** menu.

   b) You can change drives, directories and files using the standard Windows single/double click methods. The default search pattern looks for files with the ".dat" extension. To look for all files, enter *.* in the file area.

4. To print a data file, click on **Print File** in the **File** menu.
5. To create a new graph, select **New** from the **File** menu.
6. To exit the plotting program choose **Quit** in the **File** menu.

APPENDIX B
Answers to Exercises

1.1 (a) sodium, chlorine, chromium

(b) Zn, Ni, K

1.2 100. g (1 cm^3/1.12 × 10^{-3} g) = 8.93 × 10^4 cm^3

1.3 77 K - 273.15 = - 196 °C

1.4 Area of sheet = $(2.50\ cm)^2$ = 6.25 cm^2

Volume = 1.656 g (1 cm^3/21.45 g) = 0.07720 cm^3

Thickness = volume/area = 0.0124 cm

0.0124 cm (10 mm/cm) = 0.124 mm

2.1 (a) A (for Cu) = 34 n + 29 p = 63

(b) Nickel-59 has 28 protons, 28 electrons, and (59 - 28) = 31 neutrons.

(c) ${}^{28}_{14}Si$, ${}^{29}_{14}Si$, ${}^{30}_{14}Si$

2.2 (0.7577)(34.96885 amu) + (0.2423)(36.96590 amu) = 35.45 amu

2.3 (a) 2.5 mol Al (26.98 g/mol) = 67 g

(b) 454 g S (1 mol/32.07 g) = 14.2 mol

(c) 1.0 × 10^{24} atoms (1 mol/6.02 × 10^{23} atoms)(195 g/mol) = 320 g Pt

Volume = 320 g (1 cm^3/21.45 g) = 15 cm^3

3.1 (1) K^+; (2) Se^{2-}; (3) Be^{2+}; (4) V^{2+}; (5) Co^{2+} or Co^{3+}; (6) Cs^+.

3.2 Part (a): (a) 1 Na^+ and 1 F^- ion. (b) 1 Cu^{2+} and 2 NO_3^- ions. (c) 1 Na^+ and 1 $CH_3CO_2^-$ ion.

Part (b): $FeCl_2$ and $FeCl_3$

Part (c): Na_2S, Na_3PO_4, BaS, $Ba_3(PO_4)_2$.

3.3 Part 1:

(a) NH_4NO_3

(b) $CoSO_4$

(c) $Ni(CN)_2$

(d) V_2O_3

(e) $Ba(CH_3CO_2)_2$

(f) $Ca(OCl)_2$

Part 2:

(a) magnesium bromide

(b) lithium carbonate

(c) potassium hydrogen sulfite

(d) potassium permanganate

(e) ammonium sulfide

(f) copper(I) chloride and copper(II) chloride

3.4 Part (a): (a) $CaCO_3$ = 100.1 g/mol; (b) caffeine = 194.2 g/mol

Part (b): 454 g (1 mol/100.1 g) = 4.54 mol

Part (c): 2.50×10^{-3} mol (194.2 g/mol) = 0.486 g

3.5 1. NaCl has 23.0 g of Na (39.3 %) and 35.5 g of Cl (60.7 %) in 1.00 mol.

2. C_8H_{18} has 96.1 g of C (84.1 %) and 18.1 g of H (15.9 %) in 1.00 mol.

3. $(NH_4)_2SO_4$ has a molar mass of 132.15 g/mol. It has 28.0 g of N (21.2 %), 8.06 g of H (6.10 %), 32.1 g of S (24.3 %), and 64.0 g of O (48.4 %) in 1.00 mol.

3.6 Part (a):

78.14 g B (1 mol/10.811 g) = 7.228 mol B

21.86 g H (1 mol/1.008 g) = 21.69 mol H

Ratio of H to B = 21.69 mol H / 7.228 mol B = 3.000 H to 1.000 B

Empirical formula = BH_3

The molecular weight (27.7 g/mol) is twice the empirical formula weight (13.8 g/formula unit), so the molecular formula is $(BH_3)_2$ or B_2H_6.

Part (b):

Mass of chlorine (Cl) = 2.108 g - 0.532 g = 1.576 g Cl

0.532 g Ti (1 mol/47.88 g) = 0.0111 mol Ti

1.576 g Cl (1 mol Cl/35.453 g) = 0.04445 mol Cl

Ratio of Cl to Ti = 0.04445 mol Cl/0.0111 mol Ti = 4 Cl to 1 Ti

Empirical formula = $TiCl_4$

4.1 (a) Stoichiometric coefficients: 4 for Fe, 3 for O_2, and 2 for Fe_2O_3

(b) 8000 atoms of Fe require (3/4) × 8000 = 6000 molecules of O_2

4.2 (a) $C_5H_{12}(g) + 8\ O_2(g) \rightarrow 5\ CO_2(g) + 6\ H_2O(\ell)$

(b) $Pb(C_2H_5)_4(\ell) + 13\ O_2(g) \rightarrow PbO(s) + 8\ CO_2(g) + 10\ H_2O(\ell)$

4.3 (a) KNO_3 is soluble and gives K^+ and NO_3^- ions.

(b) $CaCl_2$ is soluble and gives Ca^{2+} and Cl^- ions.

(c) CuO is not water-soluble.

(d) $NaCH_3CO_2$ is soluble and gives Na^+ and $CH_3CO_2^-$ ions.

4.4 (a) SeO_2 is an acidic oxide (like SO_2).

(b) MgO is a basic oxide.

(c) P_4O_{10} is an acidic oxide.

4.5 (a) $BaCl_2(aq) + Na_2SO_4(aq) \rightarrow BaSO_4(s) + 2\ NaCl(aq)$

$Ba^{2+}(aq) + SO_4^{2-}(aq) \rightarrow BaSO_4(s)$

(b) $Pb(NO_3)_2(aq) + 2\ KCl(aq) \rightarrow PbCl_2(s) + 2\ KNO_3(aq)$

$Pb^{2+}(aq) + 2\ Cl^-(aq) \rightarrow PbCl_2(s)$

4.6 $2\ AgNO_3(aq) + K_2CrO_4(aq) \rightarrow Ag_2CrO_4(s) + 2\ KNO_3(aq)$

$2\ Ag^+(aq) + CrO_4^{2-}(aq) \rightarrow Ag_2CrO_4(s)$

4.7 $Mg(OH)_2(s) + 2\ HCl(aq) \rightarrow MgCl_2(aq) + 2\ H_2O(\ell)$

$Mg(OH)_2(s) + 2\ H^+(aq) \rightarrow Mg^{2+}(s) + 2\ H_2O(\ell)$

4.8 $PbCO_3(s) + HNO_3(aq) \rightarrow Pb(NO_3)_2(aq) + H_2O(\ell) + CO_2(g)$

Lead(II) carbonate + nitric acid → lead(II) nitrate + water + carbon dioxide

4.9 (a) Fe = +3

(b) S = +6

(c) C = +4

(d) C = -1

4.10 $Cr_2O_7^{2-}$ is reduced and so is the oxidizing agent because the oxidation number of Cr changes from +6 in this ion to +3 in Cr^{3+}. C_2H_5OH is oxidized and so is the reducing agent because the oxidation number of C is changed from -2 in C_2H_5OH to 0 in CH_3CO_2H.

5.1 454 g O_2 (1 mol/32.00 g) = 14.2 mol O_2

14.2 mol O_2 (2 mol C/1 mol O_2)(12.01 g C/mol) = 341 g C

14.2 mol O_2 (2 mol CO/1 mol O_2)(28.01 g CO/mol) = 795 g CO

5.2 20.0 g S_8 (1 mol/256.5 g) = 0.0780 mol S_8

160 g O_2 (1 mol/32.00 g) = 5.0 mol O_2

$S_8(s) + 8\ O_2(g) \rightarrow 8\ SO_2(g)$

0.0780 mol S_8 (8 mol O_2/1 mol S_8) = 0.624 mol O_2 required by 0.0780 mol S_8

5.0 mol O_2 (1 mol S_8 /8 mol O_2) = 0.625 mol S_8 required by 5.0 mol O_2

Sulfur is the limiting reagent, so some O_2 remains after reaction.

Moles of O_2 in excess = 5.0 mol - 0.624 mol = 4.4 mol

0.0780 mol S_8 (8 mol SO_2/1 mol S_8) = 0.624 mol SO_2 produced

0.624 mol SO_2 (64.07 g/mol) = 40.0 g SO_2

5.3 1.612 g (1 mol CO_2/44.01 g)(1 mol C/1 mol CO_2) = 0.03663 mol C

0.7425 g H_2O (1 mol H_2O/18.02 g)(2 mol H/1 mol H_2O) = 0.08245 mol H

0.08245 mol H/0.03663 mol C = 2.25 H/1 C = 9 H/4 C

The empirical formula is C_4H_9, which has a molar mass of 57 g/mol. The molecular formula is $(C_4H_9)_2$ or C_8H_{18}.

5.4 0.452 g (1 mol CO_2/44.01 g)(1 mol C/1 mol CO_2) = 0.0103 mol C

0.0103 mol C (12.01 g/mol) = 0.123 g C

0.0924 g H_2O (1 mol H_2O/18.02 g)(2 mol H/1 mol H_2O) = 0.0103 mol H

0.0103 mol H (1.008 g/mol) = 0.0103 g H

Mass of Cr in unknown = 0.178 g - mass of C - mass of H = 0.0447 g Cr

0.0447 g Cr (1 mol / 52.00 g) = 8.60×10^{-4} mol Cr

The ratio of C to H is 1 C/1 H. The ratio of C to Cr is 12 to 1. Therefore, the empirical formula is $C_{12}H_{12}Cr$.

5.5 26.3 g (1 mol $NaHCO_3$/84.01 g) = 0.313 mol $NaHCO_3$

0.313 mol $NaHCO_3$/0.200 L = 1.57 M

5.6 0.500 L (0.0200 mol/L) = 0.0100 mol $KMnO_4$ required.

0.0100 mol $KMnO_4$ (158.0 g/mol) = 1.58 g $KMnO_4$

Place 1.58 g $KMnO_4$ in a 500-mL volumetric flask and add water to the mark on the flask. (See the volumetric flask on sidebar to Screen 5.11.)

5.7 (a) Add 125 mL of water to 125 mL of 2.00 M NaOH.

(b) (0.15 M)(0.0060 L)/0.010 L = 0.090 M

5.8 (0.02833 L)(0.953 M) = 0.0270 mol NaOH

(0.0270 mol NaOH)(1 mol CH_3CO_2H/1 mol NaOH) = 0.0270 mol CH_3CO_2H

0.0270 mol CH_3CO_2H (60.05 g/mol) = 1.62 g CH_3CO_2H

0.0270 mol CH_3CO_2H / 0.0250 L = 1.08 M

6.1 $q = 24.1 \times 103$ J = (250. g)(0.902 J/g · K)(T_{final} - 5.0 °C)

T_{final} = 111.9 °C

6.2 (15.5 g)(C_{metal})(18.9 °C - 100.0 °C) = -(55.5 g)(4.184 J/g · K)(18.9 °C - 16.5 °C)

C_{metal} = 0.450 J/g · mol

6.3 (400. g iron)(0.451 J/g · K)(32.8 °C - $T_{initial}$)

= -(1000. g)(4.184 J/g · K)(32.8 °C - 20.0 °C)

$T_{initial}$ = 330. °C

6.4 (25.0 g CH_3OH)(2.53 J/g · K)(64.6 °C - 25.0 °C) = 2.50×10^3 J

(25.0 g CH_3OH)(2.00×10^3 J/g) = 5.00×10^4 J

Total heat energy = 5.25×10^4 J

6.5 (1.32 g ice)(333 J/g ice) = 440. J

440. J = -(9.85 g)(C_{metal})(0.0 °C - 100.0 °C)

C_{metal} = 0.446 J/g · K

6.6 a) (10.0 g)(1 mol I_2/253.8 g)(62.4 kJ/mol) = 2.46 kJ

b) The process is exothermic.

(3.45 g)(1 mol I_2/253.8 g)(62.4 kJ/mol) = 0.848 kJ

6.7 (12.6 g H_2O)(1 mol / 18.02 g)(285.8 kJ/mol) = 2.00×10^2 kJ

6.8 $\Delta H_{overall}$ = -413.7 kJ + 106.8 kJ = -306.9 kJ

(454 g)(1 mol PbS/239.3 g)(-306.9 kJ) = 582 kJ

6.9 $\Delta H^\circ_{rxn} = 6\ \Delta H^\circ_f[CO_2(g)] + 3\ \Delta H^\circ_f[H_2O(\ell)] - \{\Delta H^\circ_f[C_6H_6(\ell)] + 15/2\ \Delta H^\circ_f[O_2(g)]\}$

= 6 mol(-393.5 kJ/mol) + 3 mol (-285.8 kJ/mol) - 1 mol (+49.0 kJ/mol) - 0

= -3267.4 kJ

7.1 Blue light: 4.00×10^2 nm = 4.00×10^{-7} m

$\nu = (2.998 \times 10^8 \text{ m/s}) / (4.00 \times 10^{-7} \text{ m}) = 7.50 \times 10^{14}$ /s

$E = (6.626 \times 10^{-34} \text{ J} \cdot \text{s})(7.50 \times 10^{14} \text{ /s}) = 4.97 \times 10^{-19}$ J

Microwaves: $E = (6.626 \times 10^{-34} \text{ J} \cdot \text{s})(2.45 \times 10^9 \text{ /s}) = 1.62 \times 10^{-24}$ J

E (blue light)/E (microwaves) = 4.6×10^5

Blue light is almost half a million times more energetic than microwaves.

7.2
1. $\ell = 0$ and 1
2. $m_\ell = +1, 0$, and -1; subshell label = p
3. d subshell
4. $\ell = 0$ and $m_\ell = 0$ for an s subshell.
5. 3 orbitals in a p subshell
6. f subshell has 7 values of m_ℓ and 7 orbitals.

7.3

Orbital	n	ℓ
6s	6	0
4p	4	1
5d	5	2
4f	4	3

8.1
1. 4s ($n + \ell = 4$) filled before 4p ($n + \ell = 5$)
2. 6s ($n + \ell = 6$) filled before 5d ($n + \ell = 7$)
3. 5s ($n + \ell = 5$) filled before 4f ($n + \ell = 7$)

8.2
1. Cl
2. S has the spectroscopic notation $1s^2 2s^2 2p^4 3s^2 3p^4$ and the following configuration in the box notation.

	3s	3p		
[Ne]	↑↓	↑↓	↑	↑

3. See electron configurations given on *Periodic Table* tool.

8.3 Electron configurations for V^{2+}, V^{3+}, and Co^{3+} are given here. All three ions are paramagnetic.

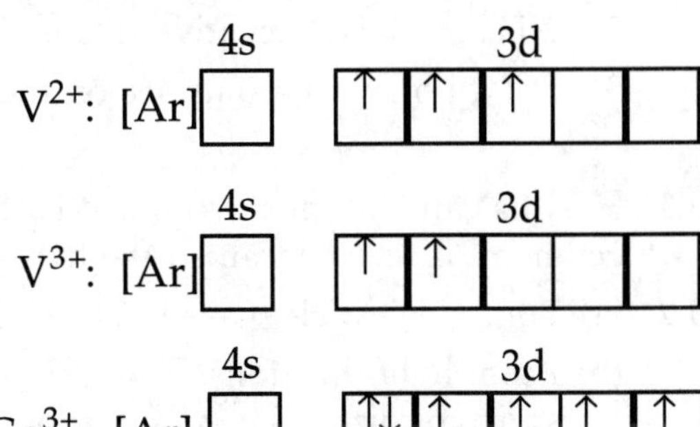

8.4 Radii are in the order $C < Si < Al$.

8.5
1. Radii are in the order $C < Si < Al$.
2. Ionization energies are in the order $Al < Si < C$.
3. Al should have a less negative EA than Si.

9.1
1.

$[\ddot{O}=N=\ddot{O}]^{\oplus}$ $\quad$ $:\ddot{Cl}—C(=\ddot{O})—\ddot{Cl}:$

2.

$:\ddot{O}—C(=O)—\ddot{O}:$ $\ominus$ $\quad$ $\ddot{O}=C(—\ddot{O}:)—\ddot{O}:$ $\ominus$ $\quad$ $:\ddot{O}—C(—\ddot{O}:)=\ddot{O}$ $\ominus$

9.2 $CH_4(g) + 2\ O_2(g) \rightarrow CO_2(g) + 2\ H_2O(g)$

Break 4 C—H bonds and 2 O=O bonds = 4 mol (414 kJ/mol) + 2 mol (498 kJ/mol) =2652 kJ

Make 2 C=O bonds and 4 H—O bonds = 2 mol (803 kJ/mol + 4 mol (464 kJ/mol) = 3462 kJ

ΔH°_{rxn} = 2652 kJ - 3462 kJ = -810 kJ (From ΔH°_f, the value of ΔH°_{rxn} is -802.3 kJ)

9.3
1. The H atom is the positive atom in each case. H—F ($\Delta\chi = 1.9$) is more polar than H—I ($\Delta\chi = 0.4$).
2. B—F ($\Delta\chi = 2.0$) is more polar than B—C ($\Delta\chi = 0.5$).
3. C—Si is more polar ($\Delta\chi = 0.7$) than C—S (which is not polar at all, $\Delta\chi = 0$)

9.4
1. The CH_2Cl_2 molecule has a tetrahedral molecular shape with a Cl—C—Cl angle of about 109°.
2. BF_3 is planar and trigonal, while adding F^- forms the tetrahedral BF_4^- ion.

9.5
1. $BFCl_2$, polar, negative side is the F atom because F is the most electronegative atom in the molecule.
2. NH_2Cl, polar, negative side is the Cl atom.
3. SCl_2, polar, Cl atoms are on the negative side.

10.1 SCl_2 has an S atom surrounded by two lone pairs and two bond pairs. The electron pair geometry is tetrahedral, so the hybridization of the S atom is sp^3.

10.2 (a) Bond angles: H—C—H = 109°; H—C—C = 109°; C—C—N = 180°

(b) Atom hybridizations: The CH_3 carbon has a tetrahedral electron-pair geometry and is sp^3 hybridized. The CN carbon has a linear electron pair geometry and is sp hybridized. The N atom has a linear electron pair geometry (the triple bond and lone pair are 180° apart) and can be considered sp hybridized.

10.3 O_2^+ [core electrons]$(\sigma_{2s})^2(\sigma^*_{2s})^2(\pi_{2p})^4(\sigma_{2p})^2(\pi^*_{2p})^1$
The net bond order is 2.5, a higher bond order than in O_2 and so O_2^+ has a stronger bond. The ion is paramagnetic to the extent of one electron.

12.1 0.83 bar (0.82 atm) > 75 kPa (0.74 atm) > 0.63 atm > 250. mm Hg (0.329 atm)

12.2 1. P_1 = 55 mm Hg and V_1 = 125 mL; P_2 = 78 mm Hg and V_2 = ?
$V_2 = P_1V_1/P_2$ = 88 mL
2. T_1 = 298 K and V_1 = 45 L; T_2 = 263 K and V2 = ?
$V_2 = V_1(T_2/T_1)$ = 40. L
3. 22.4 L CH_4 (2 L O_2/1 L CH_4) = 44.8 L O_2 required
44.8 L of H_2O and 22.4 L of CO_2 are produced.

12.3 n = 1300 mol, P = (750/760) atm, T = 293 K; V = nRT/P = 3.2×10^4 L

12.4 Molar mass = 28.96 g/mol, P = 1.00atm, T = 288 K; d = PM/RT = 1.23 g/L

12.5 P = 0.737 atm, V = 0.125 L, T = 296.2 K; n = PV/RT = 3.79×10^{-3} mol
Molar mass = 0.105 g/3.79×10^{-3} mol = 27.7 g/mol

12.6 180 g (1 mol N_2H_4/32.0 g) = 5.6 mol N_2H_4
5.6 mol N_2H_4(1 mol O_2/1 mol N_2H_4) = 5.6 mol O_2
$V(O_2)$ = nRT/P = 140 L when n = 5.6 mol, T = 294 K, and P = 0.99 atm

12.7 $P_{halothane}$ = (0.0760 mol)(0.08206 L · atm/K · mol)(298.2 K)/5.00 L
$P_{halothane}$ = 0.372 atm (or 283 mm Hg)
$P(O_2)$ = 3.59 atm (or 2730 mm Hg)
P_{total} = 3.96 atm (or 3010 mm Hg)

13.1 1. O_2 interactions occur by induced dipole-induced dipole forces, the weakest of all intermolecular forces.
2. $MgSO_4$ consists of the ions Mg^{2+} and SO_4^{2-}, so there are ion-dipole forces involved when the salt dissolves in water. The common, hydrated salt $MgSO_4 \cdot 7\,H_2O$ is widely used in agriculture and medicine.
3. Dipole-induced dipole forces exist between H_2O and O_2.
Order of strength is O_2—O_2 < O_2—H_2O < $MgSO_4$—H_2O.

13.2 $(1.00 \times 10^3$ g)(1 mol/32.04 g)(35.21 kJ/mol) = 1.10×10^3 kJ

13.3 Glycerol has three —OH groups per molecule that can be used in hydrogen bonding, as compared with only one for ethanol. The viscosity of glycerol should be greater than that of ethanol.

14.1 10.0 g sugar = 0.0292 mol and 250. g water = 13.9 mol
X_{sugar} = 0.0292 mol / (0.0292 mol + 13.9 mol) = 0.00210
Molality sugar = 0.0292 mol sugar/0.250 kg = 0.117 molal

Percentage sugar = (10.0 g / 260. g solution)100% = 3.85%

14.2 For AgCl: Enthalpy of solution = 912 kJ/mol + (-851 kJ/mol) = +61 kJ/mol

For RbF: Enthalpy of solution = 789 kJ/mol + (-792 kJ/mol) = -3 kJ/mol

AgCl, an insoluble salt, has a large, positive enthalpy of solution, whereas the soluble salt RbF has a negative enthalpy of solution.

14.3 Solution consists of sucrose (0.0292 mol) and water (12.5 mol).

X_{water} = 12.5 mol H_2O/ (12.5 mol H_2O + 0.0292 mol sucrose) = 0.998

P_{water} = (0.998)(149.4 mm Hg) = 149 mm Hg

Even with 10.0 g of sugar, the vapor pressure of water has changed very little.

14.4 Concentration (m) = $\Delta t/K_{bp}$ = 1.0 °C/(+0.512 °C/m) = 2.0 m

(2.0 mol/kg) · (0.100 kg) = 0.20 mol

0.20 kg (62.07 g/mol) = 12 g glycol

14.5 Δt_{bp} = 80.23 °C - 80.10 °C = 0.13 °C

Concentration (m) = Δt_{bp} / K_{bp} = 0.13 °C/(2.53 °C/molal) = 0.051 m

0.051 mol/kg (0.100 kg) = 0.0051 mol

The molar mass is 0.640 g / 0.0051 mol = 130 g/mol

14.6 Molality of $C_2H_4(OH)_2$ (ethylene glycol) = 8.06 mol/3.00 kg = 2.69 m

$\Delta t_{fp} = K_{fp} \cdot m$ = (-1.86 deg/molal)(2.69 molal) = -4.99 deg

500. g of glycol is not sufficient to keep water from freezing at -25 °C.

14.7 $c_M = \Pi/RT$ = (364 mm Hg/760. mm Hg atm^{-1})/(0.0821 L · atm/K · mol)(298 K)

$c_M = 1.96 \times 10^{-2}$ mol/L

(1.96×10^{-2} mol/L)(0.0250 L) = 4.89×10^{-4} mol

Molar mass = (144×10^{-3} g)/(4.89×10^{-4} mol) = 294 g/mol

15.1 The plot is not a straight line.

Rate of change over first 2 hours

= - Δ[sucrose]/Δt = (0.034 - 0.050)mol/L / 2 h = -0.0080 mol/L · h

Rate of change over last 2 hours = -0.0025 mol/L · h

Instantaneous rate at 4 hours = -0.0045 mol/L · h

15.2 1. Second order with respect to NO and first order with respect to H_2

2. Increases by factor of 4

3. The reaction rate is halved.

15.3 Rate of reaction = (0.090/h)(0.020 mol/L) = 0.0018 mol/L · h

Cl^- appears at a rate of 0.0018 mol/L · h.

15.4 (a) m = 1. The initial rates and initial concentrations are directly proportional. For example, as the concentration is doubled, the rate is doubled.

(b) Taking the data from Experiment 1, we have

$k = \text{Rate}/[\text{Reactant}] = 1.3 \times 10^{-7}$ mol/(L · min) / 1.0×10^{-3} mol/L

$k = 1.3 \times 10^{-4}$ /min

15.5 $t_{1/2} = 0.693/\ 5.40 \times 10^{-2}/\text{h} = 12.8$ h

51.2 h = 4.00 half-lives. After 4.00 half-lives the fraction remaining is 1/16.

ln(fraction remaining) = -kt = $-(5.40 \times 10^{-2}/\text{h})(18.0\ \text{h}) = -0.97$.

Fraction remaining after 18.0 h = $[R]/[R]_o = 0.38$

15.6 All three steps are bimolecular. N_2O_2, the product of the first step, is used in the second step, and N_2O, a product of the second step, is consumed in the third step. Therefore, adding the three reactions gives the equation for the overall process.

16.1 1. $K = [PCl_3][Cl_2] / [PCl_5]$

2. $K = [Cu^{2+}][OH^-]^2$

3. $K = [Cu^{2+}][NH_3]^4 / [Cu(NH_3)_4{}^{2+}]$

4. $K = [CH_3CO_2{}^-][H_3O^+] / [CH_3CO_2H]$

16.2. 1. (a) $K_{new} = [K_{old}]^2 = 6.3 \times 10^{-58}$

(b) $K = 1/(6.3 \times 10^{-58}) = 1.6 \times 10^{57}$

2. $K = (7.9 \times 10^{11})(1/4.8 \times 10^{-41})(1/2.2 \times 10^{-15}) = 7.5 \times 10^{66}$

16.3 $Q = 1.4 \times 10^{-4}$. Because Q < K, the reaction is not at equilibrium. The reaction consumes N_2 and O_2 and produces NO on proceeding to equilibrium.

16.4

	$C_6H_{10}I_2$	C_6H_{10}	I_2
Initial (M)	0.050	0	0
Change (M)	-0.035	+0.035	+0.035
Equilibrium (M)	0.015	0.035	0.035

K = 0.082

16.5 1. The equilibrium shifts to the left, and $[SO_3]$ decreases.

2. [Isobutane] = 1.93 M and [Butane] = 0.77 M.

3. Added H_2 shifts the equilibrium right, and added NH_3 shifts it left. Increasing the volume decreases all the concentrations. The equilibrium shifts to the left, toward the side with the greater number of molecules.

17.1 1. $H_3PO_4(aq) + H_2O(\ell) \rightarrow H_3O^+(aq) + H_2PO_4{}^-(aq)$

2. $H_2O(\ell) + CN^-(aq) \rightarrow OH^-(aq) + HCN(aq)$; CN^- is a Brønsted base.

3. $H_2C_2O_4(aq) + H_2O(\ell) \rightarrow H_3O^+(aq) + HC_2O_4{}^-(aq)$

$HC_2O_4{}^-(aq) + H_2O(\ell) \rightarrow H_3O^+(aq) + C_2O_4{}^{2-}(aq)$

4. $HSO_4{}^-$ is an acid and $SO_4{}^{2-}$ is its conjugate base. $CO_3{}^{2-}$ is a base and $HCO_3{}^-$ is its conjugate acid.

17.2 pH = 4.32; $[H_3O^+] = 10^{-pH} = 10^{-4.32} = 4.8 \times 10^{-5}$ M

pOH = 14.00 - 4.32 = 9.68; $[OH^-] = 10^{-pOH} = 10^{-9.68} = 2.1 \times 10^{-10}$ M

17.3 Lactic acid is a weaker acid than formic acid but stronger than benzoic acid.
$K_b = K_w/K_a = 1.0 \times 10^{-14} / 1.4 \times 10^{-4} = 7.1 \times 10^{-11}$.

17.4 $[H_3O^+] = 10^{-pH} = 10^{-2.94} = 1.1 \times 10^{-3}$ M
$K_a = [H_3O^+][CH_3CH_2CO_2^-] / [CH_3CH_2CO_2H] = (1.1 \times 10^{-3})^2 / (0.10 - 1.1 \times 10^{-3})$
$K_a = 1.2 \times 10^{-5}$

17.5 $K_a = [H_3O^+]^2 / (0.10 - [H_3O^+])$; $[H_3O^+] = 1.3 \times 10^{-3}$ M; pH = 2.87

17.6 $K_b = [OH^-]^2 / (0.025 - [OH^-])$; $[OH^-] = 6.7 \times 10^{-4}$ M; pOH = 3.17 and pH = 10.83

17.7 $K_a = [H_3O^+]^2 / (0.50 - [H_3O^+])$; $[H_3O^+] = 1.7 \times 10^{-5}$ M; pH = 4.78

17.8 1. LiBr(aq) is neutral; pH = 7
2. $FeCl_3$(aq) is acidic; pH < 7
3. NH_4NO_3(aq) is acidic; pH < 7
4. Na_2HPO_4(aq) is basic; pH > 7

18.1 Moles of HCl = (0.050 L)(0.20 M) = 0.010 mol
Moles of aniline = (0.93 g)(1 mol/93.1 g) = 0.010 mol
Because 1 mol of aniline requires 1 mol of HCl, and because they were present initially in equal molar quantities, the reaction completely consumes the acid and base. The solution after reaction contains 0.010 mol of $C_6H_5NH_3^+$ in 0.050 L of solution, so its concentration is 0.20 M. It is the conjugate acid of the weak base aniline, so

$C_6H_5NH_3^+(aq) + H_2O(\ell) \rightarrow C_6H_5NH_2(aq) + H_3O^+(aq)$

$$K_a = 2.4 \times 10^{-5} = \frac{[H_3O^+][C_6H_5NH_2]}{[C_6H_5NH_3^+]} = \frac{x^2}{0.20}$$

$x = [H_3O^+] = 2.2 \times 10^{-3}$ M, which gives a pH of 2.66.

18.2 (0.976 g)(1 mol/122.1 g) = 0.00799 mol benzoic acid
Moles of NaOH required = 0.00799 mol
(0.00799 mol NaOH)(1 L/0.100 mol) = 0.0799 L or 79.9 mL NaOH
Reaction gives 0.00799 mol of benzoate ion, $C_6H_5CO_2^-$.
Concentration benzoate ion = 0.00799 mol/0.0799 L = 0.100 M
Hydrolysis of benzoate ion:
$C_6H_5CO_2^-(aq) + H_2O(\ell) \rightarrow C_6H_5CO_2H(aq) + OH^-(aq)$
$K_b = 1.6 \times 10^{-10} = [C_6H_5CO_2H][OH^-]/[C_6H_5CO_2^-]$
$[OH^-] = 4.0 \times 10^{-6}$ M; pOH = 5.40 and pH = 8.60

18.3 (a) pH of 0.30 M formic acid (Obtain K_a for formic acid from Screen 17.7.)
$K_a = 1.8 \times 10^{-4} = [H_3O^+][HCO_2^-] / [HCO_2H] = x^2/0.30$
$x = [H_3O^+] = 7.3 \times 10^{-3}$ M, which gives a pH of 2.14.
(b) pH of 0.30 M formic acid + 0.010 M $NaHCO_2$

Equation	HCO_2H	→	H_3O^+ +	HCO_2^-
Initial concentrations (M)	0.30		0	0.10
Change (M)	-x		+x	+x
At equilibrium (M)	0.30 - x		x	0.10 + x

$K_a = 1.8 \times 10^{-4} = (x)(0.10 + x) / (0.30 - x)$
Assuming that x is small compared with 0.10 or 0.30, we find that $x = [H_3O^+] = 5.4 \times 10^{-4}$ M, which gives a pH of 3.27.

18.4 K_a for $HCO_3^- = 4.8 \times 10^{-11}$ so $pK_a = 10.32$
15.0 g $NaHCO_3$ = 0.179 mol and 18.0 g Na_2CO_3 = 0.170 mol
pH = 10.32 + log (0.170 mol/0.179 mol) = 10.30

18.5 A pH of 5.00 corresponds to $[H_3O^+] = 1.0 \times 10^{-5}$ M. This means that
$[H_3O^+] = 1.0 \times 10^{-5}\ M = \{[CH_3CO_2H] / [CH_3CO_2^-]\}(1.8 \times 10^{-5})$
The ratio $[CH_3CO_2H] / [CH_3CO_2^-]$ must be 1/1.8 to achieve the correct hydronium ion concentration. Therefore, 1.0 mol of CH_3CO_2H is mixed with 1.8 mol of a salt of $CH_3CO_2^-$ (say $NaCH_3CO_2$) in some amount of water. (The volume of water is not critical; only the relative amounts of acid and conjugate base are important.)

19.1 1. $BaSO_4(s) \rightleftharpoons Ba^{2+}(aq) + SO_4^{2-}(aq)$ $K_{sp} = [Ba^{2+}][SO_4^{2-}] = 1.1 \times 10^{-10}$
2. $BI_3(s) \rightleftharpoons Bi^{3+}(aq) + 3\ I^-(aq)$ $K_{sp} = [Bi^{3+}][I^-]^3 = 8.1 \times 10^{-19}$
3. $Ag_2CO_3(s) \rightleftharpoons 2\ Ag^+(aq) + CO_3^{2-}(aq)$ $K_{sp} = [Ag^+]^2[CO_3^{2-}] = 8.1 \times 10^{-12}$

19.2 $[Ba^{2+}] = 7.5 \times 10^{-3}$ M and $[F^-] = 2 \times [Ba^{2+}] = 1.5 \times 10^{-2}$ M
$K_{sp} = [Ba^{2+}][F^-]^2 = (7.5 \times 10^{-3})(1.5 \times 10^{-2})^2 = 1.7 \times 10^{-6}$

19.3 K_{sp} for CuI $= 5.1 \times 10^{-12} = [Cu^+][I^-] = (x)(x)$
$x = (\ 5.1 \times 10^{-12})^{1/2} = 2.3 \times 10^{-6}$ M
K_{sp} for $Mg(OH)_2 = 1.5 \times 10^{-11} = [Mg^{2+}][OH^-]^2 = (x)(2x)^2$
$x = 1.6 \times 10^{-4}$ M

19.4 1. AgCl is more soluble than AgI
2. $CaCO_3$ is more soluble than $SrCO_3$
3. CaF_2 is more soluble than $CaCO_3$

19.5 If $[Pb^{2+}] = 1.1 \times 10^{-3}$ M, then $[I^-] = 2\ (1.1 \times 10^{-3}\ M) = 2.2 \times 10^{-3}$ M
$Q = [Pb^{2+}][I\text{-}]^2 = (1.1 \times 10^{-3}\ M)(2.2 \times 10^{-3}\ M)^2 = 5.3 \times 10^{-9}$
The value of Q is less than K_{sp} (8.7×10^{-9}), so PbI_2 can dissolve to a greater extent.

19.6 $Q = [Sr^{2+}][SO_4^{2-}] = (2.5 \times 10^{-4}\ M)(2.5 \times 10^{-4}\ M) = 6.3 \times 10^{-8}$
Q is less than K_{sp} so the solution is not yet saturated, and no precipitation occurs.

19.7 $[I^-] = \{K_{sp}/[Pb^{2+}]\}^{1/2} = [(8.7 \times 10^{-9})/(0.050\ M)]^{1/2} = 4.2 \times 10^{-4}$ M
The I^- concentration needed to precipitate the lead(II) ion is 4.2×10^{-4} M. The lead(II) ion concentration remaining in solution when I^- reaches 0.0015 M.
$[Pb^{2+}] = K_{sp}/[I^-]^2 = (8.7 \times 10^{-9})/(0.0015)^2 = 3.9 \times 10^{-3}$ M

19.8 Solubility in pure water: $[Ba^{2+}] = [SO_4^{2-}] = (K_{sp})^{1/2} = 1.0 \times 10^{-5}$ M
Solubility with added Ba^{2+} ion: $[SO_4^{2-}] = K_{sp}/(x + 0.010\ M) \approx K_{sp}/(0.010\ M)$
Solubility $= 1.1 \times 10^{-8}$ M

19.9 K for the overall reaction

$= K_{sp}$ for AgCl · ($1/K_{sp}$ for AgBr) = $1.8 \times 10^{-10} / 3.3 \times 10^{-13} = 550$

AgCl can be converted to AgBr.

20.1 Entropy change for liquid to vapor = (30,900 J/K) / (353.3 K) = +87.5 J/K · mol

ΔS° for vapor to liquid = -87.5 J/K · mol

20.2 1. S° for solid CO_2 is less than for CO_2 vapor. For a given compound, molecules in the vapor state always have higher entropy because the vapor state is more disordered than the solid state.

2. KCl dissolved in water forms ions separated by water molecules, a more highly disordered state than solid KCl. Thus, the entropy of the dissolved KCl (S° = 159.0 J/K · mol) is larger than that of solid KCl (S° = 82.6 J/K · mol).

3. In this reaction a mole of solid has given rise to a mole of solid compound and a mole of gas. The entropy has increased.

20.3 $2\ NO(g) + O_2(g) \rightarrow 2\ NO_2(g)$

ΔS°_{rxn} = (2 mol NO_2)(240.1 J/K · mol)

- [(2 mol NO)(210.8 J/K · mol) + (1 mol O_2)(205.1 J/K · mol)]

= -146.5 J/K or -73.25 J/K for the formation of 1 mol of NO_2.

Notice that the sign of the entropy change is negative. This is largely due to the fact that the chemical reaction began with 3 mol of gaseous reactants and ended with 2 mol of gaseous products.

20.4 $\Delta H^\circ_{rxn} = 2\ \Delta H^\circ_f$ [HCl(g)] = -184.6 kJ

$\Delta S^\circ_{surroundings} = -[-184.6 \times 10^3$ J/298 K] = +620. J/K

ΔS°_{rxn} = 2 S°[HCl(g)] - {S°[H_2(g)] + S°[Cl_2(g)]} = 2 mol (186.9 J/K · mol)

- [1 mol (130.7 J/K · mol) + 1 mol (223.1 J/K · mol]

= +20.0 J/K

$\Delta S^\circ_{universe}$ = 640. J/K

20.5 $\Delta H^\circ_{rxn} = \Delta H^\circ_f$ [NH_3(g)] - { 1/2 ΔH°_f [N_2(g)] + 3/2 ΔH°_f [H_2(g)]}

= (1 mol)(- 46.11 kJ/mol) - [1/2 mol(0) + 3/2 mol(0)] = - 46.11 kJ

ΔS°_{rxn} = S°[NH_3(g)] - {1/2 S°[N_2(g)] + 3/2 S°[H_2(g)]} = - 99.38 J/K

$\Delta G^\circ_{rxn} = \Delta H^\circ_{rxn} - T\Delta S^\circ_{rxn}$ = - 46.11 kJ - (298 K)(-99.38 J/K)(1 kJ/1000 J) = -16.5 kJ

20.6 $C_6H_6(\ell) + 15/2\ O_2(g) \rightarrow 6\ CO_2(g) + 3\ H_2O(\ell)$

$\Delta G^\circ_{rxn} = 6\ \Delta G^\circ_f$ [CO_2(g)] + 3 ΔG°_f [$H_2O(\ell)$] - {ΔG°_f [$C_6H_6(\ell)$] + 15/2 ΔG°_f [O_2(g)]}

= (6 mol)(-394.359 kJ/mol) + (3 mol)(-237.129 kJ/mol)

- [(1 mol)(124.5 kJ/mol) + (15/2 mol)(0)]

ΔG°_{rxn} = - 3202.0 kJ

0.7 $MgO(s) + C(graphite) \rightarrow Mg(s) + CO(g)$

$\Delta H^\circ_{rxn} = +491.18$ kJ and $\Delta S^\circ_{rxn} = 197.67$ J/K

$T = \Delta H^\circ_{rxn}/\Delta S^\circ_{rxn} = 491.18$ kJ/(0.198 kJ/K) = 2480 K (or about 2200 °C)

0.8 $\Delta G^\circ_{rxn} = +130.4$ kJ/mol

(+130.4 kJ)(1000 J/kJ) = -(8.314510 J/K · mol)(298 K) ln K_p

ln $K_p = -52.6$ and so $K_p = 1.39 \times 10^{-23}$

1.1 Oxidizing agent: $I_2(s) + 2e^- \rightarrow 2\ I^-(aq)$

Reducing agent: $2\ [Cr^{2+}(aq) \rightarrow Cr^{3+}(aq) + e^-]$

$2\ Cr^{2+}(aq) + I_2(s) \rightarrow Cr^{3+}(aq) + 2\ I^-(aq)$

Cr^{2+} is oxidized by I_2 and I_2 is reduced by Cr^{2+}.

1.2 Oxidizing agent: $3\ [NO_3^-(aq) + 2\ H^+(aq) + e^- \rightarrow NO_2(g) + H_2O(\ell)]$

Reducing agent: $Co(s) \rightarrow Co^{3+}(aq) + 3e^-$

$Co(s) + 3\ NO_3^-(aq) + 6\ H^+(aq) \rightarrow Co^{3+}(aq) + 3\ NO_2(g) + 3\ H_2O(\ell)$

Cobalt metal is oxidized by nitric acid, and nitric acid is reduced by cobalt metal.

1.3 Oxidation at the anode: $Ni(s) \rightarrow Ni^{2+}(aq) + 2e^-$

Reduction at the cathode: $e^- + Ag^+(aq) \rightarrow Ag(s)$

Electrons flow from the anode (Ni) to the cathode (Ag), while NO_3^- ions flow from the cathode compartment (containing a declining concentration of Ag^+) to the anode compartment (where there is an increasing concentration of Ni^{2+}).

1.4 $\Delta G^\circ_{rxn} = -$ (2.00 mol e^-)(9.65 × 10^4 J/V · mol)(-0.76 V)(1.0 kJ/1000 J) = +150 kJ

The reaction is not product-favored as written.

1.5 $2\ Al(s) \rightarrow 2\ Al^{3+}(aq) + 6\ e^-$ $\quad E^\circ = +1.66$ V

$3\ Zn^{2+}(aq) + 6e^- \rightarrow 3\ Zn(s)$ $\quad E^\circ = -0.76$ V

$2\ Al(s) + 3\ Zn^{2+}(aq) \rightarrow 2\ Al^{3+}(aq) + 3\ Zn(s)$ $\quad E^\circ_{net} = +0.90$ V

The electrons flow through the wire from the Al electrode (anode) to the Zn electrode (cathode). Nitrate ions flow through the salt bridge from the Zn/Zn^{2+} compartment to the Al/Al^{3+} compartment.

1.6 $2\ [Ag^+(aq) + e^- \rightarrow Ag(s)]$ $\quad E^\circ = +0.800$ V

$Hg(\ell) \rightarrow Hg^{2+}(aq) + 2e^-$ $\quad E^\circ = -0.855$ V

$2\ Ag^+(aq) + Hg(\ell) \rightarrow 2\ Ag(s) + Hg^{2+}(aq)$ $\quad E^\circ_{net} = -0.055$ V

Reaction is not product-favored under standard conditions, but it is product-favored under the nonstandard conditions.

$$E_{net} = E^{\circ}_{net} - \frac{0.0257}{n} \ln \frac{[Hg^{2+}]}{[Ag^{+}]^2}$$

$$E_{net} = -0.055\ V - \frac{0.0257}{2} \ln \frac{[0.0010]}{[0.80]^2} = +0.028$$

21.7 a) Quantity of lead available = 2.41 mol

b) Moles of electrons passed through cell

= 2.41 mol Pb (2 mol e^-/1 mol Pb) = 4.83 mol Pb

c) Charge carried by 4.83 mol e^- = 4.83 mol e^- (9.65×10^4 C/mol e^-) = 4.66×10^5 C

d) Time: 4.66×10^5 C = 1.50 amp · time

Time = 3.10×10^5 s (or 86.2 h)

APPENDIX C

Answers to Study Questions

1.1 (a) carbon (b) sodium (c) chlorine (d) phosphorus (e) magnesium (f) calcium

1.2 (a) Li (b) Ti (c) Fe (d) Si (e) Co (f) Zn

1.3 557 g

1.4 498 g; 0.498 kg

1.5 0.911 g/cm^3

1.6 298 K

1.7 (a) 289 K (b) 1.0×10^2 °C (c) 230 K

1.8 Liquid; body temperature (37 °C) is above the melting point of gallium.

1.9 190 mm; 0.19 mL

1.10 5.3 cm^3; 5.3×10^{-4} m^2

1.11 800. cm^3; 0.800 L; 8.00×10^{-4} m^3

1.12 0.00563 kg; 5630 mg

1.13

Milligrams	Grams	Kilograms
693	0.693	6.93×10^{-4}
156	0.156	1.56×10^{-4}
2.23×10^6	2.23×10^3	2.23

1.14 22 cm × 28 cm; 6.0×10^2 cm^2

1.15 0.178 nm^3; 1.78×10^{-22} cm^3

1.16 0.995 g of platinum

1.17 Nickel with a density of 8.91 g/cm^3

1.18 A reasonable estimate is 0.99211 g/cm^3

2.1 (a) 9 (b) 48 (c) 70

2.2 (a) $^{23}_{11}$Na (b) $^{39}_{18}$Ar (c) $^{70}_{31}$Ga

2.3

	Electrons	Protons	Neutrons
(a)	20	20	20
(b)	50	50	69
(c)	94	94	150

2.4

Symbol	^{45}Sc	^{33}S	^{17}O	^{56}Mn
Number of protons	21	16	8	25
Number of neutrons	24	17	9	31
Number of electrons in the neutral atom	21	16	8	25

2.5 95 protons, 95 electrons, 146 neutrons

2.6 (^{6}Li mass)(% abundance) + (^{7}Li mass)(% abundance) = atomic weight of Li
(6.015121 amu)(0.0750) + (7.016003 amu)(0.9250) = 6.94 amu

2.7 ^{69}Ga abundance is 60.12%, ^{71}Ga abundance is 39.88%

2.8 (a) 27 g B (b) 0.48 g O_2 (c) 6.98×10^{-2} g Fe (d) 2.61×10^3 g He

2.9 (a) 1.9998 mol Cu (b) 0.499 mol Ca (c) 0.6208 mol Al
(d) 3.1×10^{-4} mol K (e) 2.1×10^{-5} mol Am

2.10 2.19 mol Na; 1.32×10^{24} atoms Na

2.11 4.131×10^{23} atoms Cr

2.12 1.0552×10^{-22} g/Cu atom

2.13 ^{39}K must be more abundant because the average atomic weight of K is 39.0983, which is closest to the mass number of ^{39}K.

2.14 3.40 mol Cu; 2.0×10^{24} atoms Cu

3.1 Aluminum: Al^{3+}; selenium: Se^{2-}

3.2 (a) Mg^{2+} (b) Zn^{2+} (c) Fe^{2+} and Fe^{3+} (d) Ga^{3+}

3.3 (a) Sr^{2+} (b) Al^{3+} (c) S^{2-} (d) Co^{2+}
(e) Ti^{4+} (f) HCO_3^- (g) ClO_4^- (h) NH_4^+

3.4 CoO and Co_2O_3

3.5 (a) $AlCl_3$ (b) NaF (c) correct (d) correct

3.6 Magnesium oxide, MgO; charges of the ions are greater and ionic radii are smaller; this results in stronger attractions among the ions and a higher melting temperature.

3.7 (a) potassium sulfide (b) nickel(II) sulfate
(c) ammonium phosphate (d) calcium hypochlorite

3.8 (a) $(NH_4)_2CO_3$ (b) CaI_2 (c) $CuBr_2$
(d) $AlPO_4$ (e) $AgCH_3CO_2$

3.9

K_2CO_3	potassium carbonate
KBr	potassium bromide
KNO_3	potassium nitrate
$BaCO_3$	barium carbonate
$BaBr_2$	barium bromide
$Ba(NO_3)_2$	barium nitrate
$(NH_4)_2CO_3$	ammonium carbonate
NH_4Br	ammonium bromide
NH_4NO_3	ammonium nitrate

3.10 (a) 159.7 g/mol (b) 67.81 g/mol (c) 44.02 g/mol
(d) 197.9 g/mol (e) 176.1 g/mol

3.11 (a) 0.0312 mol CH_3OH (b) 0.0101 mol Cl_2CO
(c) 0.0125 mol NH_4NO_3 (d) 4.06×10^{-3} mol $MgSO_4 \cdot 7\,H_2O$

3.12 47.1 mol C_2H_3CN

3.13 (a) 1.80×10^{-3} mol $C_9H_8O_4$; 2.266×10^{-2} mol $NaHCO_3$; 5.205×10^{-3} mol $C_6H_8O_7$
(b) 1.08×10^{21} molecules $C_9H_8O_4$

3.14 12.5 mol SO_3; 7.52×10^{24} molecules SO_3; 7.52×10^{24} S atoms; 2.26×10^{25} O atoms

3.15 (a) 86.59% Pb, 13.41% S (b) 81.71% C, 18.29% H
(c) 24.77% Co, 29.80% Cl, 5.08% H, 40.35% O
(d) 35.00% N, 5.04% H, 59.96% O

3.16 (a) 62.50 g/mol (b) 38.44% C, 4.84% H, 56.72% Cl (c) 174 g C

3.17 The molar mass of the empirical formula = 59.04 g/mol, therefore the molecular formula is $C_4H_6O_4$.

3.18 The empirical formula is CH, molecular formula is C_2H_2.

3.19 The empirical formula is N_2O_3.

3.20 The empirical formula is $C_8H_8O_3$, molecular formula is also $C_8H_8O_3$.

3.21 The empirical formula is C_2H_6As, molecular formula is $C_4H_{12}As_2$.

3.22 There are 6 F atoms for every S atom; the formula is SF_6.

4.1 (a) $4\ Cr(s) + 3\ O_2(g) \rightarrow 2\ Cr_2O_3(s)$
(b) $Cu_2S(s) + O_2(g) \rightarrow 2\ Cu(s) + SO_2(g)$
(c) $C_6H_5CH_3(\ell) + 9\ O_2(g) \rightarrow 4\ H_2O(\ell) + 7\ CO_2(g)$

4.2 (a) $3\ MgO(s) + 2\ Fe(s) \rightarrow Fe_2O_3(s) + 3\ Mg(s)$
iron(III) oxide, magnesium
(b) $AlCl_3(s) + 3\ H_2O(\ell) \rightarrow Al(OH)_3(s) + 3\ HCl(aq)$
aluminum hydroxide, hydrogen chloride
(c) $2\ NaNO_3(s) + H_2SO_4(\ell) \rightarrow Na_2SO_4(s) + 2\ HNO_3(g)$
sodium sulfate, nitric acid

4.3 (a) $CO_2(g) + 2\ NH_3(g) \rightarrow CO(NH_2)_2(s) + H_2O(\ell)$
(b) $UO_2(s) + 4\ HF(aq) \rightarrow UF_4(s) + 2\ H_2O(aq)$
$UF_4(s) + F_2(g) \rightarrow UF_6(s)$
(c) $TiO_2(s) + 2\ Cl_2(g) + 2\ C(s) \rightarrow TiCl_4(\ell) + 2\ CO(g)$
$TiCl_4(\ell) + 2\ Mg(s) \rightarrow Ti(s) + 2\ MgCl_2(s)$

4.4 (a) $FeCl_2$ is soluble (b) $AgNO_3$ is soluble (c) NaCl and $KMnO_4$ are soluble

4.5 (a) $NaCH_3CO_2$ (b) Fe_2S_3 (c) KOH (d) $PbCl_2$

4.6 (a) K^+ and I^- ions (b) K^+ and SO_4^{2-} ions
(c) K^+ and HSO_4^- ions (d) K^+ and CN^- ions

4.7 $HNO_3(aq) \rightarrow H^+(aq) + NO_3^-(aq)$

4.8 $H_2C_2O_4(aq) \rightarrow H^+(aq) + HC_2O_4^-(aq)$
$HC_2O_4^-(aq) \rightarrow H^+(aq) + C_2O_4^{2-}(aq)$

4.9 $SO_3(g) + H_2O(\ell) \rightarrow H_2SO_4(aq)$

4.10 (a) $Zn(s) + 2\ HCl(aq) \rightarrow H_2(g) + ZnCl_2(aq)$
$Zn(s) + 2\ H^+(aq) \rightarrow H_2(g) + Zn^{2+}(aq)$

(b) $Mg(OH)_2(s) + 2\ HCl(aq) \rightarrow MgCl_2(aq) + 2\ H_2O(\ell)$
$Mg(OH)_2(s) + 2\ H^+(aq) \rightarrow Mg^{2+}\ (aq) + 2\ H_2O(\ell)$
(c) $2\ HNO_3(aq) + CaCO_3(s) \rightarrow Ca(NO_3)_2(aq) + H_2O(\ell) + CO_2(g)$
$2\ H^+(aq) + CaCO_3(s) \rightarrow Ca^{2+}(aq) + H_2O(\ell) + CO_2(g)$

4.11 (a) $Ba(OH)_2(s) + 2\ HNO_3(aq) \rightarrow Ba(NO_3)_2(aq) + 2\ H_2O(\ell)$
$Ba(OH)_2(s) + 2\ H^+(aq) \rightarrow Ba^{2+}(aq) + 2\ H_2O(\ell)$
(b) $BaCl_2(aq) + Na_2CO_3(aq) \rightarrow BaCO_3(s) + 2\ NaCl(aq)$
$Ba^{2+}(aq) + CO_3^{2-}(aq) \rightarrow BaCO_3(s)$
(c) $2\ Na_3PO_4(aq) + 3\ Ni(NO_3)_2(aq) \rightarrow Ni_3(PO_4)_2(s) + 6\ NaNO_3(aq)$
$2\ PO_4^{3-}(aq) + 3\ Ni^{2+}(aq) \rightarrow Ni_3(PO_4)_2(s)$

4.12 $CdCl_2(aq) + 2\ NaOH(aq) \rightarrow Cd(OH)_2(s) + 2\ NaCl(aq)$
$Cd^{2+}(aq) + 2\ OH^-(aq) \rightarrow Cd(OH)_2(s)$

4.13 (a) $NiCl_2(aq) + (NH_4)_2S(aq) \rightarrow NiS(s) + 2\ NH_4Cl(aq)$
(b) $3\ Mn(NO_3)_2(aq) + 2\ Na_3PO_4(aq) \rightarrow Mn_3(PO_4)_2(s) + 6\ NaNO_3(aq)$

4.14 $Pb(NO_3)_2(aq) + 2\ KOH(aq) \rightarrow Pb(OH)_2(s) + 2\ KNO_3(aq)$
lead(II) nitrate, potassium hydroxide, lead(II) hydroxide, potassium nitrate

4.15 (a) $2\ CH_3CO_2H(aq) + Mg(OH)_2(s) \rightarrow Mg(CH_3CO_2)_2(aq) + 2\ H2O(\ell)$
acetic acid, magnesium hydroxide, magnesium acetate, water
(b) $HClO_4(aq) + NH_3(aq) \rightarrow NH_4ClO_4(aq)$
perchloric acid, ammonia, ammonium perchlorate

4.16 $Ba(OH)_2(s) + 2\ HNO_3(aq) \rightarrow Ba(NO_3)_2(aq) + 2\ H_2O(\ell)$

4.17 $MnCO_3(s) + 2\ HCl(aq) \rightarrow MnCl_2(aq) + CO_2(g) + H_2O(\ell)$
manganese(II) carbonate, hydrochloric acid, manganese(II) chloride, carbon dioxide, water

4.18 (a) Br is +5 and O is –2
(b) C is +3 and O is –2
(c) F is 0
(d) Ca is +2 and H is –1
(e) H is +1, Si is +4, and O is –2
(f) S is +6 and O is –2

4.19 Reactants: Na is +1, I is –1, H is +1, S is +6, O is –2, Mn is +4
Products: Na is +1, I is 0, H is +1, S is +6, O is –2, Mn is +2

4.20 (a) Precipitation reaction. (b) Oxidation-reduction reaction. The oxidation number of Ca changes from 0 to +2, while that of O changes from 0 to –2. (c) acid-base reaction

4.21 (a) Mg is oxidized and is the reducing agent. O_2 is reduced and is the oxidizing agent.
(b) C is oxidized and C_2H_4 is the reducing agent. O_2 is reduced and is the oxidizing agent
(c) Si is oxidized and is the reducing agent. Cl_2 is reduced and is the oxidizing agent.

4.22 $MgCO_3(s) + 2\ H^+(aq) \rightarrow Mg^{2+}(aq) + CO_2(g) + H_2O(\ell)$
This is a gas-forming reaction where Cl^- is a spectator ion.

4.23 (a) $(NH_4)_2S(aq) + Hg(NO_3)_2(aq) \rightarrow HgS(s) + 2\ NH_4NO_3(aq)$
(b) ammonium sulfide, mercury(II) nitrate, mercury(II) sulfide, ammonium nitrate
(c) precipitation reaction

4.24 (a) $MnCl_2(aq) + Na_2S(aq) \rightarrow MnS(s) + 2\ NaCl(aq)$
$Mn^{2+}(aq) + S^{2-}(aq) \rightarrow MnS(s)$; precipitation reaction
(b) $K_2CO_3(aq) + ZnCl_2(aq) \rightarrow ZnCO_3(s) + 2\ KCl(aq)$

$CO_3^{2-}(aq) + Zn^{2+}(aq) \rightarrow ZnCO_3(s)$; precipitation reaction

(c) $K_2CO_3(aq) + 2\ HClO_4(aq) \rightarrow 2\ KClO_4(aq) + CO_2(g) + H_2O(\ell)$

$CO_3^{2-}(aq) + 2\ H^+(aq) \rightarrow CO_2(g) + H_2O(\ell)$; gas-forming reaction

.25 C is oxidized and $C_6H_8O_6$ is the reducing agent. Br_2 is reduced and Br_2 is the oxidizing agent.

.26 Precipitation reaction: $BaCl_2(aq) + Na_2SO_4(aq) \rightarrow BaSO_4(s) + 2\ NaCl(aq)$

Gas-forming reaction: $BaCO_3(s) + H_2SO_4(aq) \rightarrow BaSO_4(s) + CO_2(g) + H_2O(\ell)$

.1 4.5 mol O_2; 310 g Al_2O_3

.2 5.64 g $CoCl_2$; 0.0876 g H_2

.3 1.1 mol O_2; 35 g O_2

.4 (a) 318 g Fe; (b) 239 g CO

.5 (a) Cl_2 is limiting reactant (b) 5.08 g $AlCl_3$ (c) 1.67 g Al left unreacted

.6 CO is the limiting reactant. 1.3 g of H_2 will be left unreacted. Theoretical yield of CH_3OH is 85.2 g.

.7 CaO is the limiting reactant; the maximum possible yield of NH_3 is 68.0 g; 10.3 g of NH_4Cl will be left unreacted.

.8 73.5%

.9 Theoretical yield is 74.0 g; 88.1%

.10 1.14 g $CuSO_4 \cdot 5H_2O$; 91.6%

.11 83.9% $Al(C_6H_5)_3$

.12 Empirical formula is CH

.13 Empirical formula is $C_3H_6O_2$

.14 Empirical formula is SiH_4

.15 0.254 M Na_2CO_3; 0.508 M Na^+; 0.254 M CO_3^{2-}

.16 0.494 g $KMnO_4$

.17 5.08×10^3 mL

.18 0.0100 M

.19 (a) 0.0450 M H_2SO_4 (b) 0.125 M H_2SO_4

(c) 0.150 M H_2SO_4 (d) 0.250 M H_2SO_4

.20 (a) 0.12 M Ba^{2+}; 0.24 M Cl^- (b) 0.0125 M Cu^{2+}; 0.0125 M SO_4^{2-}

(c) 0.146 M Al^{3+}; 0.438 M Cl^- (d) 1.000 M K^+; 0.500 M $Cr_2O_7^{2-}$

.21 0.205 g Na_2CO_3

.22 60. g NaOH; 53 g Cl_2

.23 193 mL

.24 40.9 mL

.25 42.5 mL

.26 0.0219 g citric acid/100. mL

6.1 6-Pack has potential energy, which is converted to kinetic energy as it falls. When striking the ground the kinetic energy is converted to heat.

6.2 (a) ΔH°_f refers to formation of 1 mol of product. (b) ΔH°_f for O_2 is 0. (c) q has a negative value, and is –3330 J for 10.0 g of ice.

6.3 399 Calories

6.4 0.235 J/g · K

6.5 Ethylene glycol requires more heat (1.51×10^4 J).

6.6 104 kJ

6.7 6.2 °C

6.8 0.24 J/g · K

6.9 905 kJ

6.10 48.2 kJ

6.11 Exothermic; 2.38 kJ evolved.

6.12 Exothermic; 3.29×10^4 kJ evolved.

6.13 ΔH°_{rxn} = 619 kJ; molar heat of combustion = 310 kJ.

6.14 –434.6 kJ; 260 kJ of heat evolved from 250 g of lead.

6.15 $2\ Cr(s) + 3/2\ O_2(g) \rightarrow Cr_2O_3(s)$

6.16 Reaction is exothermic; ΔH°_{rxn} = –98.89 kJ

6.17 Reaction is exothermic; ΔH°_{rxn} = –905.2 kJ; ΔH°_{rxn} = –133 kJ for 10.0 g NH_3

6.18 6.62 kJ evolved

6.19 394 kJ evolved per mol C

6.20 ΔH°_f = –56 kJ/mol

7.1 (a) red, orange, yellow (b) blue (c) blue

7.2 $6.0 \times 10^{14}\ s^{-1}$; 4.0×10^{-19} J/photon; 2.4×10^5 J/mol

7.3 $7.5677 \times 10^{14}\ s^{-1}$; 5.0143×10^{-19} J/photon; 3.0197×10^5 J/mol

7.4 (a) 837.761 nm (b) $3.46407 \times 10^{-14}\ s^{-1}$, E = 2.29532×10^{-19} J

7.5 Increasing energy: radar < microwaves < red light < UV < γ-rays

7.6 (a) n = 3 to n = 2 (b) n = 4 to n = 1

7.7 The atom must absorb 2.093×10^{-18} J.

7.8 The 4p orbital is larger. The 4p orbital has one planar node; the 2s orbital has none.

7.9 (a) ℓ = 0, 1, 2, 3 (b) m_ℓ = –2, –1, 0, 1, 2 (c) n = 4, ℓ = 0, m_ℓ = 0; (d) n = 4, ℓ = 3, m_ℓ = –3, –2, –1, 0, 1, 2, 3

7.10

n	ℓ	m_ℓ
4	1	–1
4	1	0
4	1	1

7.11 4 subshells; 4s, 4p, 4d, 4f

7.12 (a) 7 (b) 25 (c) None; ℓ cannot have a value equal to n (d) 1

7.13 (a) size and energy; shape (b) 0, 1, 2 (d) 4; 2; –2 (e) f (f) 2d, 4g
(g) n = 2, ℓ = 1, m_ℓ = 2 is not valid (h) 3, 9, none, 1

7.14 Bohr's model has the electron located in a definite orbit.

7.15 (a) green light (b) 680 nm corresponds to red light (c) green light

7.16 n, orbital size; m_ℓ, relative orbital orientation; ℓ, orbital shape

7.17 (a) Group 7 B, period 5 (b) n = 5, ℓ = 0, m_ℓ = 0
(c) $3.41 \times 10^{19}\ s^{-1}$; 8.79×10^{-12} m
(d) $HTcO_4(aq) + NaOH(aq) \rightarrow NaTcO_4(aq) + H_2O(\ell)$
8.5×10^{-3} g $NaTcO_4$; 1.8×10^{-3} g NaOH

8.1 Mg $1s^22s^22p^63s^2$

[↑↓] [↑↓] [↑↓|↑↓|↑↓] [↑↓]
1s 2s 2p 3s

Cl $1s^22s^22p^63s^23p^5$

[↑↓] [↑↓] [↑↓|↑↓|↑↓] [↑↓] [↑↓|↑↓|↑]
1s 2s 2p 3s 3p

8.2 V $1s^22s^22p^63s^23p^63d^34s^2$

8.3 Ge $1s^22s^22p^63s^23p^63d^{10}4s^24p^2$
$[Ar]3d^{10}4s^24p^2$

8.4 (a) Sr $[Kr]5s^2$
(b) Zr $[Kr]4d^25s^2$
(c) Rh $[Kr]4d^85s^1$
(d) Sn $[Kr]4d^{10}5s^25p^2$

8.5 (a) Na^+

[↑↓] [↑↓] [↑↓|↑↓|↑↓]
1s 2s 2p

(b) Al^{3+}

[↑↓] [↑↓] [↑↓|↑↓|↑↓]
1s 2s 2p

(c) Cl^-

[↑↓] [↑↓] [↑↓|↑↓|↑↓] [↑↓] [↑↓|↑↓|↑↓]
1s 2s 2p 3s 3p

8.6 (a) Ti

[Ar] [↑ |↑ | | |] [↑↓]
3d 4s

(b) Ti^{2+} The titanium(II) ion is paramagnetic.

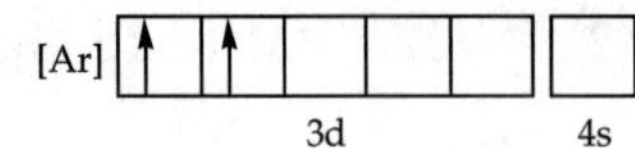

(c) Ti^{4+}, no electrons remain in 3d or 4s orbitals.

8.7 (a) Manganese

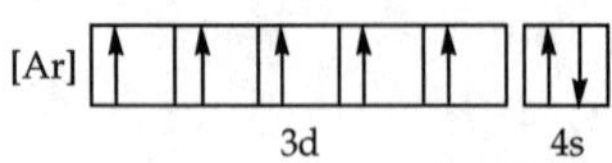

(b) Mn^{2+} ion

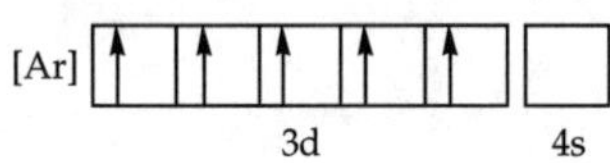

(c) There are 5 unpaired electrons in Mn^{2+}, so the ion is paramagnetic.

8.8 Magnesium

[Ne] ↑↓ (3s) $n = 3, \ell = 0, m_\ell = 0, m_s = +1/2$

$n = 3, \ell = 0, m_\ell = 0, m_s = -1/2$

8.9 (a) 6
(b) 18
(c) none; ℓ cannot have a value equal to n
(d) 1
(e) none; if $\ell = 0$, $m_\ell = 0$

8.10 C < B < Al < Na < K

8.11 K < Li < C < N

8.12 (a) K, (b) C, (c) K < Li < C < N

8.13 (a) S < O < F — IE increases moving up a group and to the right across a period.
(b) O — IE increases moving up a group
(c) Cl — The affinity for an electron increases across a period and up a group
(d) O^{2-} — Anions are larger than their neutral atoms; F^- has 10 electrons and 9 protons while O^{2-} has 10 electrons and only 8 protons.

8.14 ΔH°_f (LiCl) = –400. kJ (calculated)

8.15 Element 109 $[Rn]5f^{14}6d^7 7s^2$
Other elements in the same group are Co, Rh, Ir

8.16 (a) P (b) Be (c) N (d) Mn (e) Cl (f) Zn

8.17 (a) alkaline earth metal (b) nonmetal (halogen)
(c) element B (d) element B

8.18 (a) S (b) Cl^- (c) Na (d) O (e) N^{3-}

8.19 (a) Ti (b) Group 4B, period 4 (c) Transition element
(d) Paramagnetic, 2 unpaired electrons
(e)

Electron	n	ℓ	m_ℓ	m_s
1	3	2	−2	+1/2
2	3	2	−1	+1/2
4	4	0	0	−1/2

(f) Electrons 3 and 4 are removed; the ion is paramagnetic.

8.20 Ionization energies normally increase across the periodic table. Here the ionization energy of S is lower than expected because of electron-electron repulsions between the electrons paired in a 3p orbital.

8.21 Co^{2+}, [Ar]$3d^7$, has 3 unpaired electrons whereas Co^{3+}, [Ar]$3d^6$, has 4 unpaired electrons. The two ions have different numbers of unpaired electrons, so $Co(NO_3)_2$ would be more paramagnetic than $CoCl_2$. The magnetic moment of the reaction product could be measured to determine the identity of the product.

9.1

Atom	Group number	Number of valence electrons
N	5A	5
B	3A	3
Na	1A	1
Mg	2A	2
F	7A	7
S	6A	6

9.2

(a) :F̤—N̈—F̤: with :F̤: bonded below N

(b) [:Ö—C̈l—Ö:]⁻ with :Ö: bonded below Cl

(c) H—Ö—B̈r:

(d) [:Ö—S̈—Ö:]$^{2-}$ with :Ö: bonded below S

9.3

(a) H—C—C̈l: with :F̈: above C and :F̤: below C

(b) H—C—Ö—H with :O: double-bonded above C

(c) H—C—C≡N: with H above and H below the first C

(d) H—C—Ö—H with H above and H below C

9.4

(a)  [Lewis structures: resonance forms of SO_2]

(b) 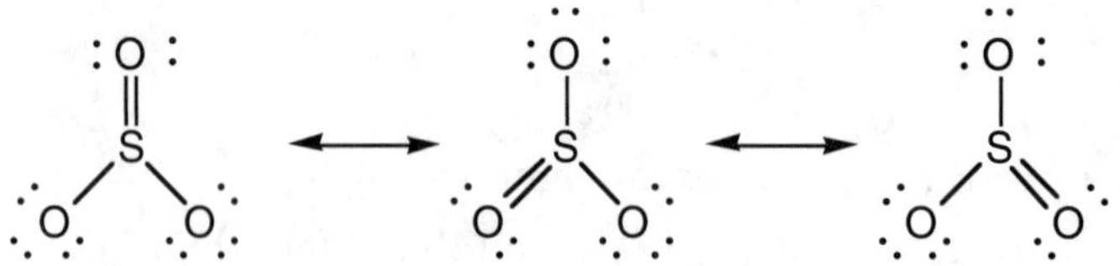 [Lewis structures: resonance forms of SO_3]

(c) 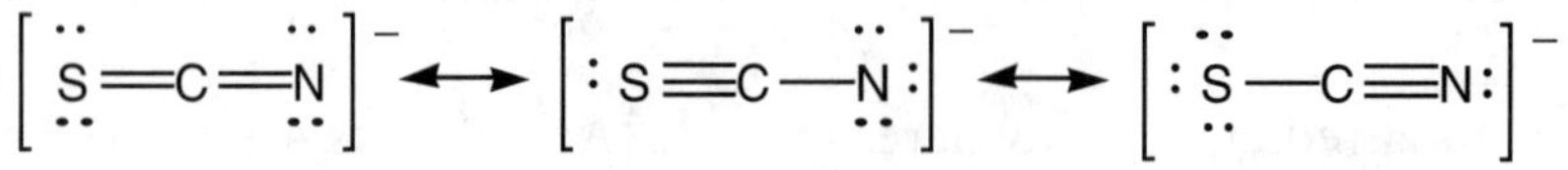 [Lewis structures: resonance forms of SCN^-]

9.5 (a) H_2CO One C=O bond, bond order = 2
Two C—H bonds, bond order = 1
(b) SO_3^{2-} Three S—O bonds, bond order = 1
(c) NO_2^+ Two N=O bonds, bond order = 2

9.6 (a) B—Cl is shorter (b) C—O is shorter
(c) P—O is shorter (d) C=O is shorter

9.7 The carbon monoxide CO triple bond is shorter and stronger than the formaldehyde C=O double bond.

9.8 The average bond order in NO_2^+ is 2, whereas it is only 1.33 in NO_3^-. Therefore, the NO bonds in NO_3^- are longer than in NO_2^+.

9.9 Enthalpy of hydrogenation = –128 kJ

9.10 $D_{O—F}$ = 195 kJ/mol

9.11 (a) The C—H and C=O bonds are polar, while the C=C and C—C bonds are nonpolar.
(b) The most polar bond is C=O. The O atom is the negative end of the bond dipole.

9.12

[Lewis structures: resonance forms of NO_2^-]

0 0 -1 -1 0 0

9.13

H—N—Cl: (with H below N)
electron-pair geometry = tetrahedral
molecular geometry = trigonal pyramid

:O—Cl: (with :Cl: below O)
electron-pair geometry = tetrahedral
molecular geometry = bent

[S=C=N]⁻
electron-pair geometry = linear
molecular geometry = linear

:O—F: (with H below O)
electron-pair geometry = tetrahedral
molecular geometry = bent or angular

9.14 As seen below, 16-electron molecules or ions are linear, 18-electron molecules or ions have a trigonal planar electron-pair geometry, and 20-electron molecules or ion have a tetrahedral electron-pair geometry.

O=C=O
electron-pair geometry = linear
molecular geometry = linear

[O=N—O:]⁻
electron-pair geometry = trigonal planar
molecular geometry = bent or angular

O=O—O:
electron-pair geometry = trigonal planar
molecular geometry = bent or angular

[:O—Cl—O:]⁻
electron-pair geometry = tetrahedral
molecular geometry = bent or angular

O=S—O:
electron-pair geometry = trigonal planar
molecular geometry = bent or angular

9.15 1 = 120°, 2 = 109°, 3 = 120°

9.16 (i) H_2O has the most polar bonds.
(ii) CO_2 and CCl_4 are not polar
(iii) In ClF the F atom is negatively charged.

9.17

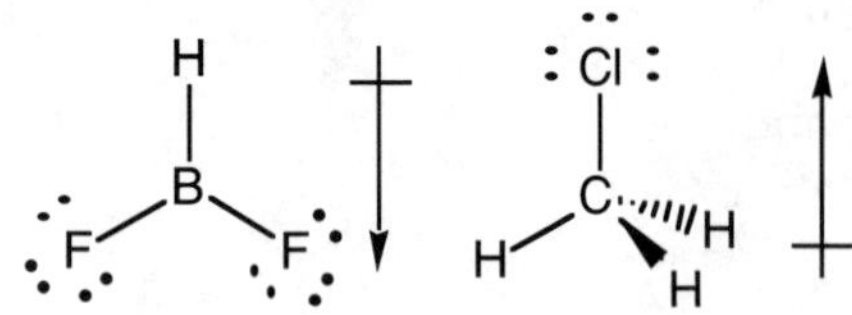

9.18 The N—O bonds in NO_2^- have a bond order of 1.5 while in NO_2^+ the bond order is 2. The shorter bonds (110 pm) are the N—O bonds with the higher bond order (2) while the N—O bonds with a bond order of 1.5 are longer (124 pm).

9.19 (a) 1 = 120°, 2 = 180°
(b) The C=C bond is shorter than the C—C bonds.
(c) The C=C bond is shorter than the C—C bonds.
(d) The CN triple bond is most polar and N is the negative end of the bond dipole.

10.1 Electron-pair geometry for OF_2 is tetrahedral; molecular geometry is bent. The O atom is sp^3 hybridized and the O—F bonds form by overlap of an O sp^3 orbital with an F p orbital.

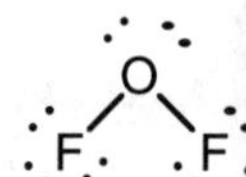

10.2 (a) sp^2 (b) sp^2 (c) sp^3 (d) sp^3
The addition of an O atom to SO_2 or SO_3^{2-} does not change the hybridization of the S atom. The addition of electrons to SO_3 to form SO_3^{2-} causes the S atom hybridization to change from sp^2 to sp^3.

10.3 The N atom is sp^2 hybridized. The N=O bond consists of a s bond resulting from overlap of a N sp^2 hybrid orbital with an O sp^2 hybrid orbital, and a p bond formed from overlap of unhybridized 2p orbitals on N and O.

10.4 (a) Carbon 1: sp^2 Carbon 2: sp^2
(b) A = 120°, B = 120°, and C = 120°

10.5 (a) 1 p bond and 11 s bonds.
(b) C1 = sp^3, C2 = sp^2, O3 = sp^3
(c) The O—H bond attached to the center carbon is shorter.
(d) A = 109°, B = 109°, C = 120°

10.6 H_2^+, $(\sigma_{1s})^1(\sigma^*_{1s})^0$. Bond order = 1/2. The H—H bond in H_2^+ is weaker than in H_2 (where the bond order is 1).

10.7 The C_2^{2-} ion has 1 σ bond and 2 π bonds for a carbon-carbon bond order of 3. On adding two electrons to C_2 the bond order increases by 1. The C_2^{2-} ion is diamagnetic.

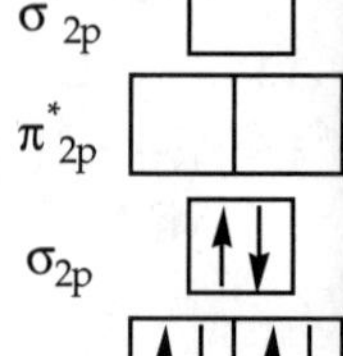

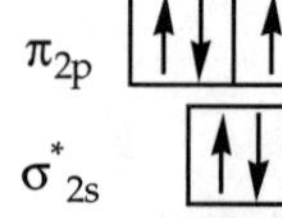

C_2^{2-} MO diagram.

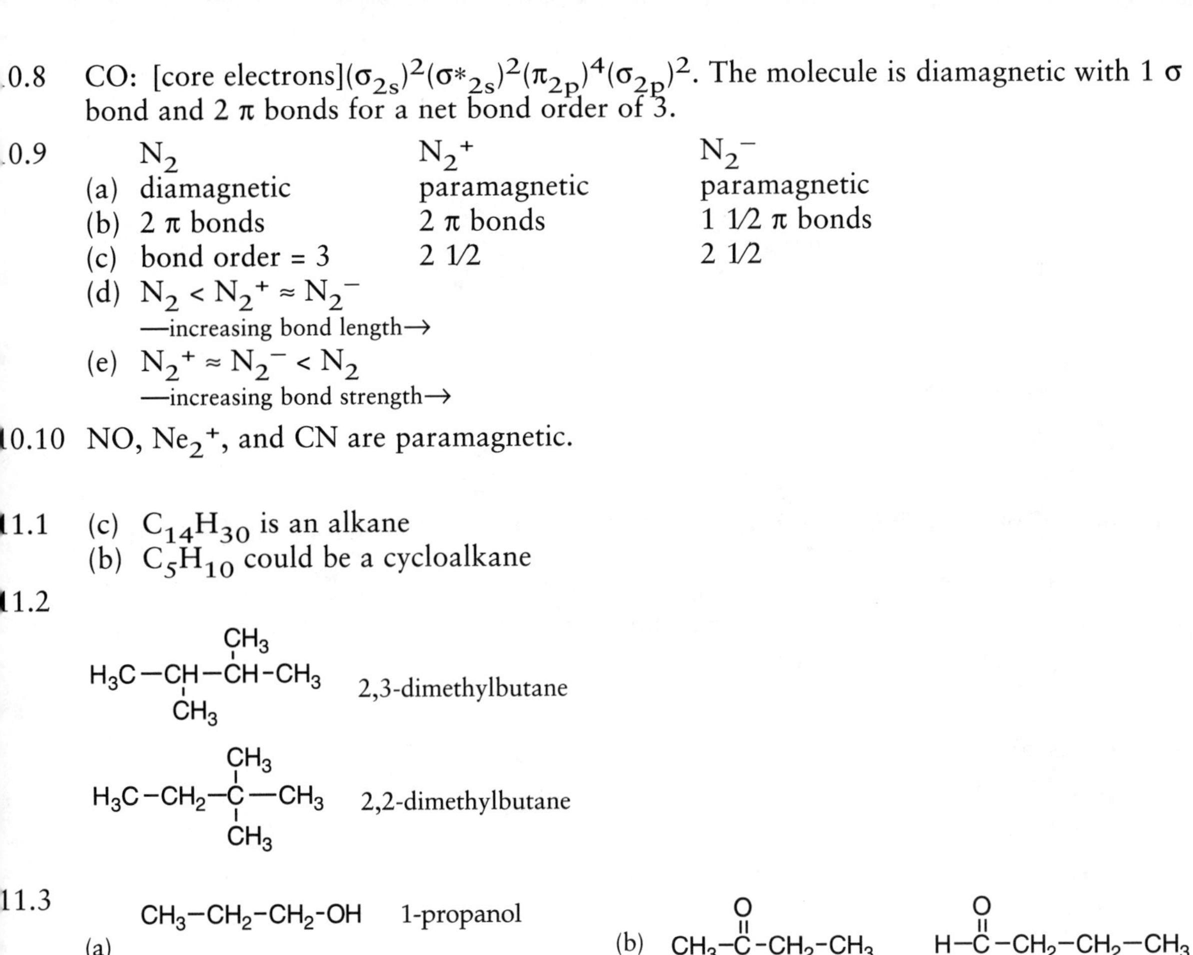

10.8 CO: [core electrons]$(\sigma_{2s})^2(\sigma^*_{2s})^2(\pi_{2p})^4(\sigma_{2p})^2$. The molecule is diamagnetic with 1 σ bond and 2 π bonds for a net bond order of 3.

10.9

	N_2	N_2^+	N_2^-
(a)	diamagnetic	paramagnetic	paramagnetic
(b)	2 π bonds	2 π bonds	1 1/2 π bonds
(c)	bond order = 3	2 1/2	2 1/2

(d) $N_2 < N_2^+ \approx N_2^-$
—increasing bond length→

(e) $N_2^+ \approx N_2^- < N_2$
—increasing bond strength→

10.10 NO, Ne_2^+, and CN are paramagnetic.

11.1 (c) $C_{14}H_{30}$ is an alkane
(b) C_5H_{10} could be a cycloalkane

11.2

$H_3C-CH(CH_3)-CH(CH_3)-CH_3$ 2,3-dimethylbutane

$H_3C-CH_2-C(CH_3)_2-CH_3$ 2,2-dimethylbutane

11.3

(a)
$CH_3-CH_2-CH_2-OH$ 1-propanol

$CH_3-CH(OH)-CH_3$ 2-propanol

$CH_3-O-CH_2-CH_3$ methylethyl ether

(b)
$CH_3-C(=O)-CH_2-CH_3$ butanone

$H-C(=O)-CH_2-CH_2-CH_3$ butanal

11.4

(a) $H_3C-CH_2-CH=CH_2 + Br_2 \longrightarrow H_3C-CH_2-CH(Br)-CH_2(Br)$

1-butene → 1,2-dibromobutane

(b) $CH\equiv C-CH_3 + H_2 \xrightarrow{\text{catalyst}} CH_2=CH-CH_3$

propyne → propene

(c) $CH\equiv C-CH_3 + 2\,H_2 \xrightarrow{\text{catalyst}} CH_3-CH_2-CH_3$

propyne → propane

(d) $H_3C-CH=CH-CH_3 + H_2O \longrightarrow H_3C-CH_2-CH(OH)-CH_3$

2-butene → 2-butanol

11.5 The alkene should react with elemental bromine to give colorless products. The cyclopentane would not react with elemental bromine.

12.1 16.9 mm Hg

12.2 3.7 L

12.3 0.50 L O_2 gas

12.4 0.811 atm

12.5 2.9 L

12.6 0.337 mol N_2

12.7 57.5 g/mol

12.8 (a) Empirical formula is C_2H_5
(b) Molar mass = 58.2 g/mol
(c) Molecular formula = C_4H_{10}

12.9 $d = 3.7 \times 10^{-4}$ g/L

12.10 10.2 g O_2

12.11 22.2 mm Hg

12.12 58 g NaN_3

12.13 1.7 atm of O_2 gas

12.14 $Fe(CO)_5$

12.15 0.047 g H_2

12.16 (a) 2.0 g He
(b) 0.54 atm O_2

12.17 $P(Cl_2)$ = 250 mm Hg

12.18 (a) The average kinetic energies of the gases are the same because they depend only on temperature.
(b) H_2 molecules are moving 4.7 times faster, on average, than CO_2 molecules.
(c) Because T and V are the same for both gases, the number of molecules is proportional to P. Therefore, there are twice as many CO_2 molecules as H_2 molecules.
(d) Because there are more molecules present of CO_2 than of H_2, the mass of CO_2 is greater than the mass of H_2.

12.19 $CH_2F_2 < Ar < N_2 < CH_4$

13.1 (a) Dipole-dipole forces (and hydrogen bonds) in water.
(b) Ion-ion forces in solid KCl.
(c) Ion-dipole forces assist in dissolving the KCl in water.

13.2 (a) Induced dipole-induced dipole (b) induced dipole-induced dipole
(c) dipole-dipole (d) dipole-dipole (and H-bonds)

13.3 No, there is no significant hydrogen bonding in H_2S.

13.4 (c) HF (b) CH_3CO_2H (f) CH_3OH

13.5 Propane can be liquefied as long as T does not exceed T_c, 96.7 °C.

13.6 Assume that black = A and white = B. The following pattern would lead to a formula of AB because there is a net of 2 A (= 1 + 4 · 1/4) and 2 B (= 4 · 1/2) inside the dashed line.

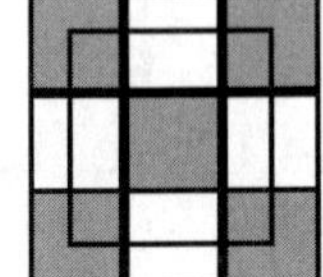

13.7 Acetone is a polar molecule. It can interact with the dipole of water. In addition, the electronegative O atom can be involved in H-bonding with water.

13.8 $HOCH_2CH_2OH$ will be more viscous than C_2H_5OH because the glycol has more sites for hydrogen bonding.

13.9 (a) Ethanol does not have as great a capacity for H-bonding as water does. Ethanol has only one O—H bond, whereas H_2O has two.
(b) A possible structure for the ion involves H-bonding between HF and an F^- ion. $[:\ddot{F}-H-\ddot{F}:]^-$
(c) A mixture of ethanol and water involves strong H-bonding between the molecules. It is stronger than in ethanol alone and causes the volume to decrease.

13.10 A unit cell of the NaCl can only have a 1:1 cation to anion ratio.

13.11 There are 8 Na^+ ions in the unit cell and 4 O^{2-} ions. (The O^{2-} ions are arranged as a face-centered cube.) Thus, the ratio is 8 Na^+ to 4 O^{2-} ions, giving a formula of Na_2O.

14.1 (a) Assume total volume is 0.500 L, so molarity is 0.0382 M.
(b) 0.0382 molal
(c) mole fraction malic acid = 6.88×10^{-4}
(d) 0.509%-

14.2 (a) 4.25 g $NaNO_3$; mole fraction $NaNO_3$ = 0.00359

14.3 16.2 molal; 37.1%

14.4 Option (c)

14.5 1.13×10^{-3} mm Hg

14.6 Vapor pressure of water over the solution = 35.0 mm Hg

14.7 1.04×10^3 g glycol

14.8 100. g/mol

14.9 100.456 °C

14.10 62.51 °C

14.11 180 g/mol

14.12 Molar mass = 180 g/mol; molecular formula = $C_{14}H_{10}$

14.13 (a) 8.60 mol/kg (b) 28.4%

14.14 170 g/mol

14.15 (a) −0.3 °C
(b) 100.10 °C
(c) Π = 4.58 atm. The osmotic pressure is the easiest to measure in the laboratory.

14.16 6.0×10^3 g/mol

14.17 (a) $BaCl_2(aq) + Na_2SO_4(aq) \rightarrow BaSO_4(s) + 2\ NaCl(aq)$
(b) Small particles of insoluble $BaSO_4$ are formed rapidly. The $BaSO_4$ particles can accumulate a charge, form a colloid, and resist forming larger particles that can precipitate.
(c) As the $BaSO_4$ particles "age," they lose their charge, form larger particles, and precipitate.

14.18 55.3 M; 55.5 molal

14.19 404 g/mol-

14.20 The -OH group in both CH_3OH and C_2H_5OH forms hydrogen bonds with water, and this overcomes the lack of interaction of the CH-containing group. In the case of alcohols with long carbon chains, however, the hydrocarbon chains are hydrophobic and dominate the interaction with water, making such alcohols only poorly soluble.

15.1 (a) Rate for 0 to 10 s = 0.167 mol/L · s
Rate for 10 to 20 s = 0.119 mol/L · s
Rate for 20 to 30 s = 0.0089 mol/L · s
Rate for 30 to 40 s = 0.0070 mol/L · s
The rate decreases because there is less A available in each successive time period.
(b) Rate of appearance of B is twice the rate of disappearance of A.
(c) Rate when [A] = 0.75 mol/L is 0.0113 mol/L · s

15.2 (a) Rate = k $[NO_2][O_3]$
(b) Tripling $[NO_2]$ triples the reaction rate.
(c) Halving $[O_3]$ halves the reaction rate.

15.3 (a) Rate = k $[NO]^2 [O_2]$
(b) k = 25 $L^2/mol^2 \cdot s$
(c) 1.3×10^{-3} mol/L · s
(d) NO is disappearing at a rate of 10.0×10^{-4} mol/L · s and NO_2 is appearing at the same rate.

15.4 5.57×10^{-3} min^{-1} or 9.28×10^{-5} sec-1

15.5 2.94 seconds

15.6 (a) $t_{1/2} = 2.08 \times 10^5$ h (b) 4.8×10^5 hr

15.7 75.8 s

15.8

	P(total)	P(HOF)
After 30. min	125 mm Hg	50. mm Hg
After 45 min	132 mm Hg	35 mm Hg

15.9 (a) 3.0% of the original amount (b) 0.14% of the original amount

15.10 (a) A graph of ln [sucrose] versus time is a straight line so the reaction is first order in sucrose.
(b) Rate = k [sucrose] where k = 3.68×10^{-3} min^{-1} (as calculated from the slope of the ln [sucrose] versus time plot).
(c) [sucrose] = 0.166 mol/L

15.11 (a) A graph of ln $[N_2O]$ versus time is a straight line so the reaction is first order in N_2O.
(b) k = 1.28×10^{-2} min^{-1}, so Rate = 4.5×10^{-4} mol/L · min

15.12 Reaction is exothermic, releasing 26 kJ/mol

15.13 (a) 2nd step (the slow step)
(b) Rate = k $[O_3][O]$
(c) Step 1 is unimolecular and step 2 is bimolecular

15.14 (a) 0.59 mg of NO_x (b) 75 hr

15.15 $2\ O_3(g) \rightarrow 3\ O_2(g)$. Ozone decomposes to oxygen molecules, resulting in a net loss of ozone in the stratosphere. Cl is a catalyst, and ClO is an intermediate.

15.16 19 hours

16.1 (a) $K_c = [H_2O]^2[O_2] / [H_2O_2]^2$
(b) $K_c = [PCl_5] / [PCl_3][Cl_2]$
(c) $K_c = [CO_2] / [CO][O_2]^{1/2}$
(d) $K_c = [CO]^2 / [CO_2]$
(e) $K_c = [CO_2] / [CO]$

16.2 $K_2 = 1/K_1^{\ 2}$

16.3 K = (1.6)(1.5) = 2.4

16.4 Q = 5.0×10^{-4}, so Q < K_c, and the reaction shifts toward products.

16.5 $K_c = 279$

16.6 (a) $K_c = 1.6$
(b) Mol CO = mol H_2O = 0.064 mol

16.7 $K_c = 0.035$

16.8 [isobutane] = 0.024 M and [butane] = 0.010 M

16.9 $[CO] = [Br_2] = 0.034$ M

16.10 0.0730 mol NO_2; 16.7% dissociated

16.11 [I] = 0.0424 M and $[I_2]$ = 0.479 M

16.12 (a) left (b) left (c) left (d) right

16.13 $K_c = 0.089$

17.1 (a) CN^-, cyanide ion
(b) SO_4^{2-}, sulfate ion
(c) F^-, fluoride ion
(d) NO_2^-, nitrite ion
(e) CO_3^{2-}, carbonate ion

17.2 (a) Products: H_3O^+ and NO_3^-
Acid is HNO_3 and conjugate base is NO_3^-
Base is H_2O and conjugate acid is H_3O^+
(b) Products: H_3O^+ and SO_4^{2-}
Acid is HSO_4^- and conjugate base is SO_4^{2-}
Base is H_2O and conjugate acid is H_3O^+
(c) Products: H_2O and HF
Acid is H_3O^+ and conjugate base is H_2O
Base is F^- and conjugate acid is HF

17.3 $CO_3^{2-}(aq) + H_2O(\ell) \rightarrow HCO_3^-(aq) + OH^-(aq)$

17.4

	Acid (A)	Base (B)	Conjugate of A	Conjugate of B
(a)	HCO_2H	H_2O	HCO_2^-	H_3O^+
(b)	H_2S	NH_3	HS^-	NH_4^+
(c)	HSO_4^-	OH^-	SO_4^{2-}	H_2O

17.5 (a) HF is the strongest acid and NH_4^+ is the weakest. (b) F^- (c) HF (d) NH_4^+

17.6 HClO has the strongest conjugate base because HClO is the weakest acid of the group. See the ionization constant table, Screen 17.6

17.7 Wine is acidic; $[H_3O^+] = 4.0 \times 10^{-4}$ M

17.8 pH = 2.89; $[OH^-] = 7.7 \times 10^{-12}$ M

17.9 pH = 11.48

17.10 (a) 1.6×10^{-4} M (b) moderately weak

17.11 $K_b = 6.6 \times 10^{-9}$

17.12 $[H_3O^+] = [A^-] = 1.3 \times 10^{-5}$ M; [HA] = 0.040 M

17.13 $[H_3O^+] = [CN^-] = 3.2 \times 10^{-6}$ M; [HCN] = 0.025 M; pH = 5.50

17.14 (a) 4-chlorobenzoic acid is the stronger acid.
(b) Benzoic acid, the weaker acid, has the higher pH.

17.15 Highest pH, Na_2S. Lowest pH, $AlCl_3$. ($AlCl_3$ gives the $Al(H_2O)_6^{3+}$ ion in solution, and this is a stronger acid than the $H_2PO_4^-$ ion.)

17.16 $[OH^-] = [HCN] = 3.3 \times 10^{-3}$ M; $[Na^+] = 0.441$ M; $[H_3O^+] = 3.0 \times 10^{-12}$ M; pH = 11.52.

17.17 (a) Lewis acid (b) Lewis base (c) Lewis base (d) Lewis base (e) Lewis acid

17.18 (a) CH_3CO_2H, NH_4Cl
(b) NH_3, Na_2CO_3, $NaCH_3CO_2$
(c) CH_3CO_2H

17.19 $HCl < CH_3CO_2H < NaCl < NH_3 < NaCN < NaOH$

18.1 $[H_3O^+] = 1.89 \times 10^{-9}$ M and pH = 8.72

18.2 (a) greater than 7 (b) less than 7 (c) equal to 7

18.3 (a) 25 mL of 0.45 M H_2SO_4 with 25 mL of 0.90 M NaOH
All of the acid and base are consumed to leave a solution of Na_2SO_4. Because SO_4^{2-} is a very weak base, the solution will be neutral to very slightly basic. pH ≥ 7
(b) 15 mL of 0.050 M formic acid with 15 mL of 0.050 M NaOH
The titration produces a solution of Na^+ and HCO_2^-. Because the latter is the conjugate base of a weak acid, the solution will be basic. pH > 7
(c) 25 mL of 0.15 M $H_2C_2O_4$ (oxalic acid) with 25 mL of 0.30 M NaOH
The titration produces a solution of Na^+ and $C_2O_4^{2-}$. The latter is the conjugate base of a weak acid, so the solution will be basic. pH > 7

18.4 (a) decrease pH (b) increase pH (c) pH stays the same

18.5 (a) decreases pH (b) increases pH

18.6 pH = 9.25

18.7 4.8 g

18.8 (a) pH = 3.90
(b) Diluting the solution with 500 mL of pure water does not change the pH.

18.9 $HCl < CH_3CO_2H < NH_4Cl < CH_3CO_2H/NaCH_3CO_2 < NaCl < NH_3$
—increasing pH→

18.10 $[H_2PO_4^-]/[HPO_4^{2-}] = 0.65$

18.11 (a) HB is the stronger acid. The pH at the equivalence point is controlled by the conjugate base of the acid. The stronger the acid, the weaker the conjugate base, and the lower the pH.
(b) A^- is the stronger conjugate base.

18.12 When pH = 7.4, $[H_3O^+] = 3.98 \times 10^{-8}$ M
$[H_3O^+] = 3.98 \times 10^{-8} = \{[\text{lactic acid}] / [\text{lactate ion}]\}K_a$
Because $K_a = 1.4 \times 10^{-4}$, the only way the above can be an equality is if [lactate ion] >> [lactic acid]. That is, the lactate ion predominates.

18.13 (a) The pH at the beginning of the titration is about 11 (calculated value is 11.12), appropriate for a weak base with $K_b = 1.8 \times 10^{-5}$.

(b) The pK_b is equivalent to the pH at the midpoint in the titration. Here pK_b is about 9.3. (The calculated value is 9.26.)

(c) At the equivalence point the solution contains NH_4^+ and Cl^- ions. The solution is acidic owing to the weak acid, NH_4^+.

(d) pH at the equivalence point is about 5.5. (Computed value = 5.20.)

(e) The curves differ because in the acetic acid/NaOH titration one begins with a weak acid. This means that the pH begins at a low value and increases as the titration proceeds.

18.14 (a) When [acid] = [conjugate base], $[H_3O^+] = K_a$, and $pH = pK_a$. This occurs when the lines cross in an alpha plot. Here this point occurs at pH of about 3.8. (Calculated value = 3.75.)

(b) A pH of 4.00 occurs when formic acid represents about 36% of the species in solution and the formate ion is about 64%. This means we could make a pH = 4.00 buffer by mixing 0.36 mol of formic acid with 0.64 mol of a formate salt.

18.15 (a) Initial pH is 2.76.

(b) pH at the midpoint is about 4.2, in agreement with the value calculated from K_a.

(c) The calculated pK_a is 4.2, which is also the value of the pH at the midpoint in the titration.

(d) pH at the equivalence point = 8.2.

19.1 $K_{sp} = 3.2 \times 10^{-6}$

19.2 $K_{sp} = 7.9 \times 10^{-6}$

19.3 $K_{sp} = 8.1 \times 10^{-12}$

19.4 $K_{sp} = 7.9 \times 10^{-6}$

19.5 (a) 1.1×10^{-8} M (b) 1.5×10^{-6} g/L

19.6 (a) AgSCN (b) $SrSO_4$ (c) MgF_2 (d) AgI

19.7 $BaCO_3$ (Solubility = 9.0×10^{-5} M) < Ag_2CO_3 (Solubility = 1.3×10^{-4} M) < BaF_2 (Solubility = 7.5×10^{-3} M)

19.8 (a) $Q < K_{sp}$, no precipitate (b) $Q > K_{sp}$, $NiCO_3$ precipitates.-

19.9 $[OH^-] = 1.6 \times 10^{-5}$ M

19.10 Q (= 1.7×10^{-8}) < K_{sp}, so no precipitate of $PbCl_2$ forms.

19.11 1.0×10^{-6} M in pure water, but only 1.0×10^{-10} M in 0.010 M NaSCN.

19.12 (a) $PbCO_3$ precipitates first. (b) $[CO_3^{2-}] = 1.7 \times 10^{-7}$ M

19.13 $AgCl(aq) \rightleftharpoons Ag^+(aq) + Cl^-(aq)$ $K = K_{sp}$ for AgCl = 1.8×10^{-10}

$Ag^+(aq) + I^-(aq) \rightleftharpoons AgI(s)$ $K = 1/K_{sp}$ for AgI = $1/1.5 \times 10^{-16}$

$AgCl(s) + I^-(aq) \rightleftharpoons AgI(s) + Cl^-(aq)$ $K_{net} = 1.2 \times 10^6$

Yes, it is possible to precipitate AgI by adding I^- ion to a precipitate of AgCl because K_{net} is much greater than 1.

19.14 If the pH is decreased (by adding acid), the OH- ions that are produced by any $Ni(OH)_2$ that dissolves are consumed. This removes a product of the equilibrium

$$Ni(OH)_2(s) \rightleftharpoons Ni^{2+}(aq) + 2\ OH^-(aq)$$

causing the equilibrium to shift to the right, dissolving more $Ni(OH)_2$.

19.15 (a) Increasing the CO_2 pressure should increase the $BaCO_3$ solubility because the equilibrium is shifted to the right.

(b) Decreasing the pH (by adding acid) should increase the $BaCO_3$ solubility.

20.1 (a) CO_2 vapor (b) dissolved sugar (c) alcohol/water mixture

20.2 (a) $AlCl_3$ (b) CH_3CH_2I (c) $NH_4Cl(aq)$

20.3 (a) $\Delta S° = +84.4$ J/K for the vaporization of liquid diethyl ether.

(b) $\Delta S° = -84.4$ J/K for the condensation of the ether.

20.4 $\Delta S° = -270.1$ J/K

20.5 (a) $\Delta S° = -252.2$ J/K (b) $\Delta S° = +217.8$ J/K

20.6 $\Delta H°_{rxn}$ is positive. $\Delta S°_{rxn}$ is positive because 1 molecule (H_2O) is being split into two, gaseous molecules (H_2 and O_2). Cannot decide if reaction is product- or reactant-favored; it depends on T.

20.7 $\Delta S°_{rxn} = 46.37$ J/K; $\Delta H°_{rxn} = -1427.8$ kJ; and $\Delta S°_{surr} = 4790$ J/K

$\Delta S°_{universe} = 46.37\text{ J/K} + 4790\text{ J/K} = 4840\text{ J/K}$

The reaction is product-favored, as expected for an exothermic reaction that produces 5 moles of gaseous products from 4.5 moles of gaseous reactants.

20.8 (a) $\Delta H°_{rxn} = \Delta H°_f[SnCl_4(\ell)] = -511.3$ kJ; $\Delta S°_{rxn} = -239.1$ J/K; $\Delta G°_{rxn} = -440.1$ kJ/mol at 298 K

The reaction is product-favored and is enthalpy-driven. In this case $\Delta G°_{rxn} = \Delta G°_f$.

(b) $\Delta H°_{rxn} = -176.01$ kJ; $\Delta S°_{rxn} = -284.8$ J/K; $\Delta G°_{rxn} = -91.10$ kJ/mol at 298 K

The reaction is product-favored and is enthalpy-driven.

20.9 $2\ Fe(s) + 3/2\ O_2(g) \rightarrow Fe_2O_3(s)$

$\Delta G°_f[Fe_2O_3(s)] = -742.2$ kJ/mol and $\Delta G°_{rxn}$ for 454 g $= -2110$ kJ

20.10 (a) $\Delta G°_{rxn} = \Delta G°_f[CaCl_2(s)] = -748.1$ kJ; product-favored

(b) $\Delta G°_{rxn} = -301.39$ kJ; product-favored

20.11 $K_p = 7 \times 10^{-16}$. A positive $\Delta G°_{rxn}$ indicates that K is less than 1. Both are consistent with a reactant-favored reaction.

20.12 The reaction is expected to be exothermic, and so the value of $\Delta H°_{rxn}$ is expected to be negative (–296.83 kJ). The reaction produces 1 mole of a gas from a solid and 1 mole of a gas. The gaseous product has more atoms than the gaseous reactant, so one might expect the value of $\Delta S°_{rxn}$ to be positive (11.28 J/K). Both point to a product-favored reaction, which we know this to be.

20.13 (a) Splitting of water: One mole of a liquid produces 1.5 moles of gas. Therefore, ΔS°_{rxn} is expected to be positive (+163 J/K). Energy is required to split water (to break the O—H bonds), so ΔH°_{rxn} is positive (+285.83 kJ). Thus, ΔG°_{rxn} is positive, and the reaction is reactant-favored.

(b) One mole of liquid nitroglycerin gives a number of moles of gas, and the reaction evolves heat. Therefore, ΔS°_{rxn} is positive and ΔH°_{rxn} is negative. Both predict a negative ΔG°_{rxn}.

(c) Seventeen moles of gas combine to produce 34 moles of gaseous products, so ΔS°_{rxn} is expected to be positive. From experience you know the reaction is exothermic, so ΔH°_{rxn} is expected to be negative. This should lead to a product-favored reaction.

20.14 (a) ΔG°_{rxn} = +91.4 kJ at 298 K

(b) $K_p = 9.71 \times 10^{-17}$

(c) Reaction is not product-favored at 298 K. T at which ΔG_{rxn} is zero is 981.3 K.

21.1 (a) Oxidation; reducing agent $Cr(s) \rightarrow Cr^{3+}(aq) + 3\ e^-$

(b) Oxidation; reducing agent $AsH_3(g) \rightarrow As(s) + 3\ H^+(aq) + 3\ e^-$

(c) Reduction; oxidizing agent $VO_3^-(aq) + 6\ H^+(aq) + 3\ e^- \rightarrow V^{2+}(aq) + 3\ H_2O(\ell)$

21.2 (a) Reduction; oxidizing agent

$Cr_2O_7^{2-}(aq) + 14\ H^+(aq) + 6\ e^- \rightarrow 2\ Cr^{3+}(aq) + 7\ H_2O(\ell)$

(b) Oxidation; reducing agent

$CH_3CHO(aq) + H_2O(\ell) \rightarrow CH_3CO_2H(aq) + 2\ H^+(aq) + 2\ e^-$

(c) Oxidation; reducing agent

$Bi^{3+}(aq) + 3\ H_2O(\ell) \rightarrow HBiO_3(aq) + 5\ H^+(aq) + 2\ e^-$

21.3 (a) $Cl_2(aq) + 2\ e^- \rightarrow 2\ Cl^-(aq)$

$2\ Br^-(aq) \rightarrow Br_2(aq) + 2\ e^-$

$Cl_2(aq) + 2\ Br^-(aq) \rightarrow 2\ Cl^-(aq) + Br_2(aq)$

(b) $Sn(s) \rightarrow Sn^{2+}(aq) + 2\ e^-$

$2\ H^+(aq) + 2\ e^- \rightarrow H_2(g)$

$Sn(s) + 2\ H^+(aq) \rightarrow Sn^{2+}(aq) + H_2(g)$

(c) $Zn(s) \rightarrow Zn^{2+}(aq) + 2\ e^-$

$2\ [VO^{2+}(aq) + 2\ H^+(aq) + e^- \rightarrow V^{3+}(aq) + H_2O(\ell)]$

$Zn(s) + 2\ VO^{2+}(aq) + 4\ H^+(aq) \rightarrow Zn^{2+}(aq) + 2\ V^{3+}(aq) + 2\ H_2O(\ell)$

21.4 (a) $2\,[Ag^+(aq) + e^- \rightarrow Ag(s)]$

$HCHO(aq) + H_2O(\ell) \rightarrow HCO_2H(aq) + 2\,H^+(aq) + 2\,e^-$

$2\,Ag^+(aq) + HCHO(aq) + H_2O(\ell) \rightarrow 2\,Ag(s) + HCO_2H(aq) + 2\,H^+(aq)$

(b) $3\,[H_2S(aq) \rightarrow S(s) + 2\,H^+(aq) + 2\,e^-]$

$Cr_2O_7^{2-}(aq) + 14\,H^+(aq) + 6\,e^- \rightarrow 2\,Cr^{3+}(aq) + 7\,H_2O(\ell)$

$3\,H_2S(aq) + Cr_2O_7^{2-}(aq) + 8\,H^+(aq) \rightarrow 2\,Cr^{3+}(aq) + 3\,S(s) + 7\,H_2O(\ell)$

(c) $3\,[Zn(s) \rightarrow Zn^{2+}(aq) + 2\,e^-]$

$2\,[VO_3^-(aq) + 6\,H^+(aq) + 3\,e^- \rightarrow V^{2+}(aq) + 3\,H_2O(\ell)]$

$3\,Zn(s) + 2\,VO_3^-(aq) + 12\,H^+(aq) \rightarrow 3\,Zn^{2+}(aq) + 2\,V^{2+}(aq) + 6\,H_2O(\ell)$

21.5 Oxidation; anode compartment: $Cr(s) \rightarrow Cr^{3+}(aq) + 3\,e^-$
Reduction; cathode compartment: $Fe^{2+}(aq) + 2\,e^- \rightarrow Fe(s)$

21.6 ΔG°_{rxn} = -562 kJ

21.7 (a) $E^\circ_{net} = -1.298$ V; not product-favored as written
(b) $E^\circ_{net} = -0.51$ V; not product-favored as written
(c) $E^\circ_{net} = -1.021$ V; not product-favored as written

21.8 (a) Weakest oxidizing agent, V^{2+}
(b) Strongest oxidizing agent, Cl_2
(c) Strongest reducing agent, V
(d) Weakest reducing agent, Cl^-
(e) Pb will not reduce V^{2+}
(f) I_2 will not oxidize Cl^-
(g) Cl_2 and I_2 can be reduced by Pb.

21.9 (a) $Zn(s) + 2\,Ag^+(aq) \rightarrow Zn^{2+}(aq) + 2\,Ag(s)$; E° = +1.56 V
(b) Zn = anode and Ag = cathode
(c) See cell diagram
(d) Ag wire is cathode.
(e) Electrons flow from Zn to Ag.
(f) NO_3^- ions flow from Ag^+ to Zn^{2+}.

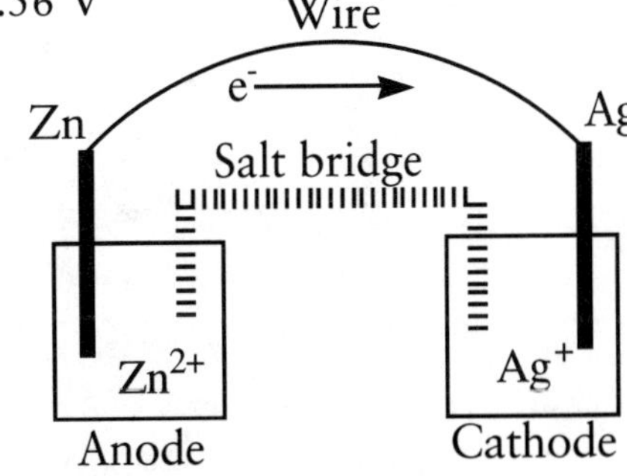

21.10 E = 0.177 V; the cell potential (E) is much less positive when the dissolved reagents are 0.10 M (E°_{net} = 0.236 V). (Note that I_2 is a solid and is not included when solving the Nernst equation.)

21.11 5.47 g Ni

21.12 270 kg Al

21.13 190 g Pb

21.14 (a) Balanced equation: $Ni^{2+}(aq) + Cd(s) \rightarrow Ni(s) + Cd^{2+}(aq)$

(b) Cd is the reducing agent and $Ni^{2+}(aq)$ is the oxidizing agent. $Ni^{2+}(aq)$ is reduced to Ni(s) and Cd(s) is oxidized to $Cd^{2+}(aq)$.

(c) Cd is the anode and Ni is the cathode. The Cd electrode is negative as this electrode supplies electrons.

(d) $E°_{net} = 0.15$ V

(e) Electrons flow from the Cd electrode (anode) to the Ni electrode (cathode).

(f) NO_3^- ions migrate from the Ni^{2+}/Ni compartment (the cathode) to the Cd/Cd^{2+} compartment (the anode).

(g) ln K = +nE° / 0.0257 = +(2.00 mol) (0.15 V) / (0.0257 V · mol)
ln K = 11.67, which gives K = 1.2 × 105

(h) For $[Cd^{2+}] = 0.010$ M and $[Ni^{2+}] = 1.0$ M, and E° = 0.15 V, we have
$E = E° - (0.0257\ V/2) \log \{[Cd^{2+}] / [Ni^{2+}]\} = 0.21$ V
The reaction is still product-favored in the direction written.

(i) The battery has, in the beginning, 1.00 mol each of Ni^{2+} and Cd^{2+}. In addition,
50.0 g Ni · 1 mol / 58.69 g = 0.852 mol Ni
50.0 g Cd · 1 mol / 112.4 g = 0.445 mol Cd
The limiting reactant here is 0.445 mol Cd. Therefore, we use this to find the lifetime of the battery.
(0.445 mol Cd) (2 mol e- / 1 mol Cd) (96500 C / mol e-) = 8.59×10^4 C
(8.59×10^4 C) (1 amp · sec / C) (1 / 0.050 amp) = 1.7×10^6 sec (about 20 days)

21.16 (a) A and C are stronger reducing agents than H_2.

(b) C is the strongest reducing agent of the series.

(c) D is a stronger reducing agent than B.
Order of strength as reducing agents: $B < D < H_2 < A < C$

APPENDIX D
Tables

TABLE 1: SELECTED THERMODYNAMIC VALUES*

Species	ΔH_f°(298.15K) kJ/mol	S°(298.15K) J/K · mol	ΔG_f°(298.15K) kJ/mol
Aluminum			
Al(s)	0	28.3	0
$AlCl_3$(s)	–704.2	110.67	–628.8
Al_2O_3(s)	–1675.7	50.92	–1582.3
Barium			
$BaCl_2$(s)	–858.6	123.68	–810.4
BaO(s)	–553.5	70.42	–525.1
$BaSO_4$(s)	–1473.2	132.2	–1362.2
Beryllium			
Be(s)	0	9.5	0
$Be(OH)_2$	–902.5	51.9	–815.0
Bromine			
Br(g)	111.884	175.022	82.396
$Br_2(\ell)$	0	152.2	0
Br_2(g)	30.907	245.463	3.110
BrF_3(g)	–255.60	292.53	–229.43
HBr(g)	–36.40	198.695	–53.45
Calcium			
Ca(s)	0	41.42	0
Ca(g)	178.2	158.884	144.3
Ca^{2+}(g)	1925.90	—	—
CaC_2(s)	–59.8	69.96	–64.9
$CaCO_3$(s; calcite)	–1206.92	92.9	–1128.79
$CaCl_2$(s)	–795.8	104.6	–748.1
CaF_2(s)	–1219.6	68.87	–1167.3
CaH_2(s)	–186.2	42	–147.2
CaO(s)	–635.09	39.75	–604.03
CaS(s)	–482.4	56.5	–477.4
$Ca(OH)_2$(s)	–986.09	83.39	–898.49
$Ca(OH)_2$(aq)	–1002.82	–74.5	–868.07
$CaSO_4$(s)	–1434.11	106.7	–1321.79
Carbon			
C(s, graphite)	0	5.740	0
C(s, diamond)	1.895	2.377	2.900
C(g)	716.682	158.096	671.257

Species	ΔH_f°(298.15K) kJ/mol	S°(298.15K) J/K · mol	ΔG_f°(298.15K) kJ/mol
CCl_4(ℓ)	−135.44	216.40	−65.21
CCl_4(g)	−102.9	309.85	−60.59
$CHCl_3$(liq)	−134.47	201.7	−73.66
$CHCl_3$(g)	−103.14	295.71	−70.34
CH_4(g, methane)	−74.81	186.264	−50.72
C_2H_2(g, acetylene)	226.73	200.94	209.20
C_2H_4(g, ethylene)	52.26	219.56	68.15
C_2H_6(g, ethane)	−84.68	229.60	−32.82
C_3H_8(g, propane)	−103.8	269.9	−23.49
C_6H_6(ℓ, benzene)	49.03	172.8	124.5
CH_3OH(ℓ, methanol)	−238.66	126.8	−166.27
CH_3OH(g, methanol)	−200.66	239.81	−161.96
C_2H_5OH(ℓ, ethanol)	−277.69	160.7	−174.78
C_2H_5OH(g, ethanol)	−235.10	282.70	−168.49
CO(g)	−110.525	197.674	−137.168
CO_2(g)	−393.509	213.74	−394.359
CS_2(g)	117.36	237.84	67.12
$COCl_2$(g)	−218.8	283.53	−204.6
Cesium			
Cs(s)	0	85.23	0
Cs^+(g)	457.964	—	—
CsCl(s)	−443.04	101.17	−414.53
Chlorine			
Cl(g)	121.679	165.198	105.680
Cl^-(g)	−233.13	—	—
Cl_2(g)	0	223.066	0
HCl(g)	−92.307	186.908	−95.299
HCl(aq)	−167.159	56.5	−131.228
Chromium			
Cr(s)	0	23.77	0
Cr_2O_3(s)	−1139.7	81.2	−1058.1
$CrCl_3$(s)	−556.5	123.0	−486.1
Copper			
Cu(s)	0	33.150	0
CuO(s)	−157.3	42.63	−129.7
$CuCl_2$(s)	−220.1	108.07	−175.7
Fluorine			
F_2(g)	0	202.78	0
F(g)	78.99	158.754	61.91
F^-(g)	−255.39	—	—
F^-(aq)	−332.63	−13.8	−278.79

Species	ΔH_f°(298.15K) kJ/mol	S°(298.15K) J/K · mol	ΔG_f°(298.15K) kJ/mol
HF(g)	–271.1	173.779	–273.2
HF(aq)	–332.63	–13.8	–278.79
Hydrogen			
H_2(g)	0	130.684	0
H(g)	217.965	114.713	203.247
H^+(g)	1536.202	—	—
$H_2O(\ell)$	–285.830	69.91	–237.129
H_2O(g)	–241.818	188.825	–228.572
$H_2O_2(\ell)$	–187.78	109.6	–120.35
Iodine			
I_2(s)	0	116.135	0
I_2(g)	62.438	260.69	19.327
I(g)	106.838	180.791	70.250
I^-(g)	–197.	—	—
ICl(g)	17.78	247.551	–5.46
Iron			
Fe(s)	0	27.78	0
FeO(s)	–272	—	—
Fe_2O_3(s, hematite)	–824.2	87.40	–742.2
Fe_3O_4(s, magnetite)	–1118.4	146.4	–1015.4
$FeCl_2$(s)	–341.79	117.95	–302.30
$FeCl_3$(s)	–399.49	142.3	–344.00
FeS_2(s, pyrite)	–178.2	52.93	–166.9
$Fe(CO)_5(\ell)$	–774.0	338.1	–705.3
Lead			
Pb(s)	0	64.81	0
$PbCl_2$(s)	–359.41	136.0	–314.10
PbO(s, yellow)	–217.32	68.70	–187.89
PbS(s)	–100.4	91.2	–98.7
Lithium			
Li(s)	0	29.12	0
Li^+(g)	685.783	—	—
LiOH(s)	–484.93	42.80	–438.95
LiOH(aq)	–508.48	2.80	–450.58
LiCl(s)	–408.701	59.33	–384.37
Magnesium			
Mg(s)	0	32.68	0
$MgCl_2$(s)	–641.32	89.62	–591.79
$MgCO_3$(s)	–1095.8	65.7	–1012.1
MgO(s)	–601.70	26.94	–569.43

Species	ΔH_f°(298.15K) kJ/mol	S°(298.15K) J/K · mol	ΔG_f°(298.15K) kJ/mol
$Mg(OH)_2(s)$	–924.54	63.18	–833.51
MgS(s)	–346.0	50.33	–341.8
Mercury			
Hg(ℓ)	0	76.02	0
$HgCl_2(s)$	–224.3	146.0	–178.6
HgO(s, red)	–90.83	70.29	–58.539
HgS(s, red)	–58.2	82.4	–50.6
Nickel			
Ni(s)	0	29.87	0
NiO(s)	–239.7	37.99	–211.7
$NiCl_2(s)$	–305.332	97.65	–259.032
Nitrogen			
$N_2(g)$	0	191.61	0
N(g)	472.704	153.298	455.563
$NH_3(g)$	–46.11	192.45	–16.45
N_2H_4(ℓ)	50.63	121.21	149.34
$NH_4Cl(s)$	–314.43	94.6	–202.87
$NH_4Cl(aq)$	–299.66	169.9	–210.52
$NH_4NO_3(s)$	–365.56	151.08	–183.87
$NH_4NO_3(aq)$	–339.87	259.8	–190.56
NO(g)	90.25	210.76	86.55
$NO_2(g)$	33.18	240.06	51.31
$N_2O(g)$	82.05	219.85	104.20
$N_2O_4(g)$	9.16	304.29	97.89
NOCl(g)	51.71	261.69	66.08
HNO_3(ℓ)	–174.10	155.60	–80.71
$HNO_3(g)$	–135.06	266.38	–74.72
$HNO_3(aq)$	–207.36	146.4	–111.25
Oxygen			
$O_2(g)$	0	205.138	0
O(g)	249.170	161.055	231.731
$O_3(g)$	142.7	238.93	163.2
Phosphorus			
P_4(s, white)	0	164.36	0
P_4(s, red)	–70.4	91.2	–48.4
P(g)	314.64	163.193	278.25
$PH_3(g)$	5.4	310.23	13.4
$PCl_3(g)$	–287.0	311.78	–267.8
$P_4O_{10}(s)$	–2984.0	228.86	–2697.7
$H_3PO_4(s)$	–1279.0	110.5	–1119.1

Species	ΔH_f°(298.15K) kJ/mol	S°(298.15K) J/K · mol	ΔG_f°(298.15K) kJ/mol
Potassium			
K(s)	0	64.18	0
KCl(s)	–436.747	82.59	–409.14
$KClO_3$(s)	–397.73	143.1	–296.25
KI(s)	–327.90	106.32	–324.892
KOH(s)	–424.764	78.9	–379.08
KOH(aq)	–482.37	91.6	–440.50
Silicon			
Si(s)	0	18.83	0
$SiBr_4$(ℓ)	–457.3	277.8	–443.9
SiC(s)	–65.3	16.61	–62.8
$SiCl_4$(g)	–657.01	330.73	–616.98
SiH_4(g)	34.3	204.62	56.9
SiF_4(g)	–1614.94	282.49	–1572.65
SiO_2(s, quartz)	–910.94	41.84	–856.64
Silver			
Ag(s)	0	42.55	0
Ag_2O(s)	–31.05	121.3	–11.20
AgCl(s)	–127.068	96.2	–109.789
$AgNO_3$(s)	–124.39	140.92	–33.41
Sodium			
Na(s)	0	51.21	0
Na(g)	107.32	153.712	76.761
Na^+(g)	609.358	—	—
NaBr(s)	–361.062	86.82	–348.983
NaCl(s)	–411.153	72.13	–384.138
NaCl(g)	–176.65	229.81	–196.66
NaCl(aq)	–407.27	115.5	–393.133
NaOH(s)	–425.609	64.455	–379.494
NaOH(aq)	–470.114	48.1	–419.150
Na_2CO_3(s)	–1130.68	134.98	–1044.44
Sulfur			
S(s, rhombic)	0	31.80	0
S(g)	278.805	167.821	238.250
S_2Cl_2(g)	–18.4	331.5	–31.8
SF_6(g)	–1209	291.82	–1105.3
H_2S(g)	–20.63	205.79	–33.56
SO_2(g)	–296.830	248.22	–300.194
SO_3(g)	–395.72	256.76	–371.06
$SOCl_2$(g)	–212.5	309.77	–198.3

Species	ΔH_f°(298.15K) kJ/mol	S°(298.15K) J/K · mol	ΔG_f°(298.15K) kJ/mol
$H_2SO_4(\ell)$	–813.989	156.904	–690.003
$H_2SO_4(aq)$	–909.27	20.1	–744.53
Tin			
Sn(s, white)	0	51.55	0
Sn(s, gray)	–2.09	44.14	0.13
$SnCl_4(\ell)$	–511.3	258.6	–440.1
$SnCl_4(g)$	–471.5	365.8	–432.2
$SnO_2(s)$	–580.7	52.3	–519.6
Titanium			
Ti(s)	0	30.63	0
$TiCl_4(\ell)$	–804.2	252.34	–737.2
$TiCl_4(g)$	–763.2	354.9	–726.7
TiO_2	–939.7	49.92	–884.5
Zinc			
Zn(s)	0	41.63	0
$ZnCl_2(s)$	–415.05	111.46	–369.398
ZnO(s)	–348.28	43.64	–318.30
ZnS(s, sphalerite)	–205.98	57.7	–201.29

*Taken from *The NBS Tables of Chemical Thermodynamic Properties*, 1982.

TABLE 2: IONIZATION CONSTANTS FOR WEAK ACIDS AT 25 °C

Acid	Formula and Ionization Equation	K_a
Acetic	$CH_3CO_2H \rightleftharpoons H^+ + CH_3CO_2^-$	1.8×10^{-5}
Arsenic	$H_3AsO_4 \rightleftharpoons H^+ + H_2AsO_4^-$	$K_1 = 2.5 \times 10^{-4}$
	$H_2AsO_4^- \rightleftharpoons H^+ + HAsO_4^{2-}$	$K_2 = 5.6 \times 10^{-8}$
	$HAsO_4^{2-} \rightleftharpoons H^+ + AsO_4^{3-}$	$K_3 = 3.0 \times 10^{-13}$
Arsenous	$H_3AsO_3 \rightleftharpoons H^+ + H_2AsO_3^-$	$K_1 = 6.0 \times 10^{-10}$
	$H_2AsO_3^- \rightleftharpoons H^+ + HAsO_3^{2-}$	$K_2 = 3.0 \times 10^{-14}$
Benzoic	$C_6HgCO_2H \rightleftharpoons H^+ + C_6H_5CO_2^-$	6.3×10^{-5}
Boric	$H_3BO_3 \rightleftharpoons H^+ + H_2BO_3^-$	$K_1 = 7.3 \times 10^{-10}$
	$H_2BO_3 \rightleftharpoons H^+ + HBO_3^{2-}$	$K_2 = 1.8 \times 10^{-13}$
	$HBO_3^{2-} \rightleftharpoons H^+ + BO_3^{3-}$	$K_3 = 1.6 \times 10^{-14}$
Carbonic	$H_2CO_3 \rightleftharpoons H^+ + HCO_3^-$	$K_1 = 4.2 \times 10^{-7}$
	$HCO_3^- \rightleftharpoons H^+ + CO_3^{2-}$	$K_2 = 4.8 \times 10^{-11}$

Acid	Formula and Ionization Equation	K_a
Citric	$H_3C_6H_5O_7 \rightleftharpoons H^+ + H_2C_6H_5O_7^-$	$K_1 = 7.4 \times 10^{-3}$
	$H_2C_6H_5O_7^- \rightleftharpoons H^+ + HC_6H_5O_7^{2-}$	$K_2 = 1.7 \times 10^5$
	$HC_6H_5O_7^{2-} \rightleftharpoons H^+ + C_6H_5O_7^{3-}$	$K_3 = 4.0 \times 10^{-7}$
Cyanic	$HOCN \rightleftharpoons H^+ + OCN^-$	3.5×10^{-4}
Formic	$HCO_2H \rightleftharpoons H^+ + HCO_2^-$	1.8×10^{-4}
Hydrazoic	$HN_3 \rightleftharpoons H^+ + N_3^-$	1.9×10^{-5}
Hydrocyanic	$HCN \rightleftharpoons H^+ + CN^-$	4.0×10^{-10}
Hydrofluoric	$HF \rightleftharpoons H^+ + F^-$	7.2×10^{-4}
Hydrogen peroxide	$H_2O_2 \rightleftharpoons H^+ + HO_2^-$	2.4×10^{-12}
Hydrosulfuric	$H_2S \rightleftharpoons H^+ + HS^-$	$K_1 = 1 \times 10^{-7}$
	$HS^- \rightleftharpoons H^+ + S^{2-}$	$K_2 = 1 \times 10^{-19}$
Hypobromous	$HOBr \rightleftharpoons H^+ + OBr^-$	2.5×10^{-9}
Hypochlorous	$HOCl \rightleftharpoons H^+ + OCl^-$	3.5×10^{-8}
Nitrous	$HNO_2 \rightleftharpoons H^+ + NO_2^-$	4.5×10^{-4}
Oxalic	$H_2C_2O_4 \rightleftharpoons H^+ + HC_2O_4^-$	$K_1 = 5.9 \times 10^{-2}$
	$HC_2O_4^- \rightleftharpoons H^+ + C_2O_4^{2-}$	$K_2 = 6.4 \times 10^{-5}$
Phenol	$C_6H_5OH \rightleftharpoons H^+ + C_6H_5O^-$	1.3×10^{-10}
Phosphoric	$H_3PO_4 \rightleftharpoons H^+ + H_2PO_4^-$	$K_1 = 7.5 \times 10^{-3}$
	$H_2PO_4^- \rightleftharpoons H^+ + HPO_4^{2-}$	$K_2 = 6.2 \times 10^{-8}$
	$HPO_4^{2-} \rightleftharpoons H^+ + PO_4^{3-}$	$K_3 = 3.6 \times 10^{-13}$
Phosphorous	$H_3PO_3 \rightleftharpoons H^+ + H_2PO_3^-$	$K_1 = 1.6 \times 10^{-2}$
	$H_2PO_3 \rightleftharpoons H^+ + HPO_3^{2-}$	$K_2 = 7.0 \times 10^{-7}$
Selenic	$H_2SeO_4 \rightleftharpoons H^+ + HSeO_4^-$	K_1 = very large
	$HSeO_4^- \rightleftharpoons H^+ + SeO_4^{2-}$	$K_2 = 1.2 \times 10^{-2}$
Selenous	$H_2SeO_3 \rightleftharpoons H^+ + HSeO_3^-$	$K_1 = 2.7 \times 10^{-3}$
	$HSeO_3^- \rightleftharpoons H^+ + SeO_3^{2-}$	$K_2 = 2.5 \times 10^{-7}$
Sulfuric	$H_2SO_4 \rightleftharpoons H^+ + HSO4^-$	K_1 = very large
	$HSO_4^- \rightleftharpoons H^+ + SO_4^{2-}$	$K_2 = 1.2 \times 10^{-2}$
Sulfurous	$H_2SO_3 \rightleftharpoons H^+ + HSO_3^-$	$K_1 = 1.7 \times 10^{-2}$
	$HSO_3^- \rightleftharpoons H^+ + SO_3^{2-}$	$K_2 = 6.4 \times 10^{-8}$
Tellurous	$H_2TeO_3 \rightleftharpoons H^+ + HTeO_3^-$	$K_1 = 2 \times 10^{-3}$
	$HTeO_3^- \rightleftharpoons H^+ + TeO_3^{2-}$	$K_2 = 1 \times 10^{-8}$

TABLE 3: IONIZATION CONSTANTS FOR WEAK BASES AT 25 °C

Base	Formula and Ionization Equation	K_b
Ammonia	$NH_3 + H_2O \rightleftharpoons NH_4^+ + OH^-$	1.8×10^{-5}
Aniline	$C_6H_5NH_2 + H_2O \rightleftharpoons C_6H_5NH_3^+ + OH^-$	4.0×10^{-10}
Dimethylamine	$(CH_3)_2NH + H_2O \rightleftharpoons (CH_3)_2NH_2^+ + OH^-$	7.4×10^{-4}
Ethylenediamine	$(CH_2)_2(NH_2)_2 + H_2O \rightleftharpoons (CH_2)_2(NH_2)_2H^+ + OH^-$	$K_1 = 8.5 \times 10^{-5}$
	$(CH_2)_2(NH_2)_2H^+ + H_2O \rightleftharpoons (CH_2)_2(NH_2)_2H_2^{2+} + OH^-$	$K_2 = 2.7 \times 10^{-8}$
Hydrazine	$N_2H_4 + H_2O \rightleftharpoons N_2H_5^+ + OH^-$	$K_1 = 8.5 \times 10^{-7}$
	$N_2H_5^+ + H_2O \rightleftharpoons N_2H_6^{2+} + OH^-$	$K_2 = 8.9 \times 10^{-16}$
Hydroxylamine	$NH_2OH + H_2O \rightleftharpoons NH_3OH^+ + OH^-$	6.6×10^{-9}
Methylamine	$CH_3NH_2 + H_2O \rightleftharpoons CH_3NH_3^+ + OH^-$	5.0×10^{-4}
Pyridine	$C_5H_5N + H_2O \rightleftharpoons C_5H_5NH^+ + OH^-$	1.5×10^{-9}
Trimethylamine	$(CH_3)_3N + H_2O \rightleftharpoons (CH_3)_3NH^+ + OH^-$	7.4×10^{-5}

TABLE 4: SOLUBILITY PRODUCT CONSTANTS FOR SOME INORGANIC COMPOUNDS AT 25 °C

Substance	K_{sp}	Substance	K_{sp}
Aluminum compounds			
$AlAsO_4$	1.6×10^{-16}	$BiPO_4$	1.3×10^{-23}
$Al(OH)_3$	1.9×10^{-33}	Bi_2S_3	1.6×10^{-72}
$AlPO_4$	1.3×10^{-20}	**Cadmium compounds**	
Antimony compounds		$Cd_3(AsO_4)_2$	2.2×10^{-32}
Sb_2S_3	1.6×10^{-93}	$CdCO_3$	2.5×10^{-14}
Barium compounds		$Cd(CN)_2$	1.0×10^{-8}
$Ba_3(AsO_4)_2$	1.1×10^{-13}	$Cd_2[Fe(CN)_6]$	3.2×10^{-17}
$BaCO_3$	8.1×10^{-9}	$Cd(OH)_2$	1.2×10^{-14}
$BaC_2O_4 \cdot 2H_2O$*	1.1×10^{-7}	CdS	3.6×10^{-29}
$BaCrO_4$	2.0×10^{-10}	**Calcium compounds**	
BaF_2	1.7×10^{-6}	$Ca_3(AsO_4)_2$	6.8×10^{-19}
$Ba(OH)_2 \cdot 8H_2O$*	5.0×10^{-3}	$CaCO_3$	3.8×10^{-9}
$Ba_3(PO_4)_2$	1.3×10^{-29}	$CaCrO_4$	7.1×10^{-4}
$BaSeO_4$	2.8×10^{-11}	$CaC_2O_4 \cdot H_2O$*	2.3×10^{-9}
$BaSO_3$	8.0×10^{-7}	CaF_2	3.9×10^{-11}
$BaSO_4$	1.1×10^{-10}	$Ca(OH)_2$	7.9×10^{-6}
Bismuth compounds		$CaHPO_4$	2.7×10^{-7}
$BiOCl$	7.0×10^{-9}	$Ca(H_2PO_4)_2$	1.0×10^{-3}
$BiO(OH)$	1.0×10^{-12}	$Ca_3(PO_4)_2$	1.0×10^{-25}
$Bi(OH)_3$	3.2×10^{-40}	$CaSO_3 \cdot 2H_2O$*	1.3×10^{-8}
BiI_3	8.1×10^{-19}	$CaSO_4 \cdot 2H_2O$*	2.4×10^{-5}

olubility Product Constants for Some Inorganic Compounds at 25 °C *(Continued)*

Substance	K_{sp}	Substance	K_{sp}
Chromium compounds		$PbCO_3$	1.5×10^{-13}
$CrAsO_4$	7.8×10^{-21}	$PbCl_2$	1.7×10^{-5}
$Cr(OH)_3$	6.7×10^{-31}	$PbCrO_4$	1.8×10^{-14}
$CrPO_4$	2.4×10^{-23}	PbF_2	3.7×10^{-8}
Cobalt compounds		$Pb(OH)_2$	2.8×10^{-16}
$Co_3(AsO_4)_2$	7.6×10^{-29}	PbI_2	8.7×10^{-9}
$CoCO_3$	8.0×10^{-13}	$Pb_3(PO_4)_2$	3.0×10^{-44}
$Co(OH)_2$	2.5×10^{-16}	$PbSeO_4$	1.5×10^{-7}
CoS (α)	5.9×10^{-21}	PbSO4	1.8×10^{-8}
$Co(OH)_3$	4.0×10^{-45}	PbS	8.4×10^{-28}
Copper compounds		**Magnesium compounds**	
CuBr	5.3×10^{-9}	$Mg_3(AsO_4)_2$	2.1×10^{-20}
CuCl	1.9×10^{-7}	$MgCO_3 \cdot 3H_2O$*	4.0×10^{-5}
CuCN	3.2×10^{-20}	MgC_2O_4	8.6×10^{-5}
Cu_2O ($Cu^+ + OH^-$)†	1.0×10^{-14}	MgF_2	6.4×10^{-9}
CuI	5.1×10^{-12}	$Mg(OH)_2$	1.5×10^{-11}
Cu_2S	1.6×10^{-48}	$MgNH_4PO_4$	2.5×10^{-12}
CuSCN	1.6×10^{-11}	**Manganese compounds**	
$Cu_3(AsO_4)_2$	7.6×10^{-36}	$Mn_3(AsO_4)_2$	1.9×10^{-11}
$CuCO_3$	2.5×10^{-10}	$MnCO_3$	1.8×10^{-11}
$Cu_2[Fe(CN)_6]$	1.3×10^{-16}	$Mn(OH)_2$	4.6×10^{-14}
$Cu(OH)_2$	1.6×10^{-19}	MnS	5.1×10^{-15}
CuS	8.7×10^{-36}	$Mn(OH)_3$	$\approx 1 \times 10^{-36}$
Gold compounds		**Mercury compounds**	
AuBr	5.0×10^{-17}	Hg_2Br_2	1.3×10^{-22}
AuCl	2.0×10^{-13}	Hg_2CO_3	8.9×10^{-17}
AuI	1.6×10^{-23}	Hg_2Cl_2	1.1×10^{-18}
$AuBr_3$	4.0×10^{-36}	Hg_2CrO_4	5.0×10^{-9}
$AuCl_3$	3.2×10^{-25}	Hg_2I_2	4.5×10^{-29}
$Au(OH)_3$	1×10^{-53}	$Hg_2O \cdot H_2O$ ($Hg_2^{2+} + 2OH^-$)*†	1.6×10^{-23}
AuI_3	1.0×10^{-46}	Hg_2SO_4	6.8×10^{-7}
Iron compounds		Hg_2S	5.8×10^{-44}
$FeCO_3$	3.5×10^{-11}	$Hg(CN)_2$	3.0×10^{-23}
$Fe(OH)_2$	7.9×10^{-15}	$Hg(OH)_2$	2.5×10^{-26}
FeS	4.9×10^{-18}	HgI_2	4.0×10^{-29}
$Fe_4[Fe(CN)_6]_3$	3.0×10^{-41}	HgS	3.0×10^{-53}
$Fe(OH)_3$	6.3×10^{-38}	**Nickel compounds**	
Fe_2S_3	1.4×10^{-88}	$Ni_3(AsO_4)_2$	1.9×10^{-26}
Lead compounds		$NiCO_3$	6.6×10^{-9}
$Pb_3(AsO_4)_2$	4.1×10^{-36}	$Ni(CN)_2$	3.0×10^{-23}
$PbBr_2$	6.3×10^{-6}	$Ni(OH)_2$	2.8×10^{-16}

Solubility Product Constants for Some Inorganic Compounds at 25 °C (Continued)

Substance	K_{sp}	Substance	K_{sp}
NiS (α)	3.0×10^{-21}	$SrC_2O_4 \cdot 2H_2O$*	5.6×10^{-8}
NiS (β)	1.0×10^{-26}	$SrCrO_4$	3.6×10^{-5}
NiS (γ)	2.0×10^{-28}	$Sr(OH)_2 \cdot 8H_2O$*	3.2×10^{-4}
Silver compounds		$Sr_3(PO_4)_2$	1.0×10^{-31}
Ag_3AsO_4	1.1×10^{-20}	$SrSO_3$	4.0×10^{-8}
AgBr	3.3×10^{-13}	$SrSO_4$	2.8×10^{-7}
Ag_2CO_3	8.1×10^{-12}	**Tin compounds**	
AgCl	1.8×10^{-10}	$Sn(OH)_2$	2.0×10^{-26}
Ag_2CrO_4	9.0×10^{-12}	SnI_2	1.0×10^{-4}
AgCN	1.2×10^{-16}	SnS	1.0×10^{-28}
$Ag_4[Fe(CN)_6]$	1.6×10^{-41}	$Sn(OH)_4$	1.0×10^{-57}
Ag_2O ($Ag^+ + OH^-$)†	2.0×10^{-8}	SnS_2	1.0×10^{-70}
AgI	1.5×10^{-16}	**Zinc compounds**	
Ag_3PO_4	1.3×10^{-20}	$Zn_3(AsO_4)_2$	1.1×10^{-27}
Ag_2SO_3	1.5×10^{-14}	$ZnCO_3$	1.5×10^{-11}
Ag_2SO_4	1.7×10^{-5}	$Zn(CN)_2$	8.0×10^{-12}
Ag_2S	1.0×10^{-49}	$Zn_3[Fe(CN)_6]$	4.1×10^{-16}
AgSCN	1.0×10^{-12}	$Zn(OH)_2$	4.5×10^{-17}
Strontium compounds		$Zn_3(PO_4)_2$	9.1×10^{-33}
$Sr_3(AsO_4)_2$	1.3×10^{-18}	ZnS	1.1×10^{-21}
$SrCO_3$	9.4×10^{-10}		

* Since $[H_2O]$ does not appear in equilibrium constants for equilibria in aqueous solution in general, it doe *not* appear in the K_{sp} expressions for hydrated solids.

† Very small amounts of oxides dissolve in water to give the ions indicated in parentheses. Solid hydroxide are unstable and decompose to oxides as rapidly as they are formed.